Immobilized Enzymes for Food Processing

Editor

Wayne H. Pitcher, Jr., Sc.D.
Engineering Supervisor
Corning Glass Works
Corning, New York

CRC Press
Taylor & Francis Group
Boca Raton London New York

CRC Press is an imprint of the
Taylor & Francis Group, an informa business

First published 1980 by CRC Press
Taylor & Francis Group
6000 Broken Sound Parkway NW, Suite 300
Boca Raton, FL 33487-2742

Reissued 2018 by CRC Press

Library of Congress Cataloging in Publication Data

Main entry under title:

Immobilized enzymes for food processing.

Bibliography: p.
Includes index.
1. Immobilized enzymes—Industrial applications.
2. Food industry and trade. I. Pitcher, Wayne H.
TP456.E58I45 664'.01 79-25738
ISBN 0-8493-5345-9

A Library of Congress record exists under LC control number: 79025738

ISBN 13: 978-1-315-89430-0 (hbk)
ISBN 13: 978-1-351-07340-0 (ebk)

Visit the Taylor & Francis Web site at http://www.taylorandfrancis.com and the
CRC Press Web site at http://www.crcpress.com

FOREWORD

Although enzymes have been used in the food processing industry for a number of years, enzyme immobilization, allowing continuous processing, has been applied to food processing only in the past decade. Much has been written about immobilized enzymes during this period of time. So much, in fact, that it can become difficult even for those involved in developing new enzymatic food processing operations to bridge the gap between the field of immobilized enzymes and their specific requirements. It is the purpose of this book to assist those engaged in this difficult task. Perhaps an equally important goal is to bring to the researcher in enzyme immobilization an appreciation for the requirements of the food processing industry. After all, most of the commercial applications of immobilized enzymes, have been in the area of food processing.

The organization of the book was designed to give the reader background in enzyme immobilization, engineering and economic factors, and the unique requirements of the food processing industry in the first three chapters as a prelude to the consideration of specific applications. Chapters dealing with applications include potential as well as already commercialized processes. In so young a field the potential applications still far outnumber those that have reached practical use.

As editor, I am extremely pleased that such a well-qualified group from industry and the academic community consented to contribute to this work. My deepest thanks go to them. Thanks should also go to the editors of CRC Press for their patience and assistance.

THE EDITOR

Wayne H. Pitcher, Jr. received his B.S. in chemical engineering from the California Institute of Technology in 1966. He continued his education in the field of chemical engineering at the Massachusetts Institute of Technology where he received his S.M. degree in 1968 and his Sc.D. in 1972.

Since joining Corning Glass Works in 1972, Dr. Pitcher has been primarily involved in developing immobilized enzyme systems for industrial applications. Among other responsibilities he has headed projects to develop immobilized glucose isomerase and immobilized lactase systems.

Dr. Pitcher is a member of various professional organizations including the American Institute of Chemical Engineers and the American Chemical Society. He has chaired sessions at several national and international technical conferences, authored numerous articles, and holds several patents.

CONTRIBUTORS

Marvin Charles, Ph.D.
Department of Chemical Engineering
Lehigh University
Bethlehem, Pennsylvania

Robert W. Coughlin, Ph.D., P.E.
Professor of Chemical Engineering
University of Connecticut
Storrs, Connecticut

Robert V. MacAllister, Ph.D.
Director
Scientific and Technological Evaluation
Clinton Corn Processing Company
Clinton, Iowa

Ralph A. Messing, M.S.
Senior Research Associate
Head
Department of Fundamental Life Sciences
Corning Glass Works
Corning, New York

Peter J. Reilly, Ph.D.
Professor of Chemical Engineering
Department of Chemical Engineering
Iowa State University
Ames, Iowa

John F. Roland, B.S.
Senior Group Leader
Kraft, Inc.
Research and Development
Glenview, Illinois

Bhavender P. Sharma, Ph.D.
Senior Project Engineer-Chemical
Technical Staffs Division
Corning Glass Works
Corning, New York

Howard H. Weetall, M.S.
Manager
Biomedical Research
Corning Glass Works
Corning, New York

TABLE OF CONTENTS

Chapter 1
Introduction to Immobilized Enzymes 1
Wayne H. Pitcher, Jr.

Chapter 2
Immobilized Enzyme Engineering 15
Wayne H. Pitcher, Jr.

Chapter 3
Requirements Unique to the Food and Beverage Industry.......................... 55
J. F. Roland

Chapter 4
Manufacture of High Fructose Corn Syrup Using Immobilized Glucose Isomerase .. 81
R. V. MacAllister

Chapter 5
Potential and Use of Immobilized Carbohydrases.............................. 113
P. J. Reilly

Chapter 6
Applications of Lactose and Immobilized Lactase 153
R. W. Coughlin and Marvin Charles

Chapter 7
Immobilized Proteases — Potential Applications 175
H. H. Weetall

Chapter 8
Application and Potential of Other Enzymes in Food Processing: Aminoacylase, Aspartase, Fumarase, Glucose Oxidase-Catalase 185
B. P. Sharma and R. A. Messing

Index 211

Chapter 1

INTRODUCTION TO IMMOBILIZED ENZYMES

Wayne H. Pitcher, Jr.

TABLE OF CONTENTS

I. Introduction 2

II. Adsorption 2

III. Covalent Bonding 6

IV. Entrapment 10

References 13

I. INTRODUCTION

The purpose of this chapter on enzyme immobilization is to introduce the reader to the diversity of immobilization techniques available and some of the variables that affect the actual immobilization procedures. No effort has been made to judge the suitability of these methods for immobilizing enzymes for food processing. These considerations are discussed in Chapter 3. It is not intended that this section necessarily be comprehensive or contain detailed descriptions of immobilization procedures. Since enzyme immobilization can be accomplished in numerous ways under so many sets of conditions, these details would, at best, only provide a starting point. What is important is that the reader understand the various options known to be available in order that he may be able to select a reasonable approach to his particular problem. Thus, the emphasis in this chapter is on the attributes of the various methods of immobilization and the ways in which workers in this field have attempted to vary them.

Although there are hundreds of immobilization procedures that have been categorized in various ways, for the purposes of this treatment of the subject they are placed into three general groups. These three types of immobilization are adsorption, covalent bonding, and entrapment. Such a classification is convenient, if somewhat arbitrary. There are cases where two methods are combined, a common example being adsorption and cross-linking (a form of covalent bonding). For more detailed information on enzyme immobilization, other references should be consulted.[1,2] Some additional examples are also given in Chapter 4.

II. ADSORPTION

Adsorption is the oldest of the techniques used to immobilize enzymes, dating back to 1916 when Nelson and Griffin[3] used both charcoal and aluminum hydroxide to adsorb invertase. Since that time a wide range of organic and inorganic substances have been utilized as supports for adsorbed enzymes. Both organic materials such as charcoal, various cellulose derivatives, and ion exchange resins and inorganic materials including silica, alumina, titania, glass, and various naturally occurring minerals have been used.

Although adsorption has had a dubious reputation in the past, probably as a result of problems with desorption and inactivation upon adsorption, commercially it has seen relatively frequent usage. Clinton Corn Processing Company has reported using DEAE-cellulose to adsorb glucose isomerase.[4] Tanabe Seiyaku Company has been immobilizing aminoacylase on DEAE-sephadex and other ion exchange resins for use in a process to racemize mixtures of D- and L-isomers of amino acids.[5] CPC-International is adsorbing glucose isomerase to porous alumina beads via a process developed by Corning Glass Works.

Development of a useful adsorbed enzyme derivative depends on many factors. Perhaps the most important, or at least the first to be encountered, is the interaction between the enzyme and the surface of the carrier.

The same enzyme will be adsorbed on different carriers to varying degrees and exhibit different levels of activity as a function of the support material properties. Similarly, a carrier which is effective for one enzyme may be totally useless for another.

Several striking examples of the effect of carrier composition on adsorbed enzyme activity have been reported. Pitcher and Ford[6] adsorbed β-galactosidase (lactase) onto various porous ceramic beads with widely varying results as shown in Table 1. Stanley and Palter[7] reported adsorbing this same enzyme on a wide range of phenol resins. They found that several materials including Duolite® A-1 (Diamond Shamrock Co.),

Table 1
LACTASE ADSORPTION

Carrier composition	Ave. pore diameter (Å)	Pretreatment and adsorption pH	Percent activity at pH 4.5 relative to lactase chemically bound to silica
SiO_2	370	3	35
SiO_2	370	7	24
Al_2O_3	230	3	53
Al_2O_3	230	7	32
TiO_2	380	3	117
TiO_2	380	7	127

Note: Enzyme: Wallerstein Co. Lactase LP, 0.1 g enzyme per gram carrier offered. Carriers: 30/45 mesh, porous particles.

From Pitcher, W. H., Jr. and Ford, J. R., *Enzyme Engineering*, Vol. 3, Pye, E. K. and Weetall, H. H., Eds., Plenum Press, New York, 1978.

charcoal, *p*-hydroxybenzaldehyde, salicylaldehyde, and *o*-cresol-phenol-formaldehyde polymer failed to retain any enzyme activity. Oxidized catechol, humic acid, phloroglucinol-formaldehyde polymer, and catechol-formaldehyde polymer exhibited good initial activity but exhibited significantly lower activity after treatment with salt solution. Duolite® S-30 (Diamond Shamrock Co.), resorcinol-formaldehyde polymer, and Duolite® S-30 formylated with DMF-$POCl_3$, showed relatively high activity retention even after soaking in 2*M* NaCl or 4*M* urea for several hours. The authors did eventually resort to glutaraldehyde cross-linking to stabilize the composite after adsorbing the enzyme to the S-30 resin.

Caldwell et al.[8] reported affecting the activity of a β-amylase-Sepharose 6B derivative by varying the hexyl-group substitution levels in the sepharose. Maximum activity was observed at a hexyl to galactose residue molar ratio of 0.5.

A patent granted to Eaton and Messing[9] describes the effect of varying amounts of magnesia in a porous alumina carrier on the activity of adsorbed glucose isomerase. Addition of magnesia to the alumina support composition affects pH which may, in turn, influence enzyme activity. From Table 2, it can be seen that enzyme activity falls off at low and high MgO levels. Other data in the patent indicates the optimal magnesia level to be in the 1 to 4% range.

Some of the effects of carrier composition in this case may be attributable to pH shifts. Examples of pH effects on enzyme adsorption are well known. Kennedy and Kay[10] reported optimal adsorption of dextranase on porous titanium oxide spheres at pH 5.0. This derivative was evidently at least fairly stable when used in a column reactor. However, when the pH of the dextran solution feed was raised from 5.0 to 7.3, enzyme was rapidly eluted from the titania bed.

Boudrant and Cheftel[11] reported on the stability of invertase adsorbed to several macroreticular ion exchange resins. Enzyme desorption was more pronounced at pH 5.9 to 6.3 than at lower pH (2.4 to 3). However, other conditions such as temperature and ionic strength more strongly influenced enzyme desorption.

High ionic strength solutions of electrolytes do tend to cause the desorption of adsorbed proteins. The extent of this problem depends on the specific enzyme and support material involved, pH, temperature, and perhaps other variables. Boudrant and Cheftel,[11] and Stanley and Palter,[17] and Baratti et al.[12] all reported the effect of ionic concentration on the adsorption of enzymes as had many others before them. Each of

Table 2
GLUCOSE ISOMERASE ADSORPTION[9]

MgO in carrier (wt %)	pH of carrier slurry 1g/10g water	Activity[a] (IGIU/g)
0	4.4	387
0.84	7.0	805
6.65	8.8	916
12.0	8.9	768
28.6	9.3	200

Note: Enzyme: Glucose isomerase.
Carriers: 30/45 mesh, porous alumina particles containing various levels of MgO.

[a]International Glucose Isomerase Units.

these groups, however, also reported other factors having even greater influence on the susceptibility of the immobilized enzyme to desorption. Boudrant and Cheftel[11] observed marked variation in enzyme adsorption with pH. Stanley and Palter[7] reported wide variations in enzyme adsorption as a function of the support material. Baratti et al.[12] observed that different sodium chloride concentrations affected the adsorption of two different enzymes on the same DEAE-cellulose support. Less than 20% of catalase was adsorbed from 0.1 *M* NaCl solution, while more than 90% methanol oxidase was adsorbed. A patent by Cayle[13] describes specific adsorption of β-galactosidase from a crude enzyme preparation onto aluminum silicate at a pH between 3 and 6 and desorption at a pH between 7 and 8. In other words, knowledge of the enzyme-support interaction as a function of adsorption conditions can be utilized for enzyme purification as well as enzyme immobilization.

Other factors are also important in the adsorption of enzymes to support materials. These include the duration of enzyme-support contact, enzyme to support ratio, and enzyme concentration. Temperature can also influence adsorption but this usually is not readily apparent in systems of interest where adsorption is normally strong and temperature ranges are relatively small.

Enzyme adsorption seems to require a longer time than would be expected solely from diffusion rate considerations. Typical enzyme uptake curves are shown in Figure 1. It can be seen that the adsorption rate can be affected by the support properties and adsorption conditions, further evidence that the adsorption reaction itself is the rate-limiting step. In the absence of enzyme adsorption rate data, it is usually prudent to allow 16 to 24 hr contacting time for the preparation of an adsorbed enzyme derivative.

Another commonly observed phenomenon is illustrated by Table 3 based on data from Caldwell et al.[8] From this table it is apparent that the higher the enzyme or protein concentration, the higher the loading. Concentration, not amount, was the critical factor since excess protein was available in all cases. It is also clear that at high enzyme concentration the support material becomes saturated and little additional protein can be adsorbed even at higher concentrations. Finally, the specific activity decreases with increasing enzyme loading, even to the extent that lower total activity is observed for the derivative. This phenomenon has been observed, even more dramatically in other cases. It is apparent that simply adsorbing a large amount of enzyme on a support material is not sufficient to produce a high activity derivative. Kennedy and Kay[10] reported a higher protein loading on titania when they attempted to adsorb dextranase in the presence of ammonia. However, when they used no ammonia, the titania

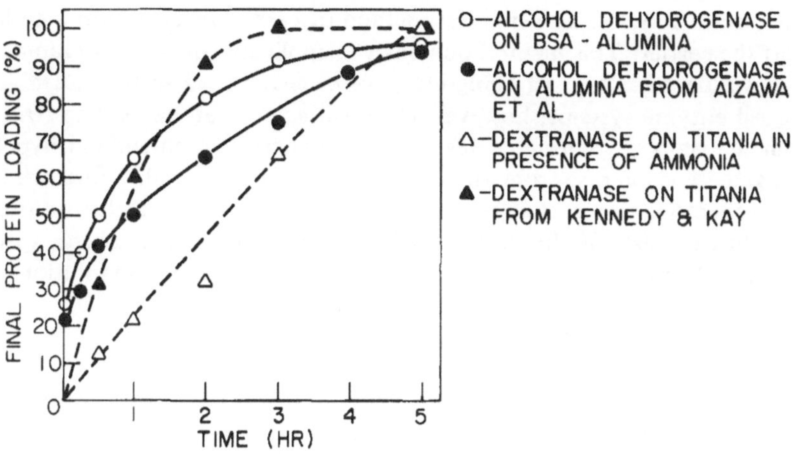

FIGURE 1. Enzyme adsorption as a function of time.

Table 3
AMYLOGLUCOSIDASE ADSORPTION ON HEXYL-SEPHAROSE

Protein concentration in coupling solution (mg/mℓ)		Protein loading on support	Derivative activity	
Initial	Final	(mg/g)	(units/g)	Specific activity (%)
5	3.4	7.8	560	96
10	7.0	15.1	975	86
40	36.7	16.5	845	68

Based on data from Caldwell, K. D., Axen, R., Bergwall, M., and Porath, J., *Biotechnol. Bioeng.*, 18, 1589, 1976.

derivative had lower protein loading (3.2 vs. 25.1 mg/g matrix), but much higher activity (9.6 vs. 2.2 units/g).

The apparent activity of an adsorbed enzyme derivative has frequently been observed to rise as smaller support particles are used. Usually this phenomenon is attributed to internal diffusion limitations and indeed that is frequently the case. However, at least in the case of adsorbed enzymes there is another effect that is sometimes overlooked. Enzyme loading itself can vary with particle size.

One might expect to observe higher enzyme loading on small vs. large particles from a sample of various size particles contacted with enzyme at the same time. Diffusion effects would favor the smaller particles. However, at Corning Glass Works, enzyme was adsorbed on two separate porous alumina samples, one 30/45 mesh and the other 50/60, under identical conditions. The protein loading on the 30/45 mesh material was 20.8 mg/g and on the 50/60 mesh material, 30.5 mg/g. A similar result is discussed in the section on covalent bonding.

Another interesting feature of adsorption as an immobilization technique is that it lends itself to various methods of regenerating the carrier for reuse after the enzyme has lost its useful activity. The simplest method of carrier regeneration is to burn the enzyme off the carrier. This method obviously is applicable only to inorganic carriers, but does have the advantage of being effective even for covalently bonded enzymes. For adsorbed enzyme systems, however, chemical agents can be used to strip the enzyme from the support material. Gregory and Pitcher[14,15] used sodium hypochlorite to remove glucose isomerase from an alumina carrier. Fujita et al.[16] patented a procedure for regenerating ion exchange resins used for immobilizing glucose isomerase. Their procedure consists of alternating washing with strong acid and base, specifically with sulfuric acid and a 0.1 *N* sodium hydroxide, 0.5 *M* sodium sulfate solution.

III. COVALENT BONDING

Probably the most frequently reported coupling methods in the literature are the various types of covalent attachment of enzymes to support materials. The purpose of this section is not to describe each of the dozens of covalent bonding techniques, but to review some of the important attributes of this immobilization method. Covalent bonding has not found as extensive use in the food processing industry as one might expect from the numerous articles describing different chemical coupling schemes. One reason is that many of the reagents used in the coupling processes would not be suitable for applications where they would contact food. Other chemical compounds or at least the resulting derivatives are probably quite safe, but the testing program to establish this fact, in light of the regulatory situation, may be prohibitively expensive. Finally many covalent bonding techniques are complicated and expensive.

Undoubtedly the greatest advantage of covalent coupling is the durability of the derivative. Variations in pH, ionic strength, and temperature will not normally cause the enzyme elution problem that can plague certain adsorbed enzyme systems.

Of course there are exceptions and variations in stability with the coupling method used. Epton et al.[17] report coupling thermolysin to a polyacrylamide copolymer via diazo coupling at pH 7.5, diazo coupling at pH 10.0, and acyl azide coupling. The acrylazide derivative demonstrated the best stability at 80°C as well as the highest activity per gram of enzyme and per gram of derivative. The pH 10 diazo coupling derivative was the least stable at 80°C, but retained 96% of its original activity when stored at 0 to 2°C for 6 months. Under the same storage conditions the pH 7.5 diazo coupling and acylazide derivatives had activity losses of 30% and 56%, respectively.

In some cases it can be the support material and not the binding method that is responsible for enzyme activity loss. Weetall and Havewala[18] observed activity decay rates for glucoamylase bonded to controlled pore glass via silane glutaraldehyde. The decay rates were flow dependent. This fact, along with the detection of silica in the reactor eluent, indicated a carrier dissolution problem. In some cases, the percentage weight loss of carrier exceeded the percentage enzyme activity loss. A switch to a zirconia-coated porous glass particle eliminated the flow rate decay dependence and substantially lengthened the immobilized enzyme half-life.

The activity of a covalently bonded enzyme derivative depends on a number of factors including size and shape of carrier material, the nature of the coupling method, the composition of the carrier material, and the specific conditions during coupling. The following examples of the influence of these factors on enzyme immobilization will serve to illustrate the range of considerations necessary to arrive at an optimal system.

An excellent example of the effect of the type of chemical coupling on the activity of the derivative is the study of enzyme immobilization on porous cellulose beads by Chen and Tsao.[19,20] Without elaborating on the details of the specific coupling procedures, it can be seen from Table 4 that the activity of these derivatives varies greatly. It is also significant to note that the different enzymes do not respond uniformly to the same coupling method. In other words, the best bonding technique for one enzyme may not be the best for another. As shown by Weetall and Havewala,[18] pH optima can be shifted by using a different coupling method. They reported bonding glucoamylase to alklyamine porous glass via azo- and glutaraldehyde methods, the former yielding a pH optimum of 5.0 and the latter 4.0, as compared with a broad optimum between pH 3.5 and 5.0 for the soluble enzyme.

Weetall et al.[21] have reported the results of a study of glucoamylase immobilized on various materials via silane and glutaraldehyde. Table 5 shows the more than five-fold variation in activity over the range of carrier materials.

Even though a support material and a coupling method have been decided upon, variations in the conditions or details of the bonding procedure may cause substantial changes in the level of enzyme activity of the derivative.

Earlier reference was made to the work of Epton et al.[17,22] which included the diazo coupling of thermolysin to a polyacrylamide copolymer. The procedure was carried out at pH 7.5 and at pH 10.0, with the former pH resulting in 75% higher protein binding than at pH 10.0. Depending on the assay, however, the caseinolytic activity of the pH 10.0 preparation was about 12 times that of the pH 7.5 preparation.

Lai and Cheng[23] have provided some interesting examples of the influence of coupling procedure on enzyme loading. They made a styrene-maleic anhydride (SMA) copolymer support material which they prepared by cross-linking with hexamethylenediamine (HMDA) and then activated with ethylenediamine (EDA) followed by a glutaraldehyde cyanogen bromide or diazotization process. As the molar ratio of SMA anhydride to HMDA was decreased from 30:1 to 10:1, the amount of α-chymotrypsin that could be bound via glutaraldehyde decreased from 73 to 36 mg/g. The effects of EDA solution concentration are shown in Figure 2. The optimum H_2O to EDA molar ratio appears to be in the 3 to 5:1 range.

The same authors also reported enzyme loading as a function of coupling time, another important parameter, as shown in Figure 3.

Brillouet et al.[24] covalently bound amyloglucosidase to collagen films using a three-step process involving acidic methylation, hydrazine treatment, and azidation by nitrous acid. Although the authors do not make the point clear, apparently an excess of enzyme was available during the immobilization. They reported a tenfold increase in immobilized enzyme activity when the concentration of the enzyme solution contacted with the membrane during the coupling procedure was increased from 0.5 to 27 mg/mℓ.

Another interesting sidelight in the area of chemical coupling is the article by Gray and Livingstone[25] which describes the immobilization of a number of enzymes including dextranase to various supports, including nylon, cellulose, celite, and glass, via the diazotized *m*-diaminobenzene method (formation of Bismarck brown). The unusual aspect of their work was the reuse of the carrier after the enzyme had been inactivated. Enzyme was coupled to this used carrier utilizing only half the original amount of reagent. This second preparation exhibited levels of activity comparable to the original preparations, with glass and cellulose achieving activities equal to the originals.

Similar to the case with adsorption, Maeda et al.[26] observed an effect of particle size on enzyme loading when coupling β-amylase to porous cellulose beads by the diisocyanate method. The smallest particles (65 to 100 mesh) bound four times as much enzyme as an equal weight of 20- to 28-mesh particles. The observed activity of the derivative was eight times greater for the smaller particles than for the larger ones. Diffusion limitations obviously caused the difference in enzyme efficiency between the

Table 4
ENZYME LOADING AS A FUNCTION OF COUPLING METHOD ON POROUS CELLULOSE BEADS[19,20]

Method of immobilization	Enzymes			
Covalent bonding	**Glucoamylase (*Asperigillis oryzae*)**	Glucoamylase (*A. niger*)	Glucose isomerase (*Streptomyces albus*)	Invertase (*Candida utilis*)
Cyanogen bromide	1800			
Diisocyanate	550	2000	90	1140
Diisocyanatediazo	275			
Diisocyanateglutaraldehyde	530			
Glutaraldehyde	190			
Silane-glutaraldehyde	200	200		
Adsorption				
DEAE-cellulose		9000	300	2000
Guanidino cellulose		1000	160	1840

Table 5
IMMOBILIZED GLUCOAMYLASE ACTIVITY AS A FUNCTION OF SUPPORT MATERIAL

Support composition	Support mesh size (U.S. standard sieve)	Average pore diameter (Å)	Pore volume (mℓ/g)	Initial activity (IU/g)
75% SiO_2, 25% TiO_2	30/45	465	0.76	1750
90% SiO_2, 10% ZrO_2	30/60	435	0.76	658
100% SiO_2	30/60	435	0.76	2400
84.3% SiO_2, 15.7% ZrO_2	30/60	235	1.30	423
75% SiO_2, 25% Al_2O_3	30/60	435	0.89	1729
98% TiO_2, 2% MgO	30/45	410	0.53	1607

From Weetall, H. H., Vann, W. P., Pitcher, W. H., Jr., Lee, D. D., Lee, Y. Y., and Tsao, G. T., *Methods in Enzymology,* Vol. 44, Academic Press, New York, 1976.

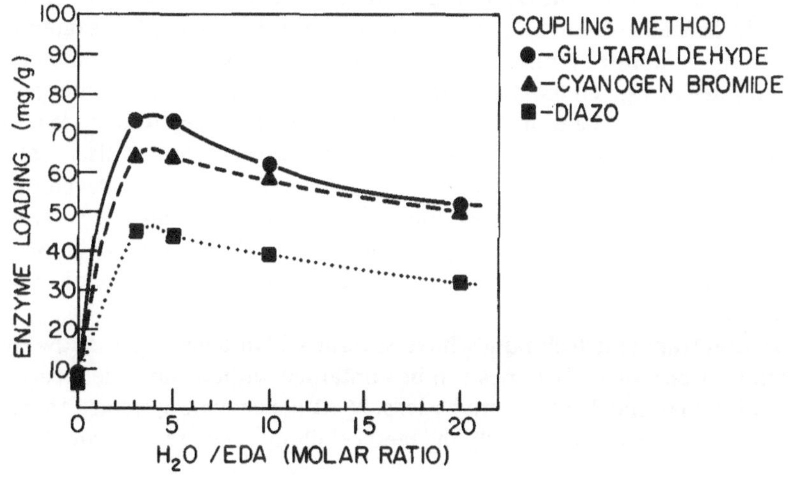

FIGURE 2. Effect of EDA concentration on enzyme loading.

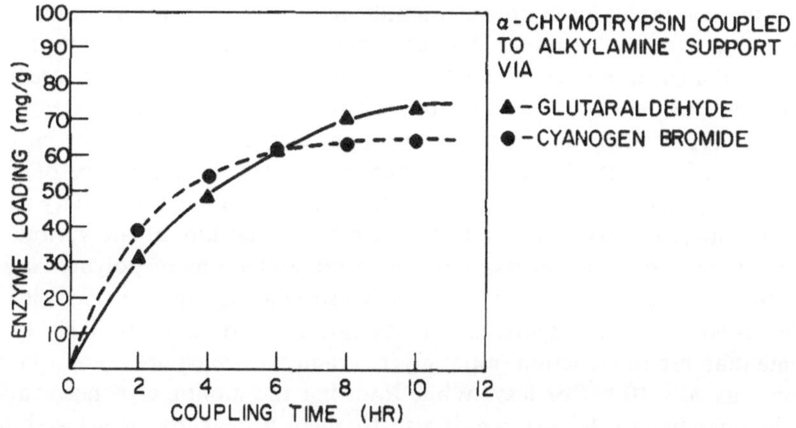

FIGURE 3. Enzyme loading as a function of coupling time.

two particle sizes. However, the difference in enzyme loading is even more dramatic and more difficult to explain. The loading difference by a factor of 4 would seem to be readily explained by the difference in particle diameters, and thus external surface area, by a factor of 4. If the loading was primarily near the surface, as would seem to be indicated, then the diffusion limitation should not have been so severe. Whatever the explanation, particle size will sometimes have an effect on enzyme loading as well as observed activity.

O'Neill et al.[27] reported superior stability for chymotrypsin bound to DEAE-cellulose particles as opposed the enzyme coupled to soluble dextran. Both immobilized enzyme derivatives exhibited a marked improvement in stability over soluble chymotrypsin.

IV. ENTRAPMENT

The term entrapment has been used to cover a range of immobilization techniques. O'Driscoll[28] has chosen to distinguish between those methods, where a gel or membrane is formed around an enzyme by polymerization and those methods where the enzyme is retained in a preformed polymer by subsequent formation of a cross-linked network. Microencapsulation and fiber entrapment are clearly other examples of this immobilization technique. Hollow fibers and membrane devices have also been used to retain enzymes in a given location without actually binding the enzyme to a support material. In some cases, the term entrapment has been extended to include copolymerization which is in essence a form of cross-linking. Entrapment has also seen frequent use in the area of whole cell immobilization. Increasing interest is developing in the immobilization of cells. In some cases, the activity of only one enzyme is retained with the whole cell being used to avoid costly purification. In other cases, a complex enzyme system or a completely functioning cell is desired to carry complex chemical or biological conversions.

In general, entrapment techniques have several advantages. They allow high local concentration of enzymes. Enzymes can be contained without any chemical modification or bonding that may lead to activity loss. Multiple enzyme systems can be handled readily since entrapment is essentially the method living cells use to retain their enzyme systems.

Unfortunately, entrapment has the reputation, justified to a certain extent, of resulting in enzyme loss by leakage. It should be noted, however, that sufficiently tight membranes and suitable polymers can lead to much better enzyme retention than observed for the commonly used polyacrylamide gels. Mass transfer limitations are also a problem with entrapped enzymes. This consideration in conjunction with the physical properties of the entrapped enzyme composites or devices places some constraints on reactor design. Furthermore, it is obvious that none of the stabilizing effects of bonding to a rigid support are available to the entrapped enzyme.

As with the other methods of enzyme immobilization, the conditions of the actual immobilization process are extremely important in determining the activity and stability of the entrapped enzyme. O'Driscoll[28] points out that any of the various components of the type of mixture necessary to gel-entrap an enzyme by polymerization have the potential to denature the enzyme. He suggests that the monomer, typically comprising 20 to 60% of the polymerization reaction mixture, is more likely to damage the enzyme than are the reaction initiating free radicals, which are present in low concentrations, usually $10^{-6} M$ or less. While lowering the monomer concentration may result in less damage to the enzyme, it also frequently weakens the gel and decreases the amount of protein entrapped. Thus a trade-off between gel strength and enzyme activity must sometimes be made. The presence of the monomer and other components of the polymerization mixture can decrease the solubility of the enzyme limiting the enzyme concentration and the final immobilized enzyme activity. While the optimum concentration of the polymerization mixture components varies from case to case, O'Driscoll[28] observed that typically the optimum cross-linking agent concentration is about 5%. Data also lead one to conclude that increasing the water content of a gel will increase the effective diffusivity of substrate and products, thus enhancing the apparent activity of the entrapped enzyme. Unfortunately, the higher water content generally also leads to a weaker gel.

While these generalizations have some value, it is perhaps more instructive to examine some of the different entrapment techniques in more detail. While by no means a comprehensive list or a complete description of the possible techniques, the following examples should give an idea of the diversity and some of the problems with various methods of entrapping enzymes.

Jaworek and co-workers[29] have used a method of immobilization by copolymerization, which is perhaps as much a technique of cross-linking as it is entrapment. Their method consists of vinylating the enzyme with acylating and alkylating monomers and then copolymerizing the vinylated enzyme with comonomers. Table 6 shows the effect of the different vinylating agents on the activity of the final immobilized enzyme which was copolymerized with acrylamide and *N,N'*-methylenebisacrylamide. From this table it can be seen that the preferred vinylating agent depends on the particular enzyme being immobilized. Copolymerization does appear to give better results than simple mechanical inclusion, at least from the standpoint of activity.

Kawashima and Umeda[30] achieved copolymerization by Co^{60} irradiation. The resulting immobilized enzyme had a spongy structure, although it could be prepared in various shapes such as beads, tubes, and membranes. Suzuki et al.[31] have also reported extensive work in the area of radiation polymerization. They found that increasing the radiation dosage resulted in a more rigid, lighter gel with less leakage. Use of electron beam irradiation allowed reduction in irradiation time from 90 hr to 5 min and resulted in higher activity immobilized enzyme composites. These were higher dosages than used by Kawashima and Umeda. Some enzymes such as urease were found to be sensitive to radiation and could not be immobilized by this method. Glucoamylase, invertase, and β-galactosidase were successfully immobilized.

Although enzymes can be copolymerized into a polymer gel structure to achieve additional stability, even enzymes immobilized by simple entrapment can be further stabilized. Change,[32] who pioneered enzyme microencapsulation, has reported stabilizing enzymes by encapsulation with 100 g/ℓ hemoglobin solution. Cross-linking with glutaraldehyde, which can be carried out after the enzyme is encapsulated was used to achieve additional stability. As Chang points out, enzymes can also be microencapsulated with various types of adsorbents such as ion exchange resins, activated charcoal, or porous ceramics. In earlier work,[33] Chang described microencapsulating magnetically active materials, such as iron filings, along with the enzymes. This allows magnetic recovery of the microcapsules from a reaction solution.

Formation of microcapsules can be accomplished in various ways. Nakamura and Mori[34] classified microencapsulation methods into three groups: interfacial polymerization, interfacial coacervation, or drying in liquid. Additional variations have been used. Miura et al.[36] described a secondary emulsion technique, involving water-in-oil-in-water emulsions with no chemical reaction.

As mentioned previously, when polymerization reactions are used, the enzyme being immobilized is susceptible to inactivation by the agents used for polymerization. Kobayashi et al.[35] described the use of chemical agents to protect the enzyme during the

Table 6
ACTIVITIES (UNITS/g) OF ENZYMES IMMOBILIZED BY COPOLYMERIZATION

Vinylating reagent	Glucose oxidase	Catalase	Acylase	Chymotrypsin	Hexokinase
Acrylic acid chloride	300	8500	0	0.12	100
1-Allyloxy-3-(N-ethylenimine-2-propanol	300	7000	160	3.3	48
Acrylic acid-2,3-epoxy-propyl ester	240	5900	70	0.03	80
Mechanical inclusion	90	5200	50	—	40

polymerization reaction. In tests involving encapsulating β-galactosidase from three sources (*Aspergillis oryzae, Klebsiellba lactis,* and *Escherichia coli*), dithiothreitol proved beneficial for all three enzymes while other protective agents including glucono-δ-lactone, galacton-γ-lactone, bovine serum albumin, and glutathione worked for some enzymes but not for others.

There are other variables to be considered in encapsulating enzymes. In general, the microcapsule size, which can effect apparent activity because of diffusional resistance, can be controlled by stirrer speed. Nakamura and Mori,[34] for example, show a correlation between microcapsule diameter and stirrer speed. Interfacial surface tension obviously affects droplet size. Kobayashi et al.[35] gave an example of SPAN® 85 percentage on particle size. Miura et al.[36] found that the interfacial surface tension also affected the yield of microcapsules.

Mass transfer effects in microcapsules have been modeled. Wadiak and Carbonell,[37] using *E. coli* β-D-galactosidase immobilized in cellulose nitrate membrane microcapsules as a model, examined the reaction kinetics taking mass transfer resistance into consideration. They used both continuous stirred tank and packed bed reactors in determining effectiveness factors (ratio of apparent to actual activity).

Dinelli[38] has described the procedure used by Snamprogetti, an Italian firm, to spin fibers containing enzyme. Using this procedure, Grazi et al.[39] entrapped sucrose phosphorylase in cellulose triacetate fibers. This enzyme in its soluble form has a sharp pH optimum at a pH of about 7.5. However, when entrapped in the fibers, the sucrose phosphorylase exhibited little variation in activity between pH 6.0 and 8.0. The authors suggest that this phenomenon is due at least partially to enzyme-polymer interaction since the magnitude of the effect is larger than expected for pH optimum broadening caused by diffusion limitations.

More common is the use of hollow fibers to contain enzymes. The enzymes are usually contained within the hollow center or lumen, but Romicon[40] has experimented with entrapping enzymes in the porous wall of the hollow fiber. A membrane sandwich reactor[41] in which enzymes are contained between semipermeable membranes is another variation of membrane containment. The idea of recovering enzymes from a batch reactor by ultrafiltration has also been advanced.

Various forms of cross-linking have also been used to immobilize enzymes, one of the most notable, from a commercial standpoint, being the cross-linked cell homogenates containing glucose isomerase made by Novo Industri.[42] The cell homogenate is spray dried, then moistened and extruded to form pellets. These pellets are then treated with glutaraldehyde followed by washing and drying.

Whole cells as well as enzymes can be immobilized by entrapment. Chibata[43,44] has reported a number of examples including *E. coli* cells for producing L-aspartic acid from fumaric acid and ammonia and for production of 6-aminopenicillanic acid from penicillin, *Pseudomonas putida* to produce L-citrulline, and *Achromobacter liquidum* to produce urocanic acid from L-histidine. Vieth and Vankatasubramanian[45] reported entrapping cells with glucose isomerase activity in collagen membranes. Kierstan and Bucke[46] used calcium alginate to immobilize cells, organelles (chloroplasts and mitochondria), and enzymes with all exhibiting activity.

REFERENCES

1. **Zaborsky, O. R.**, *Immobilized Enzymes*, CRC Press, Cleveland, 1973.
2. **Messing, R. A.**, Ed., *Immobilized Enzymes for Industrial Reactors*, Academic Press, New York, 1975.
3. **Nelson, J. M. and Griffin, E. G.**, Adsorption of invertase, *J. Am. Chem. Soc.*, 38, 1109, 1916.
4. **Schnyder, B. J. and Logan, R. M.**, Commercial application of immobilized glucose isomerase, Paper No. 9c, in 77th Natl. Meeting, AIChE, Pittsburgh, 1974.
5. **Chibata, I., Tosa, T., Sato, T., Mori, T., and Matsuo, Y.**, Preparation and industrial application of immobilized aminoacylases, in *Fermentation Technology Today, Proc. IV IFS Fermentation Technology Today*, Society of Fermentation Technology, Osaka, Japan, 1972.
6. **Pitcher, W. H., Jr. and Ford, J. R.**, Development of an adsorbed lactase immobilized enzyme system, in *Enzyme Engineering*, Vol. 3, Pye, E. K. and Weetall, H. H., Eds., Plenum Press, New York, 1978.
7. **Stanley, W. L. and Palter, R.**, Lactase immobilization on Phenolic Resins, *Biotechnol. Bioeng.*, 15, 597, 1973.
8. **Caldwell, K. D., Axen, R., Bergwall, M., and Porath, J.**, Immobilization of enzymes based on hydrophobic interaction. II. Preparation and properties of an amyloglucosidase adsorbate, *Biotechnol. Bioeng.*, 18, 1589, 1976.
9. **Eaton, D. L. and Messing, R. A.**, Support of Alumina-Magnesia for the Adsorption of Glucose Isomerase Enzymes, U.S. Patent 3,992,329, 1976.
10. **Kennedy, J. F. and Kay, I. M.**, The use of titanium (IV) oxide for the immobilization of carbohydrate-directed enzymes, *Carbohydr. Res.*, 56, 211, 1977.
11. **Boudrant, J. and Cheftel, C.**, Continuous hydrolysis of sucrose by invertase adsorbed in a tubular reactor, *Biotechnol. Bioeng.*, 17, 827, 1975.
12. **Baratti, J., Coudrerc, R., Cooney, C. L., and Wang, D. I. C.**, Preparation and properties of immobilized methanol oxidase, *Biotechnol. Bioeng.*, 20, 333, 1978.
13. **Cayle, T.**, Acid-Active Lactase, U.S. Patent 3,629,073, 1971.
14. **Gregory, J. L. and Pitcher, W. H., Jr.**, Enzyme Carrier Regeneration, U.S. Patent 4,002,576, 1977.
15. **Gregory, J. L. and Pitcher, W. H., Jr.**, Immobilization of Enzymes Using Recycled Support Materials, U.S. Patent 4,087,330, 1978.
16. **Fujita, Y., Matsumota, A., Miyachi, I., Imai, N., Kawakami, I., Itishida, T., and Kumata, A.**, Verfahren zur erneurern unlaslich gemachten Glucoseisomerase, *German Offenbach*, 2720538, Dec. 29, 1977.
17. **Epton, R., McLaren, J. V., and Thomas, T. H.**, Water-insoluble polyacrylamide-thermolysin conjugates, *Biochim. Biophys. Acta*, 328, 418, 1973.
18. **Weetall, H. H. and Havewala, N. B.**, The continuous production of dextrose from cornstarch. A study of reactor parameters necessary for commercial application, in *Enzyme Engineering*, Wingard, L. B., Jr., Ed., Interscience, New York, 1972.
19. **Chen, L. F. and Tsao, G. T.**, Chemical procedures for enzyme immobilization on porous cellulose beads, *Biotechnol. Bioeng.*, 19, 1463, 1977.
20. **Chen, L. F. and Tsao, G. T.**, Physical characteristics of porous cellulose beads as supporting material for immobilized enzymes, *Biotechnol. Bioeng.*, 18, 1507, 1976.
21. **Weetall, H. H., Vann, W. P., Pitcher, W. H., Jr., Lee, D. D., Lee, Y. Y., and Tsao, G. T.**, Scale-up studies on immobilized, purified, glucoamylase, covalently coupled to porous ceramic support, in *Methods in Enzymology*, Vol. 44, Mosbach, K., Ed., Academic Press, New York, 1976.

22. **Epton, R., Hibbert, B. L., and Thomas, T. H.**, Enzymes covalently bound to polyacrylic and polymethacrylic copolymers, in *Methods in Enzymology*, Vol. 44, Mosbach, K., Ed., Academic Press, New York, 1976.
23. **Lai, T.-S. and Cheng, P. S.**, A new type of enzyme immobilization support derived from styrene-maleic anhydride copolymer and the α-chymotrypsin with which it immobilizes, *Biotechnol. Bioeng.*, 20, 773, 1978.
24. **Brillouet, J.-M., Coulet, P. R., and Gautheron, D. C.**, Chemically activated collagen for amyloglucosidase attachment. Use in a helicoidal reactor, *Biotechnol. Bioeng.*, 19, 125, 1977.
25. **Gray, C. J. and Livingstone, C. M.**, Properties of enzymes immobilized by the diazotized m-diaminobenzene method, *Biotechnol. Bioeng.*, 19, 349, 1977.
26. **Maeda, H., Tsao, G. T., and Chen, L. F.**, Preparation of immobilized soybean β-amylase on porous cellulose beads and continuous maltose production, *Biotechnol. Bioeng.*, 20, 383, 1978.
27. **O'Neill, S. P., Wykes, J. R., Dunnill, P., and Lilly, M. D.**, An ultrafiltration-reactor system using a soluble immobilized enzyme, *Biotechnol. Bioeng.*, 13, 319, 1971.
28. **O'Driscoll, K. F.**, Techniques of enzyme entrapment in gels, in *Methods in Enzymology*, Vol. 44, Mosbach, K., Ed., Academic Press, New York. 1976.
29. **Jaworek, D., Botsch, H., and Maier, J.**, Preparation and properties of enzymes immobilized by copolymerization, in *Methods in Enzymology*, Vol. 44, Mosbach, K., Ed., Academic Press, New York, 1976.
30. **Kawashima, K. and Umeda, K.**, Immobilization of enzymes by radiation copolymerication of synthetic monomers, in *Immobilized Enzyme Technology — Research and Application*, Weetall, H. H. and Suzuki, S., Eds., Plenum Press, New York, 1975.
31. **Suzuki, H., Maeda, H., and Yamauchi, A.**, Immobilization of enzymes by radiation polymerization, in *Immobilized Enzyme Technology — Research and Applications*, Weetall, H. H. and Suzuki, S., Eds., Plenum Press, New York, 1975
32. **Chang, T. M. S.**, Microencapsulation of enzymes and biologicals, in *Methods in Enzymology*, Vol. 44, Mosbach, K., Ed., Academic Press, New York, 1976.
33. **Chang, T. M. S.**, Semipermeable aqueous microcapsules ("artificial cells"): with emphasis on experiments in an extracorporeal shunt system, *Trans. Am. Soc. Artif. Intern. Organs*, 12, 13, 1966.
34. **Nakamura, K. and Mori, Y.**, Mass transfer and reaction with multicomponent microcapsules, in *Immobilized Enzyme Technology — Research and Applications*, Weetall, H. H. and Suzuki, S., Eds., Plenum Press, New York, 1975.
35. **Kobayashi, T., Ohmiya, K., and Shimizu, S.**, Immobilization of β-galactosidase by polyacrylamide gel, in *Immobilized Enzyme Technology — Research and Applications*, Weetall, H. H. and Suzuki, S., Eds., Plenum Press, New York, 1975.
36. **Miura, Y., Miyamoto, K., Fujii, T., Takamatsu, N., and Mitsuo, O.**, Activity of enzyme immobilized by microencapsulation, in *Immobilized Enzyme Technology — Research and Applications*, Weetall, H. H. and Suzuki, S., Eds., Plenum Press, New York, 1975.
37. **Wadiak, D. T. and Carbonell, R. G.**, Kinetic behavior of microencapsulated β-galactosidase, *Biotechnol. Bioeng.*, 17, 1157, 1975.
38. **Dinelli, D.**, Fibre-entrapped enzymes, *Process Biochem.*, 9 August 1972.
39. **Grazi, E., Trombetta, G., and Morisi, F.**, Properties of free and immobilized sucrose phosphorylase, *J. Mol. Catal.*, 2, 453, 1977.
40. **Breslau, B. R. and Kilcullen, B. M.**, Hollow fiber enzymatic reactors: an engineering approach, in *Enzyme Engineering*, Vol. 3, Pye, E. K. and Weetall, H. H., Eds., Plenum Press, New York, 1978.
41. **Vorsilak, P. and McCoy, B. J.**, Enzyme immobilized in a membrane sandwich reactor, *J. Food Sci.*, 40, 431, 1975.
42. **Poulsen, P. B. and Zittan, L.**, Continuous production of high-fructose syrup by cross-linked cell homogenates containing glucose isomerase, in *Methods in Enzymology*, Vol. 44, Mosbach, K., Ed., Academic Press, New York, 1976.
43. **Chibata, I., Tosa, T., Sato, T., Mori, T., and Yamamoto, K.**, Applications of immobilized enzymes and immobilized microbial cells for L-amino acid production, in *Immobilized Enzyme Technology — Research Applications*, Weetall, H. H. and Suzuki, S., Eds., Plenum Press, New York, 1975.
44. **Chibata, I., Tosa, T., and Sato, T.**, Production of L-aspartic acid by microbial cells entrapped in polyacrylamide gels, in *Methods in Enzymology*, Vol. 44, Mosbach, K., Ed., Academic Press, New York, 1976.
45. **Vieth, W. R. and Venkatasubramanian, K.**, Process engineering of glucose isomerization by collagen-immobilized whole microbial cells, in *Methods in Enzymology*, Vol. 44, Mosbach, K., Ed., Academic Press, New York, 1976.
46. **Kierstan, M. and Bucke, C.**, The immobilization of microbial cells, subcellular organelles, and enzymes in calcium alginate gels, *Biotechnol. Bioeng.*, 19, 387, 1977.

Chapter 2

IMMOBILIZED ENZYMES FOR FOOD PROCESSING

Wayne H. Pitcher, Jr.

TABLE OF CONTENTS

I. Introduction 16

II. Reactor Types 16

III. Reaction Kinetics and Reactor Behavior 20

IV. Mass Transfer 25
A. External Mass Transfer 25
B. Internal Mass Transfer 28
C. Dispersion and Backmixing 34
D. Electrostatic Effects 35
E. Transient Behavior of Immobilized Enzyme Reactors 36

V. Heat Transfer and Other Temperature Effects 37

VI. Immobilized Enzyme Activity Loss 37

VII. Pressure Drop 42

VIII. Reactor Operating Strategy 43

IX. Multienzyme Systems 45

X. Cost Estimates 47

XI. General System Design Considerations 48

References 49

I. INTRODUCTION

Although immobilized enzyme engineering is an essential part of developing a commercial immobilized enzyme system for use in food processing or other applications, much less attention has been paid to the engineering aspects than to the details of enzyme immobilization. There is, perhaps, some justification for this situation in that much of the engineering applicable to other heterogeneous catalysis systems is also appropriate for immobilized enzyme systems. Nevertheless, much of this information is not readily available to many of the people involved in developing or evaluating immobilized enzyme systems. One of the author's aims in this chapter is to consolidate sufficient engineering information applicable to immobilized enzyme systems to allow the scientist or engineer to appreciate the role of engineering in developing commercial processes using immobilized enzymes. While detailed design of large-scale equipment is beyond the scope of this book and of little interest in the early stages of process development, an understanding of general principles and the factors that may determine commercial feasibility are vitally important.

Since generalizations are risky, with each immobilized enzyme system requiring individual analysis, heavy reliance has been placed on practical examples to illustrate the various aspects of reactor design and operation. Specific topics including reactor types, reactor performance, reaction kinetics, mass transfer, enzyme activity loss, and operating strategy are treated individually. These subjects are then put into perspective by the concluding sections on cost estimates and general design considerations. Reviews of this subject that may be of interest have been written by this author[1,2,3] and Vieth et al.[4]

II. REACTOR TYPES

Immobilized enzyme reactors, like other reactors for heterogeneous catalysis, can be separated into several different categories including batch reactors, continuous stirred tank reactors, fluidized-bed reactors, and fixed-bed reactors. The reactor types can then, in turn, be combined or additionally modified. For example, the behavior of a fixed-bed reactor can be made to approach that of a continuous stirred tank reactor by the simple addition of a recycle loop. A large body of applicable information already exists in the heterogeneous catalysis literature and can readily be applied to many immobilized enzyme systems. In the chemical and petroleum industries, similar reactor design problems have been analyzed for many years.

Probably the simplest reactor is the batch reactor. However, unless the immobilized enzyme can easily be recovered for reuse after the reaction has run its course, this type of reactor fails to take advantage of one of the fundamental assets of immobilized enzymes, long life or enzyme stability. In fact, as will be noted later, frequently immobilization will not increase enzyme stability. In such cases the only justification for immobilization is to aid in enzyme recovery from batch reactors or in enzyme retention in continuous reactor systems.

Especially in the case of the batch reactor, there are other approaches not fitting the conventional conception of immobilized enzymes. Since the important thing is to recover the enzyme for reuse, in some applications the free enzyme could be used and recovered by ultrafiltration. Butterworth et al.[5] demonstrated this concept in laboratory-scale continuous reactor experiments using α-amylase and glucoamylase in separate series of experiments to hydrolyze starch and thinned starch, respectively. The enzymes, starch, and high molecular weight dextrins were retained by the ultrafiltration membrane, while the various saccharide products passed through the membranes and out of the system.

Where relatively high molecular weight reaction products occur, such a separation of enzyme from product might be more difficult or even impossible. O'Neill et al.[6] overcame this problem by attaching the enzyme to a soluble high molecular weight polymer. Thus a membrane with a higher molecular weight cut-off could be used to retain the enzyme while letting the reaction products leave the reactor. To demonstrate this point, O'Neill et al.[6] immobilized chymotrypsin on a soluble dextran (molecular weight 2×10^6) and used it in a small ultrafiltration cell operated continuously to hydrolyze casein.

Coughlin et al.[7] used a slightly different approach. They also prepared a soluble derivative, in their case by coupling NAD to alginic acid using 1,2,7,8-diepoxyoctane. This derivative could be recovered by precipitation after the completion of the reaction by lowering the pH below 3. Raising the pH would again make the derivative soluble for the next batch.

More conventional particulate immobilized enzyme composites can, of course, be used and recovered by filtration or centrifugation. Robinson et al.[8] attached enzymes to magnetizable particles which could then be recovered magnetically using technology already developed for other industrial applications. With particulate immobilized enzyme composites, one disadvantage is that the reactor must be stirred if any significant contact of reactants with the enzyme is to be achieved. This agitation generally leads to attrition of the immobilized enzyme carrier and may make enzyme losses high or subsequent recovery difficult. One means of avoiding some of the abrasion and attrition problem is to enclose the immobilized enzyme particles in mesh containers (panels) attached to a stirrer.[9] Another option is to recycle the reaction medium repeatedly through a fixed-bed reactor to give batch reaction conditions. Snamprogetti,[10-13] an Italian company, has chosen this operating mode to exploit its unique process of entrapping enzymes within the pores of wet spun synthetic fibers (such as cellulose triacetate or γ-methyl-polyglutamate). For one application, the enzymatic hydrolysis of penicillin G for the production of 6 APA, bundles of cellulose acetate fibers containing the penicillin amidase were stretched the length of a column reactor. The buffered penicillin G solution was then recirculated through the column, with the pH maintained at 8.2 by the automatic addition of sodium hydroxide solution. This procedure illustrates another advantage of batch reactors, namely, that the conditions in the reactor can be modified as the reaction proceeds. Acids or bases can be added to maintain or vary pH. Gases such as oxygen can be added. The reaction mixture temperature can be changed or held constant by addition or removal of heat. While these operations can also readily be accomplished in continuous stirred tank reactors, it is difficult to do so in fixed-bed or fluidized-bed reactors.

As shown by the studies of Butterworth et al.[5] and O'Neill et al.,[6] many of the techniques can also be used or were originally developed for use in continuous stirred tank reactors. Closset et al.[14,15] analyzed and performed experimental studies with tubular membrane reactors for the hydrolysis of starch by β-amylase. The membrane retained the enzyme and starch substrate, but was permeable to the maltose product.

The fixed-bed reactor has been the most widely studied of the major reactor types. This attention is appropriate since this type of reactor is used in most, if not all, large-scale commercial operations. The earliest examples were Tanabe Seiyaku Company's use of aminoacylase bound to DEAE-sephadex in a process for resolving D,L-amino acids and Clinton Corn Processing Company's use of immobilized glucose isomerase for the conversion of glucose to fructose. More recently developed systems such as Novo's immobilized glucose isomerase also make use of fixed-bed reactors. Due to its high efficiency, ease of operation, and general simplicity, this reactor type will almost certainly continue to dominate large-scale commercial applications. Although there are many forms of fixed-bed reactors, the most common consists of a packed bed of particulate material to which the enzymes have been immobilized. This configuration, the packed bed, has certain unique problems. Since the particulate support materials are usually porous to provide greater available surface area for enzyme immobilization, diffusional resistance can limit the effective activity of the immobilized enzyme if the particles of carrier material are too large. On the other hand, use of very small particles in a packed bed can result in high pressure drops and plugging problems.

A number of reactor variants have been developed to circumvent or at least alleviate this problem. Generally, these reactors have more open structures and higher void volumes than the conventional packed bed. Emery et al.[16,17] and Venkatasubramanian and Vieth[18] developed rolled membrane and backing systems with the membrane containing the enzyme and the backing providing support for the membrane. This design allows increased areas of membrane per unit volume of reactor without requiring the substrate to be forced through the membrane. Monoliths, developed as the support for automotive exhaust catalysts, have been used as enzyme supports. Benoit and Kohler[19] evaluated monoliths, manufactured by Corning Glass Works, as a support for catalase. Pressure drops were found to be one to two orders of magnitude less than for typical reactors containing beds of beads. These open structures can be a significant advantage in the case of high specific enzyme activity where the short diffusional distances will result in higher observed activity than for particulate derivatives of a practical size. For expensive enzymes, this increased effective activity may be essential for acceptable process economics. These open reactor designs are also less susceptible to plugging by particulate material in the feed.

Thus far in commercial practice, high effectiveness factors have been achieved with acceptable particle size, making packed beds the usual choice. Packed beds tend to be less expensive and to have higher loadings of enzyme per unit volume of reactor than the more open reactor designs. For regenerable inorganic supports, the cost of the particulate carrier may be quite minimal.

More extreme examples of open structure fixed-bed reactors, such as tubular reactors with enzyme immobilized on or in the walls, are not generally attractive for large-scale commercial application. Such reactors may be ideal for analytical applications where efficiency or cost are not important or in biomedical applications such as extracorporeal shunts where low pressure drop and turbulence are necessary. A variation using the tubular reactor is slug flow, the introduction of inert gas along with the feed to the reactor. Horvath et al.[20] found that the secondary flow patterns within the liquid slugs increased transport to the tube wall in the case of laminar flow in small diameter tubes.

Another type of reactor, the fluidized bed,[21] has received increasing attention. Among its advantages are low pressure drop, leading to low pumping costs, and the ability to handle certain types of fine particulate feeds. The liquid velocities necessary to properly suspend the enzyme carrier particles may result in residence times insufficient to achieve the required conversion. Possible solutions to this problem include use of recycle or a series of fluidized beds. Since recycling will cause the reactor system to approach continuous stirred tank reactor performance, it will frequently result in efficiency losses. Some reduction in fluid velocity can also be achieved by using smaller carrier particles if this is practical from a manufacturing standpoint. An even less likely approach is to raise the feed concentration and thus the feed viscosity. Unfortunately, this approach will normally require an even longer residence time to achieve the same percentage conversion of the higher concentration substrate. Charles et al.,[22] using immobilized lactase, and Lee et al.,[23] using immobilized glucose isomerase, reported achieving identical performance with fixed-bed and fluidized-bed reactors under certain conditions. Their experiments seemed to indicate that the extent of bed expansion can influence the performance of fluidized-bed reactors. Emery and his group at the University of Birmingham[24,25] have paid particular attention to the dispersion characteristics of fluidized-bed reactors. Tracer studies by Emery and Cardoso[25] indicated more loss of efficiency due to backmixing in fluidized-bed vs. fixed-bed reactors. They also reported that increased fluid velocity increased dispersion effects for the fluidized bed while decreasing them for the fixed-bed reactor as predicted by the correlation of Chung and Wen[26] (further discussion in the section on backmixing). However, they also reported much larger backmixing effects than predicted and consistently superior enzyme conversion performance for the fluidized-bed vs. the fixed-bed reactor. They offered two possible explanations for this unexpected result: preferential flow pattern or covering of surface area due to particle-particle contact in the fixed bed. Allen[27] has indicated that the inlet flow distribution in a fluidized-bed reactor can have a profound influence on its flow patterns and performance.

Scott et al.[28] reported using a conically tapered (large end up) fluidized-bed reactor which was more stable than the normal vertical wall reactor. The likelihood of washing out the carrier particles can also be decreased by merely enlarging the diameter of the top of the reactor. Gelf and Boudrant[29] solved the problem of retaining carrier in a fluidized-bed reactor in another fashion.

They used magnetizable particles as carriers which were prevented from leaving the reactor by a magnetic collar around the top of the reactor column. Evidently, extremely high flow rates could be achieved without loss of carrier.

The selection of a reactor design depends on enzyme kinetics, type of immobilization, economics, and, to a lesser degree, on other factors. The frequently used fixed-bed reactor has the advantage of high efficiency in terms of approaching plug flow performance, as well as the potential for the highest loadings of enzyme per unit volume of reactor. Batch reactors that are well mixed and continuous reactors that approach plug flow performance are more efficient than ideal continuous stirred-tank reactors (backmix reactors) or other nonideal reactors when reaction rates are higher than zero order in dependence on reactant concentration. Reactions catalyzed by enzymes frequently follow Michaelis-Menten kinetics, between zero and first-order in substrate concentration dependence. The ratio of ideal plug flow to backmix reactor volume required to achieve the same level of conversion is shown in Figure 1. This ratio depends on the value of S_o/K_m (S_o = initial substrate concentration and K_m = Michaelis constant) where low values of S_o/K_m cause the kinetics to approach first order and high values of S_o/K_m result in an approach to zero order kinetics. For zero order kinetics, ideal backmix and plug flow reactors give the same performance. There are cases such as substrate inhibition, where backmix reactors are superior, but these are not common in practice. Various types of product inhibition are more frequently encountered and tend to accentuate the advantage of the plug flow reactor.

Other factors, such as mass transfer limitations, enter into the question of efficiency. In some cases diffusion limitations can be overcome by the use of smaller immobilized enzyme carrier particles practical only in fluidized beds or stirred tanks. Fixed-bed configurations, other than the packed bed, can sometimes also be used to promote rapid mass transfer.

Stirred tank reactors, batch or continuous, and to a lesser extent, fluidized beds, subject the immobilized enzymes to harsher treatment than do fixed-bed reactors. While the dangers of enzyme deactivation by shear have probably been exaggerated,[30] the enzyme carrier material can be severely affected.

Where pH or temperature control are critical, stirred tank reactors can be used. Plug flow performance can even be approached, if necessary, by using a series of continuous

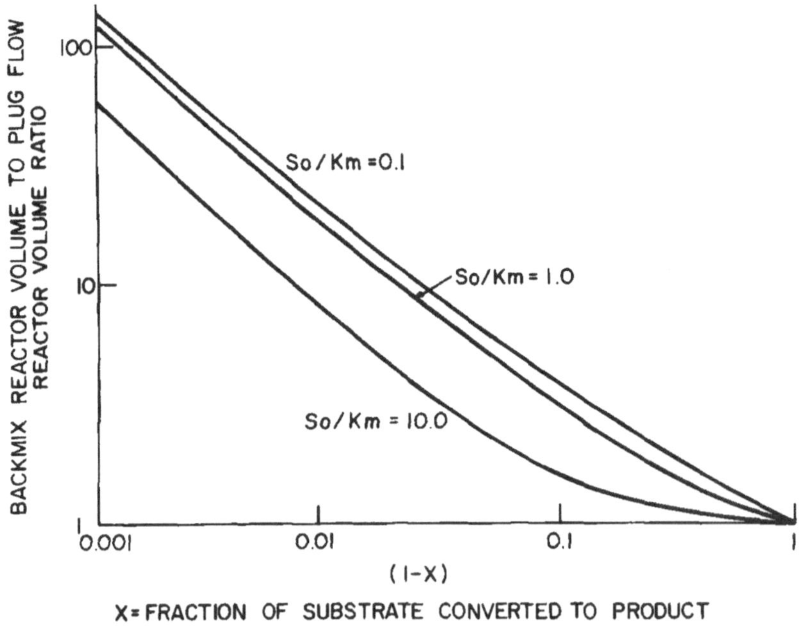

FIGURE 1. Plug flow vs. backmix reactor performance.

stirred tank reactors. Often temperature control is unnecessary since many enzyme catalyzed reactions have low heats of reaction. Operating stability and process control are perhaps easier, in general, to achieve with fixed-bed reactors. Energy costs should not normally prove a major factor in reactor choice. Continuous reactors, of course, are desirable for large volume processes to reduce labor costs.

Overall design considerations and economics will be considered at the end of this chapter.

III. REACTION KINETICS AND REACTOR BEHAVIOR

Knowledge of reaction kinetics can be used to predict the performance of certain ideal immobilized enzyme reactors. Three basic types of reactors are normally encountered: batch reactors, continuous stirred tank or backmix reactors, and plug flow reactors. The former two reactor types are assumed to be perfectly mixed at all times. The performance of real, as opposed to ideal, continuous reactors generally falls somewhere in between that of the backmix and ideal plug flow reactors. Packed-bed reactors do, at times, closely approach ideal behavior.

In the following treatment of kinetics, the simple irreversible Michaelis-Menten rate expression is used as an example, but the same mathematical operations are appropriate for other types of enzyme kinetics. For the irreversible Michaelis-Menten type kinetics, the reaction velocity can be written as

$$v = \frac{kES}{K_m + S} \tag{1}$$

where S is the substrate (reactant) concentration, K_m is the Michaelis constant, E is the amount of enzyme, k is the turnover number, and $v = -V_s(dS/dt)$ (V_s is substrate volume and t is time). Equation 1 can be integrated from time zero to time t to obtain the following expression for a batch reactor:

$$kEt/V_s = S_oX - K_m \ln(1 - X) \qquad (2)$$

where S_o is the initial substrate concentration and the fractional conversion X is equal to $(S_o-S)/S_o$. Similarly for an ideal plug flow reactor, where each volume element of fluid proceeds through the reactor behaving as an infinitesimal batch reactor, not mixing with the adjacent fluid elements, Equation 2 can be written as

$$kE/F = S_oX - K_m \ln (1 - X) \qquad (3)$$

where F is the volumetric flow rate.

For a backmix reactor, a material balance gives

$$v = F(S_o - S) \qquad (4)$$

where S_o is the feed substrate concentration and S is the concentration in the reactor and at the outlet. The right hand sides of Equations 1 and 4 can be set equal to each other to give

$$F(S-S_o) = \frac{kES}{K_m+S} \qquad (5)$$

After substitution of the fractional conversion X and rearrangement, the final equation becomes

$$\frac{kE}{F} = X\left(\frac{Km}{1-X} + S_o\right) \qquad (6)$$

Equivalent equations for reversible Michaelis-Menten, substrate inhibited, and product inhibited kinetics are presented in Table 1. Siimer[31] has taken a more general approach and gives equations for additional types of kinetic expressions.

Generally integrated rate equations or reactor operating data can be expressed graphically as conversion vs. residence time curves. In the case of reactions promoted by immobilized enzymes or other heterogeneous catalysts, the residence time should be normalized to reflect the amount of catalyst (enzyme) present. Normalized residence time can be expressed in a number of ways. For example in the case of a batch reactor, it can be written as

$$\frac{Wt}{V_s} \text{ or } \frac{E_t t}{V_s}$$

where W is the weight of the immobilized enzyme and E_t is the total enzyme activity. For a continuous reactor, normalized residence time can be expressed as

$$\frac{W}{F} \text{ or } \frac{E_t}{F}.$$

A comparison of continuous and batch reactor data for the same immobilized enzyme[32] is shown in Figure 2.

Table 1
INTEGRATED RATE EQUATIONS

Kinetic type and rate equation

Michaelis-Menten:

$$v = \frac{kES}{K_m + S}$$

$$\text{Plug flow: } \frac{kE}{F} = S_o X - K_m \ln(1 - X)$$

$$\text{Backmix: } \frac{kE}{F} = X\left\{\frac{K_m}{1 - X} + S_o\right\}$$

Reversible Michaelis-Menten:

$$v = \frac{kE\,(S - P/K)}{K_m + S - K_m P/K_p}$$

$$\text{Plug flow: } \frac{kE}{F} = X_e S_t \left\{(1 - K_m/K_p)(X_t - X_i) + \frac{K_m}{S_t} + 1 - X_e \frac{K_m K_e}{K_p} \ln \frac{X_e - X_i}{X_e - X_t}\right\}$$

$$\text{Backmix: } \frac{kE}{F} = \frac{(X_t - X_i)(K_m + S_t - X_t S_t + K_m S_t X_t/K_p)}{1 - X - X_t/K}$$

Substrate inhibition:

$$v = \frac{kE}{1 + K_m/S + S/K'_m}$$

$$\text{Plug flow: } \frac{kE}{F} = S_o X - K_m \ln(1 - X) + \frac{S_o^2 X}{K'_m} - \frac{S^2 X^2}{2\,K'_m}$$

$$\text{Backmix: } \frac{kE}{F} = XS_o \left\{1 + \frac{K_m}{S_o(1 - X)} + \frac{S_o(1 - X)}{K'_m}\right\}$$

Competitive product inhibition:

$$v = \frac{kES}{S + K_m(1 + P/K_i)}$$

$$\text{Plug flow: } \frac{kE}{F} = S_o(1 - K_m/K_i)(X_t - X_i) - \left(K_m + \frac{K_m S_t}{K_i}\right) \ln \left(\frac{1 - X_e}{1 - X_i}\right)$$

$$\text{Backmix: } \frac{kE}{F} = \frac{(X_t - X_i)\left\{S_t(1 - X_t) + \dfrac{K_m + K_m X_t S_t}{K_i}\right\}}{1 - X_t}$$

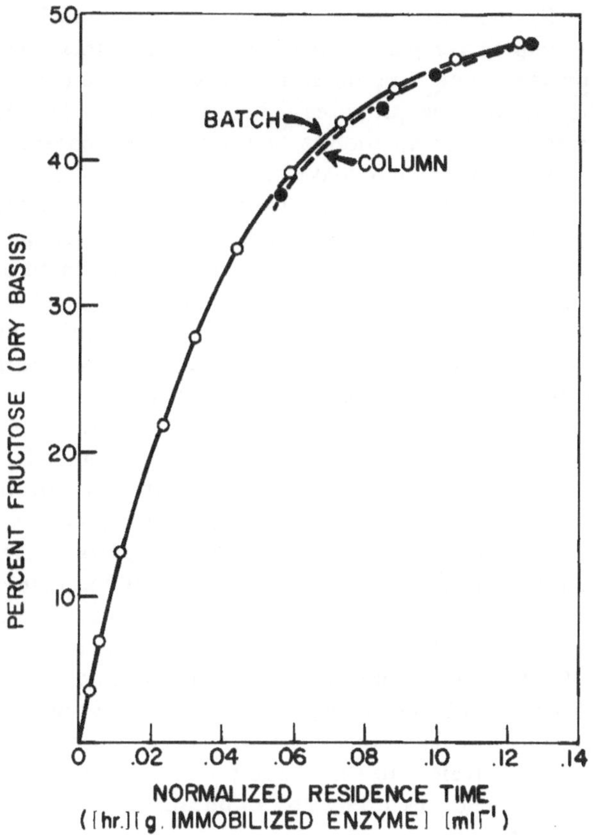

FIGURE 2. Conversion vs. normalized residence time (From Havewala, N. B. and Pitcher, W. H., Jr., in *Enzyme Engineering*, Vol. 2, Pye, E. K. and Wingard, L. B., Jr., Eds., Plenum Press, New York, 1974. With permission.)

For ideal plug flow and batch reactors with no external mass transfer limitations, the conversion vs. normalized residence time curves will coincide. This type of curve, or an equation describing the curve, can be determined from experimental data and then used to calculate enzyme activity from experimentally observed flow rate and conversion data. More explicitly, this type of curve can be used along with the observed conversion level to find the corresponding normalized residence time. From this value and the substrate flow rate (or elapsed reaction time and substrate volume in the case of a batch reaction), the enzymatic activity at the time of the observation can be calculated.

Although this pragmatic approach is often necessary for systems that are not ideal or cannot be described by simple kinetic expression, it can nevertheless be valuable to gain an understanding of the enzyme kinetics to allow use of the equations in Table 1. An understanding of the kinetics, not just at low conversions but at the levels envisioned for commercial application, can save time and effort when trying to improve or optimize a system. Since the subject of enzyme kinetics has been treated comprehensively by others,[33-36] only some recent developments and practical considerations are appropriate for inclusion here.

For many years, enzymologists have used linear plots of initial rate data to determine the constants in kinetics expressions such as Equation 1. Of the three types of these plots used, the most common is the Lineweaver-Burk plot, 1/V vs. 1/S. The Eadie or

Hanes plot, S/v vs. S, and the Hofstee plot v vs. v/S, see much less frequent use. Recently, various attempts have been made to improve upon these techniques that had been used for many years with little questioning. Among these attempts are several studies in which the authors have proposed using full time course kinetic data in conjunction with the integrated rate equation to determine kinetic constants.[37-40] Of course in industrial applications where high conversion levels are usually required, the plot of the reaction over its full time course, is probably even more valuable than the calculated constants. Nevertheless, having a reliable kinetic model can be useful when one is considering system changes. However, even predictions based on a good kinetic model should normally be verified experimentally before the operation of a large-scale plant is committed to change. Eisenthal and Cornish-Bowden[41,42] have objected on practical and statistical grounds to the use of the three traditional plots based on linear transformations of the basic rate expression. Of more practical value was their proposal of a simple new graphical procedure for estimating enzyme kinetic constants.[41] As shown by the simplified example in Figure 3, each reaction velocity point on the vertical axis is connected with its corresponding substrate concentration point on the horizontal axis. Each intersection point is an estimate of V_m (defined as kE) and K_m. The best estimates, V_m and K_m, are the medians from the appropriate sets of estimates. One important advantage of this method is the reduction of bias in the case of outliers.

More recently Cornish-Bowden and Eisenthal[43] have refined their estimation method to remove the bias encountered if any of the pairs of kinetic observations result in a negative value for V_m. They accomplish this by using $1/V_m$ and K_m/V_m as the primary parameters of the Michaelis-Menten equation. They point out that the same effect can be achieved by regarding any line intersection lying in the third quadrant (negative V and negative S) as the equivalent to very large positive values for both K_m and V_m. The authors also recommend discarding any duplicate observations on the grounds that these too can cause bias.

The use of integrated rate equations to determine kinetic constants has still not been widely accepted and has indeed encountered some criticism. Cornish-Bowden[42] was one to argue against this approach for various reasons including the claim that different kinetic models including product inhibition could not be handled. Yun and Suelter[44] developed a technique that eliminated this problem but still did not address another problem raised by Cornish-Bowden, that of enzyme inactivation during the assay. Where the enzyme is relatively stable or the course of the reaction brief, this procedure may be useful.

Several articles have indicated the increasing interest in statistically sound methods of kinetic analysis. Storer et al.[45] found that, contrary to past assumptions, variances increased with initial velocities of enzyme catalysis for three different experimental systems. Fjellstedt and Schlesselman[46] proposed a method for statistical analysis of initial velocity and inhibition data. Duggleby and Morrison[47] analyzed time course reaction data by applying nonlinear regression techniques. Somewhat later, evidently not completely satisfied with their earlier effort, the same authors[48] modified and extended their approach to more complicated systems. Using a technique based on the Gauss-Newton method for nonlinear regression, they examined various kinetic mechanisms as well as parameter standard errors for a two substrate, reversible reaction catalyzed by aspartate aminotransferase. They selected the applicable model by determining the one which best fit the data (lowest residual sums of squares). This approach appears worth pursuing in cases where complicated reactions of unknown kinetic mechanism are involved. In spite of the advantages of such techniques, it must be realized that reactants, products, and enzyme must be stable under time course assay conditions. A lesser disadvantage, pointed out by Duggleby and Morrison,[46] is that

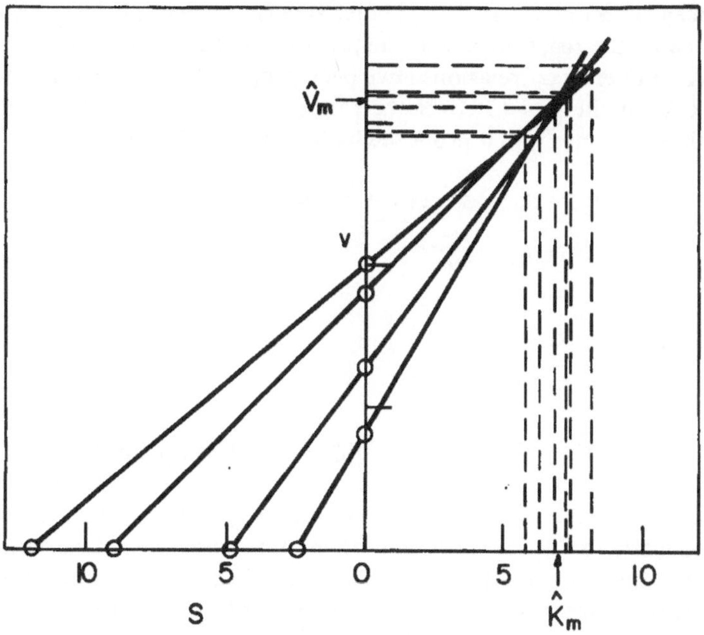

FIGURE 3. Kinetic constant estimation.

these approaches do not lend themselves readily to graphical representation of the data.

An important factor to remember in treating immobilized enzyme kinetic data by any of these methods is that mass transfer limitations may mask intrinsic kinetics and make the assumption of simple kinetic models meaningless. Unfortunately, it is usually essential to make kinetic measurements on the immobilized enzyme rather than the soluble or free enzyme. Upon immobilization, the enzyme kinetic mechanism normally remains the same, but the kinetic constants may change, although not as frequently as supposed. Many times mass transfer limitations are the cause of apparent changes in kinetic behavior. However, the conversion vs. normalized residence time plots should, with proper attention to scale-up implications, always give an accurate description of reaction behavior.

IV. MASS TRANSFER

A. External Mass Transfer

Although several methods can be used to estimate the effects of mass transfer to the immobilized enzyme carrier surface from the bulk liquid, the conditions encountered are not typical of those frequently encountered in other applications. Laminar liquid flow with a low Reynolds number falling outside the range of many mass transfer correlations is normal for immobilized enzyme systems.

The rate of mass transfer from the bulk solution to a surface can be written as

$$v = k_m a_m (S_b - S_s) \tag{7}$$

where k_m is the mass transfer coefficient (with units of cm/s), a_m is the surface area per unit volume (cm^{-1}), S_b is the bulk phase substrate concentration, and S_s is the substrate concentration at the surface. Under steady-state conditions, this mass transfer rate must equal the reaction rate at the particle surface or the apparent rate within the particle. A number of correlations have been proposed for estimating mass transfer coefficients. One of the few to cover the range of conditions normally encountered with immobilized enzymes is that proposed by Wilson and Geankoplis.[49]

For $0.0016 < N_{Re} < 55$ and $0.35 < \epsilon < 0.75$

$$J = (1.09/\epsilon)\, N_{Re}^{-2/3} \tag{8}$$

For $55 < N_{Re} < 1500$

$$J = (0.250/\epsilon)\, N_{Re}^{-0.31} \tag{9}$$

where $N_{Re} = dpG/\mu$, $J = (K_m\varrho/G)N_{Sc}^{2/3}$, $N_{Sc} = \mu/\varrho D$, is the liquid viscosity, G is the mass velocity per unit superficial bed cross-sectional area, ϱ is the liquid density, ε is the void fraction, D is the substrate diffusivity, and dp is the particle diameter. These correlations suggest, for the lower Reynold's number range usually encountered with immobilized enzyme systems, the mass transfer coefficient should be proportional to flow rate to the one third power. Rovito and Kittrell[50] found their data best fit a model with flow rate to the 0.5 power, while Raman et al.[51] observed reaction rates in tubular reactors proportional to the one third power of flow rate. Since these correlations were developed for packed beds, the question arises as to their relevance or usefulness in the case of fluidized beds.

Tournie et al.[52] found experimental agreement with the correlation developed by Vanadurongwan et al.[53] for liquid fluidized beds. These correlations, now shown valid in the 1 to 100 Reynold's number range, take the following form:

For $\epsilon < 0.815$

$$N_{Sh} = 0.215\, N_{Re}^{0.011}\, N_{Ga}^{0.309}\, N_M^{0.303}\, N_{Sc}^{0.436} \tag{10}$$

For $\epsilon > 0.815$

$$N_{Sh} = 0.115\, N_{Re}^{0.076}\, N_{Ga}^{0.390}\, N_M^{0.303}\, N_{Sc}^{0.436} \tag{11}$$

where $N_{Sh} = k_m d_p/D$, $N_{Ga} = dp^3\varrho^2/\mu^2$, $N_M = (\varrho_s - \varrho)/\varrho$, ϱ_s = particle density, and g is the gravitational constant. The mass transfer coefficients determined by Tournie et al.[52] were generally about 50 to 100% higher than predicted from Equations 8 and 9. The implications of these correlations are that for the same flow rates mass transfer from fluid to particle is better in a fluidized bed than in a packed bed and that the expansion of the fluidized bed has little effect on mass transfer rates. On the other hand, Emery and Cardoso[25] cited a number of references from which they concluded that the mass transfer coefficients for fixed beds should be higher for fixed beds than for fluidized beds. However, their experimental comparisons of immobilized glucoamylase conversion of starch in fixed and fluidized-bed reactors at Reynolds numbers of about 1 revealed better performance for the fluidized-bed reactors. Dispersion and other nonidealities that also affect reactor performance cannot be neglected and are discussed later in the subsection on backmixing.

Engasser[54] has presented the interesting concept of an external effectiveness factor, η_e, the ratio of observed reaction rate to the reaction rate without external mass transfer limitations. This external effectiveness factor can then be multiplied by the internal effectiveness factor to give an overall effectiveness factor. While Engasser gave a diagram of η_e values as a function of (v/Ak_mS_b) for various values of Sb/Km, the following modified presentation allows rapid direct calculation of η_e from $(v/AkmS_b)$ and S_b/K_m.

The rate of mass transfer per unit reactor volume to a surface can be written as

$$v = k_m a_m (S_b - S_s) \tag{12}$$

where S_b and S_s are substrate concentrations in the bulk phase and at the immobilized enzyme composite external surface, respectively. This mass transfer rate can be equated with the reaction rate at the particle surface or with the apparent rate within the particle. For the steady-state case where the reaction kinetics are simple Michaelis-Menten, the following expression holds:

$$v = k_m a_m (S_b - S_s) = \frac{kES_s}{K_m + S_s} \tag{13}$$

where E is enzyme concentration.

Rearranging this equation yields

$$\frac{S_s}{S_b} = 1 - \frac{v}{a_m k_m S_b} \tag{14}$$

The external effectiveness factor can be written as the ratio of the Michaelis-Menten rate expressions assuming S_s and S_b as the appropriate substrate concentration.

$$n_e = \frac{\dfrac{kES_s}{K_m + S_s}}{\dfrac{kES_b}{K_m + S_b}} \tag{15}$$

This equation can be rearranged into the following form:

$$n_e = \left(\frac{S_s}{S_b}\right)\left(\frac{1 + S_b/K_m}{1 + \left(\dfrac{S_s}{S_b}\right) S_b/K_m}\right) \tag{16}$$

The ratio S_s/S_b can be calculated readily from Equation 14 and then used in Equation 16 to calculate η_e.

Satterfield[55,56] suggested approaching the problem in another direction by calculating the height of bed required for the necessary mass transfer to occur assuming film diffusion to be the rate controlling step. This method is convenient since it requires only one reactor performance datapoint (flow rate and conversion) and applies to an integral reactor (effectiveness factors are usually only calculated for a given point in a reactor). The bed height, z, can be estimated from a correlation based, in part, on Equation 8.

$$z = \frac{\epsilon N_{Re}^{2/3} N_{Sc}^{2/3}}{1.09\, a_m} \ln \frac{Y_1}{Y_2} \tag{17}$$

where Y_1 is the mole fraction substrate in the feed and Y_2 is the mole fraction substrate in the product. Havewala and Pitcher[32] performed such a calculation for an immobilized glucose isomerase fixed-bed reactor of 12.8 cm bed height operated at 100 mℓ/hr feed rate (50% glucose solution) yielding 45% conversion. The bed height assuming film diffusion control, as calculated from Equation 17, was 0.23 cm. Since this height is much less than the actual bed height, it can be assumed that external diffusion resistance is not significant. Satterfield's method is particularly helpful in cases such as this where concerns about film diffusion can be dismissed with a minimum of effort.

Another practical method of experimentally testing for external mass transfer effects in fixed-bed reactors is to change the linear flow velocity while maintaining the same residence time in the reactor. This result can be accomplished by changing the bed depth, usually by a factor of 2 or 3 or more. If film diffusion effects are significant, the change in linear velocity will cause a change in conversion by the reactor even though the residence time is the same. Care must be exercised to examine a sufficient range in linear velocity to allow observation of any effect if present. Caution must also be taken to operate at realistic conversion levels. If conversion levels are too high, no effect may be observed even if present because conversion varies only slightly with activity. On the other hand, if conversion levels are too low, flow rates may be unrealistically high. Fortunately in scaling up, the linear velocities usually become sufficient to eliminate any external mass transfer limitations.

Satterfield[55] discussed the problems of mass transfer in slurry reactors, citing the correlation from Brian and Hales.[57]

$$\left(\frac{k_m d_p}{D}\right)^2 = 4.0 + 1.21\, N_{Pe}^{2/3}, \tag{18}$$

where $N_{Pe} = d_p u/D$ and μ is the fluid velocity. He also gave the correlation

$$k'_m\,(N_{Sc})^{2/3} = 0.38\left(\frac{g\mu\Delta\rho}{\rho^2}\right)^{1/3} \tag{19}$$

where k'_m is the value of k_m for a sphere settling at its terminal velocity, and $\Delta\varrho$ is the difference between particle and fluid densities. In stirred-tank reactors, particle entrainment in the moving fluid can limit the mass transfer rate even if agitation is increased. O'Neill[58] also discussed mass transfer problems in stirred batch reactors.

Limited information also exists concerning mass transfer effects in other types of reactors. Horvath and Solomon[58] reported experimental data and developed a theoretical treatment of reactions controlled by bulk diffusion for reactors consisting of tubes with enzymes immobilized on their inner walls. Bunting and Laidler[60] and Kobayashi and Laidler[61] described similar work. Experimental checks were run later by the same group.[62,63]

Unusual and unexplained results which appear to be mass transfer related appear occasionally. Nonideal flow patterns and backmixing are sometimes a problem and are considered in the subsection on backmixing.

B. Internal Mass Transfer

Since enzymes are frequently immobilized within porous bodies or retained by semipermeable membranes, the topic of internal mass transfer or pore diffusion is a relevant one. Even in cases where external mass transfer effects are not significant, pore diffusion may be important. Simultaneous reaction and diffusion is a situation commonly encountered and, as might be expected, has resulted in an extensive body of literature. Satterfield's[55] book is probably the most comprehensive treatment and review of this subject. Even the occurrence of literature concerning pore diffusion specifically in immobilized enzyme systems is frequent[64-75] and somewhat repetitive. Coverage of the subject here is intended to give an understanding of the basic approaches involved and some typical examples with emphasis on practical ways, experimental and theoretical, of evaluating internal diffusion effects.

When a reaction is promoted by an enzyme immobilized within a porous material, a substrate concentration gradient is established with the concentration decreasing as the distance from the surface into the porous body increases.

The differential equation describing diffusion in a sphere is

$$\frac{d^2S}{dr^2} + \frac{2}{r}\frac{dS}{dr} = \frac{v_i}{D_{eff}} \tag{20}$$

where r is the radial distance in the sphere, v_i is the intrinsic reaction rate (usually a function of substrate concentration), and D_{eff} is the effective diffusivity of the substrate. An analytical solution to this equation can only be found when $v_i = k_v S^m$, where m and k_v are constants. Other more complicated cases such as Michaelis-Menten kinetics must be solved numerically.

The ratio of observed or apparent reaction rate to the rate if no diffusion limitation existed, called the effectiveness factor, η, can be calculated from the solution to Equation 20. For immobilized enzyme systems, where the amount of heat generated is usually small and rapidly dispersed, the effectiveness factor will normally be unity or less.

A number of generalized plots have been developed relating the effectiveness factor to various kinds of general moduli. These moduli are of two basic types, those that depend on knowledge of the values of the kinetic constants and those that utilize observed reaction rate information.

Bischoff[76] expressed η graphically as a function of a general modulus M for various levels of S/K_m. The modulus M can be written in terms of Michaelis-Menten kinetic constants assuming D_{eff} does not vary with concentration. For flat plate geometry, this modulus is

$$M = L\left(\frac{V_m}{2\,K_m\,D_{eff}}\right)^{1/2}\left(\frac{S}{K_m + S}\right)\left(\frac{S}{K_m} - \ln\,(1 + S/K_m)\right)^{-1/2} \tag{21}$$

where for M greater than 2, $\eta = 1/M$. The effectiveness factor plot for several levels of S/K_m is shown in Figure 4.

Similar plots[1,2,55] utilizing a moduli based on observed reaction rates are shown in Figures 5 and 6. This modulus,

$$L = \frac{L^2}{D_{eff}}\left(-\frac{1}{V_c}\frac{dn}{dt}\right)\frac{1}{S} = \frac{R^2}{9D_{eff}}\left(-\frac{1}{V_c}\frac{dn}{dt}\right)\frac{1}{S} \tag{22}$$

applicable to flat plate (L = plate thickness) and spherical geometry, (R = sphere radius) assumes D_{eff} to be constant. The effective diffusivity, D_{eff}, is equal to $D\theta/\tau$ where θ = porosity (internal void fraction), τ = tortuosity (ratio of actual diffusion path to straight line distance), and D = bulk diffusivity. This type of modulus is probably more practical than moduli depending upon intrinsic kinetic constants, since

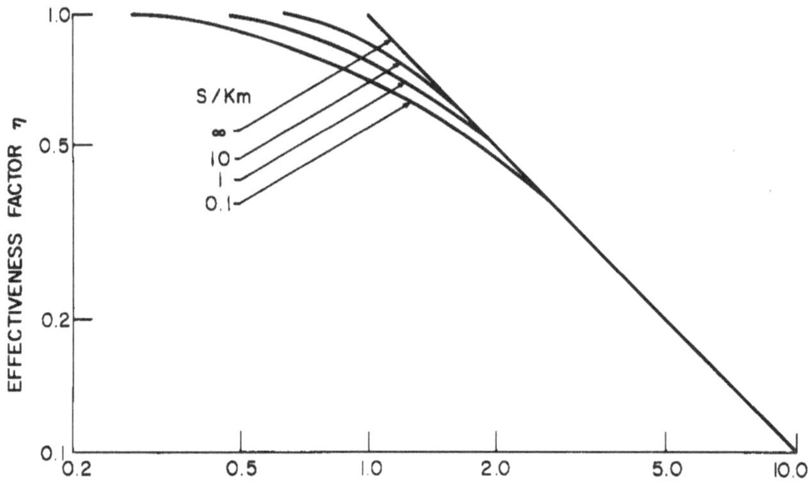

$$M = L\sqrt{\frac{V_m}{2K_mD_{eff}}}\left(\frac{S}{K_m+S}\right)\left[\frac{S}{K_m} - \ln\left(1+\frac{S}{K_m}\right)\right]^{-1/2}$$

FIGURE 4. Effectiveness factor for Michaelis-Menten kinetics. (From Bischoff, K. B., *AIChE J.*, 11, 351, 1965. With permission.)

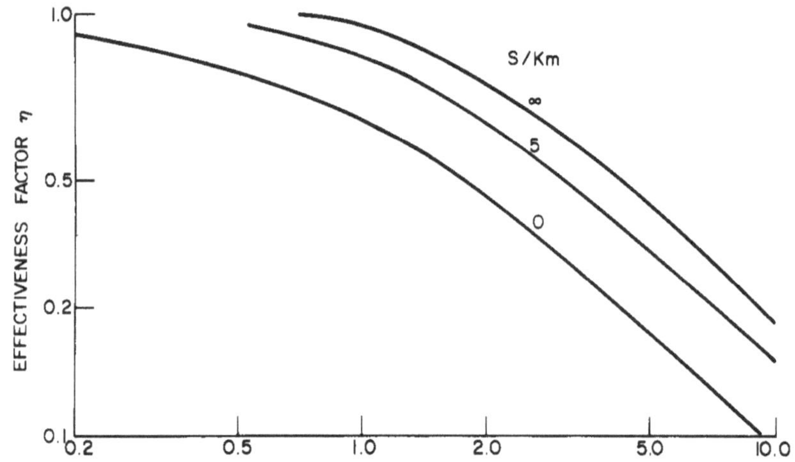

$$\Phi_L = \frac{R^2}{9D_{eff}}\left(\frac{1}{V_c}\frac{dn}{dt}\right)\frac{1}{S}$$

FIGURE 5. Effectiveness factor for Michaelis-Menten kinetics: spherical geometry.

the term, $1/V_c \cdot dn/dt$, is readily measurable, being simply the observed reaction rate per unit volume of porous carrier. Moo-Young and Kobayashi[77] have given numerical solutions for the more complex product and substrate inhibition cases. Unfortunately, calculation of their moduli is lengthy and involved, undoubtedly preventing frequent use of their approach.

Both types of moduli depend on effective diffusivity, which is difficult to estimate or measure and is probably the largest single source of error in modulus estimation. One reason for this problem is that the tortuosity, which must be found experimentally or estimated from data on similar materials, is difficult to determine accurately. A further complication is introduced if the diameter of the diffusing substrate molecule

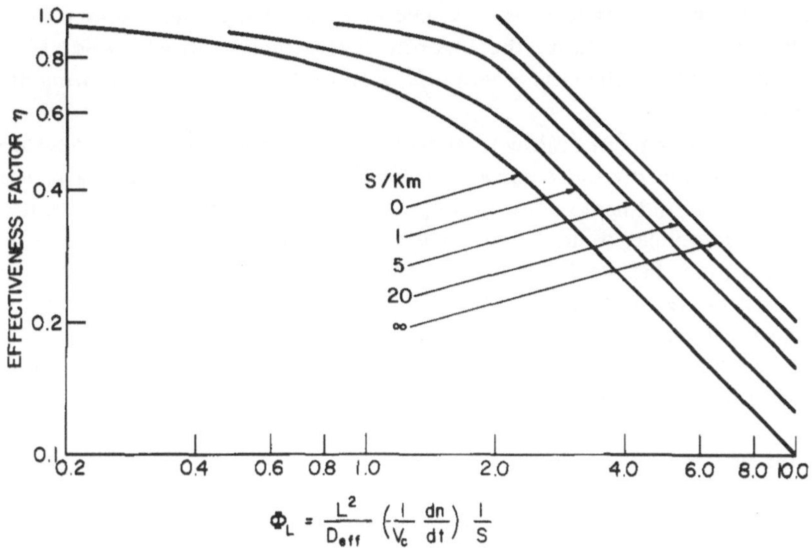

FIGURE 6. Effectiveness factor for Michaelis-Menten kinetics: flat plate geometry.

is a significant fraction of the pore diameter. In this event, the effective diffusivity will be decreased. Pitcher[7] and Satterfield et al.[79] have proposed an empirical correlation that can be used to estimate the resulting effective diffusivity

$$\log_{10}\left(\frac{D_{eff}}{D\Theta}\right) = 2.0\,\tau \qquad (23)$$

where τ is the ratio of the diffusing solute molecule critical diameter to pore diameter and D is the bulk diffusivity. The critical diameter is defined as the smallest cylinder into which the molecule will fit without distortion. This correlation does not appear to hold for polymer solutions where Colton et al.[80] found no restriction of diffusivity.

More recent theoretical treatments of diffusion have combined analysis of internal and external mass transfer effects. Hamilton et al.[81] reviewed the appropriate literature in addition to providing their own work on internal and external diffusion effects. Frouws et al.[82] devised a graphical method for calculating conversion when both intra- and extra-particle diffusion are limiting. Engasser,[54] whose approach to external mass transfer was discussed in the previous section, combined (by multiplication) external and internal effectiveness factors to obtain an overall effectiveness factor. He used the approach described by Pitcher[1,2] based on plots developed by Satterfield and others[59,83-85] using a modulus depending on observed reaction rate data. Mogensen and Vieth[73] analyzed the combined effects of external and internal diffusion on the performance of enzyme immobilized in microcapsules.

An example of recent experimental data for a diffusion-limited immobilized enzyme system is the work by Swanson et al.[86] They studied a packed-bed reactor system utilizing glucoamylase immobilized on agarose carriers. When they used the modulus proposed by Bischoff,[76] they found agreement with theoretical plots to within the limits of experimental error.

Other reactor configurations have also been analyzed theoretically. As previously noted in the section on external mass transfer, Kobayashi and Laidler[61] developed a theoretical treatment of flow systems inside tubes lined with a porous layer in which enzyme was immobilized. The group headed by Laidler[62,63] has reported testing this theoretical approach by experiment on several occasions. At least under the limiting cases examined experimentally, the theoretical treatment appears valid. Horvath et al.[87] have also analyzed the case where diffusion in the tube wall is limiting as well as external diffusion.

Specialized treatments of hollow fiber reactor performance have been developed.[88,89] One of the most recent, by Webster and Shuler,[90] consists of making effectiveness factor vs. Thiele modulus plots for limiting cases (zero and first-order reactions).

Even more complicated situations have been studied theoretically and experimentally. For example, Halwachs et al.[91] analyzed the hydrolysis of D,L-phenylalanine methylester to L-phenylalanine and D-phenylalanine methylester using immobilized α-chymotrypsin. Under their experimental conditions, the pH gradients were set up in the pores of the carrier particles due to diffusion limitations. This variation in pH across each particle in turn caused a change in enzymatic activity. The authors have described a theoretical model requiring soluble enzyme kinetic data, particle size, and effective diffusivity.

It is important to note that diffusion limitations will affect initial rate assays and thus distort the standard plots used to estimate kinetic constants. For example, the effect internal diffusion resistance on a Lineweaver-Burk plot can be readily seen in Figure 7, where β-galactosidase was trapped in polyacrylamide gel by Bunting and Laidler.[92] All the curves for the immobilized enzyme have a slope equal to ηV_m over the straightline portion of the plots, where the substrate concentration is small. The intercept ($1/V_m$) is the same for all curves but is higher than that of the soluble enzyme curve because of enzyme activity losses associated with immobilization. The higher the enzyme loading, the steeper the slope of the plot, indicating the expected increase in diffusion effect. The soluble enzyme, as would be predicted, gave the smallest slope.

Obviously, comparisons of this type with soluble enzyme can be used to estimate the effect of pore diffusion. However, this technique is limited to low conversions, which makes it only an approximation for most cases of practical interest. One must also be wary of enzyme kinetic changes caused by immobilization, although these are probably infrequent. There are several other practical methods of estimating effectiveness factors and evaluating mass transfer limitations due to internal diffusion resistance. The most useful of these avoid the drawback of the difficult-to-estimate effective diffusivity values required for use of the generalized plots. One method is to decrease membrane thickness or particle size until no additional increase in observed reaction rate results. This final and maximum rate can then be assumed to be the intrinsic rate. The effectiveness factor can then be calculated for other particle sizes by dividing the observed rate by the intrinsic rate. Several precautions must be taken to insure that this type of data is meaningful. Care should be taken to couple enzyme to different-sized particles in independent batches. If a wide size range of particles are coupled in the same batch, the risk of preferential coupling to the smaller particles is a real one. Smaller particle sizes can be obtained by crushing immobilized enzyme particles already prepared. The technique has its own risks of affecting enzyme activity by the crushing of the particles.

Marsh et al.[93] used this approach to determine effectiveness factors for glucoamylase immobilized on porous glass particles. They compared reaction rates using immobilized enzyme particles ranging in size down to finely crushed powder.

Another practical method of estimating the effect of pore diffusion is to compare the variation of reaction rate or activity with temperature for the immobilized enzyme vs. the soluble enzyme. If an Arrhenius plot (log of reaction rate vs. reciprocal absolute temperature) for the immobilized enzyme gives a straight line (in the region of interest) with a slope equal to that of a similar plot for the soluble enzyme, it usually implies

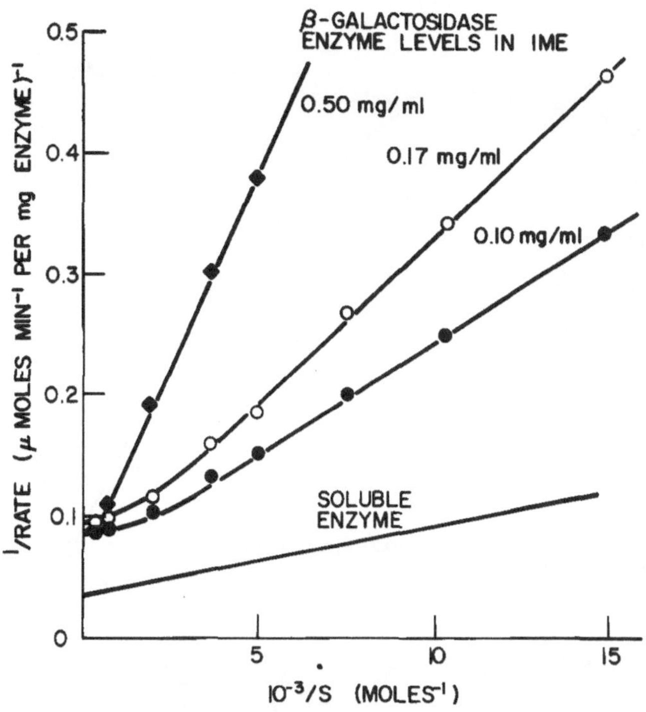

FIGURE 7. Effects of diffusion on a Lineweaver-Burk plot. (From Bunting, P. S. and Laidler, K. J., *Biochemistry*, 11, 4477, 1972. With permission.)

that no internal diffusion limitations exist. Pore diffusion limitations will normally cause a decrease in slope at higher temperatures on a plot of this nature. Wheeler[94] demonstrated that as the effectiveness factor falls below unity, the apparent activation energy also decreases, approaching the arithmetic mean of the activation energies for the chemical reaction and diffusion. Since the activation energy for diffusion is almost always small, especially under conditions suitable for enzymes, the observed activation energy will approach half the intrinsic value as diffusion becomes the limiting factor. In Figure 8 an Arrhenius plot for glucoamylase bound to 20/30 mesh porous glass particles, evidence of pore diffusion limitation can be observed at 60°C and above.[95] Buchholz and Ruth[96] have more recently reported given another example of such behavior.

Another qualitative and much less reliable indication of pore diffusion problems is the observation of substantial reductions in enzyme activity upon immobilization. Since there are other, perhaps more frequent, causes of reduced activity, caution must be used in interpreting this type of information. On the other hand, agreement between immobilized enzyme and soluble enzyme specific activities (units of activity per gram of protein) at their respective pH optima does indicate low diffusion resistance as well as an efficient coupling method. Since coupling methods are usually less than 100% efficient this case will not often occur.

There are several ways of reducing pore diffusion effects, which can be important when the enzyme is costly and must be used efficiently. Enzyme loading must be optimized considering the tradeoff between less efficient enzyme use at high loadings and larger carriers and immobilization cost at low enzyme loadings. Reactor configurations can be selected to reduce internal diffusion resistance. Certain membrane

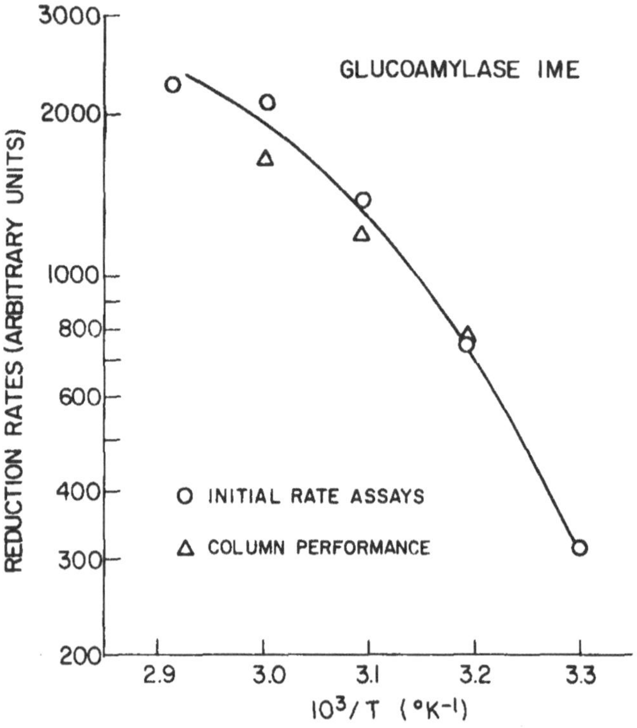

FIGURE 8. Arrhenius plot for immobilized glucoamylase.

reactors and monolith supports can be used to decrease the distance the substrate molecules must diffuse. Another approach is the pellicular immobilized enzyme described by Horvath.[97,98] This concept involves the use of solid particles coated with a porous layer to which the enzyme is immobilized. Basically, the two methods of increasing efficiency of enzyme use (where pore diffusion effects are significant) are to decrease the diffusion path for the reactants and products or to reduce the activity of the enzyme (lower loading or temperature).

C. Dispersion and Backmixing

In an ideal plug flow reactor, each element of fluid passes through the reactor behaving as a small batch reactor, not mixing with neighboring fluid. However, in real reactors there is always some mixing or dispersion into the adjoining fluid. This dispersion, when it occurs axially, forward, and backward in the direction of the flow, results in what is known as backmixing. This common source of inefficiency causes fixed and fluidized-bed reactors to tend toward stirred-tank reactor type of behavior, which in the case of most enzyme catalyzed reactions means less efficient enzyme utilization. The extent of this backmixing can be estimated from a calculated dispersion number as described by Levenspiel.[99] A dispersion number is defined as D_c/uL_c where D_c is the dispersion coefficient, u is the interstitial velocity, and L_c is the bed depth. In the literature, the reciprocal of the dispersion number, called the Bodenstein number or uL_c/D_c, is also used. The smaller the dispersion number, the less severe the backmixing. It is apparent that increasing linear velocity or bed depth will decrease the dispersion number and thus the backmixing. At higher conversion levels, the effect of backmixing on reactor performance is more severe than at lower conversions.

The dispersion number can be written in terms of the Peclet number ($N_{Pe} = ud_p/D_c$):

$$\frac{D_c}{uL_c} = \frac{d_p}{L_c N_{Pe}} \tag{24}$$

Correlations predicting value for the Peclet number can then be used to estimate the dispersion number. Chung and Wen[26] developed one such correlation, based on extensive experimental data, between the Peclet number and the Reynolds number (N_{Re}) applicable both to fixed and fluidized beds.

$$\frac{N_{Pe}\epsilon}{Z} = 0.20 + 0.011\ N_{Re}^{0.48} \tag{25}$$

where $Z = 1$ for a fixed bed, $Z = (N_{Re})_{mf}/N_{Re}$ for fluidized beds, and $(N_{Re})_{mf}$ = the minimum Reynolds number for fluidization. This minimum Reynolds number required for fluidization can either be measured experimentally or estimated from the following equation:

$$\left(N_{Re}\right)_{mf} = \left(33.7^2 + 0.0408\ N_{Ga}N_m\right)^{1/2} - 33.7 \tag{26}$$

where the $N_{Ga} = [d_p{}^3\varrho^2 g]/\mu^2$, $N_m = (p_s - \varrho)/\varrho, \varrho$ is the fluid density, ϱ_s is the particle density, g is the gravitational constant, and μ is the fluid viscosity. For Reynolds numbers less than 10 in fixed beds, the Peclet number can be approximated at 0.5. This value can also be obtained from Equation 25 by neglecting the Reynolds number term (it will be small) and by assuming a void fraction of 0.4.

At high substrate concentrations ($S >> K_m$), reactions following Michaelis-Menten kinetics become essentially zero order and thus conversions are unaffected by backmixing. At low substrate concentrations, the reaction becomes first order, a case considered by Levenspiel.[99] Kobayashi and Moo-Young[100] have prepared generalized plots comparing reactor sizes for actual vs. ideal plug flow conditions for Michaelis-Menten kinetics at various substrate concentration levels. These plots are of the same type as developed by Levenspiel.

Emery and Cardoso[25] used amyloglucosidase immobilized on solid and porous carriers to study fixed- and fluidized-bed reactor performance. Of particular interest were the glucose tracer experiments used to measure dispersion in both types of reactors at various flow rates. They found actual dispersion numbers much higher than predicted by Equation 25. Also they observed higher conversions in the fluidized beds as opposed to the fixed beds, contrary to their expectations. They were not certain of the reasons for this behavior, but cited preferential flow patterns, liquid distribution, and other similar problems as possible causes. Allen has reported vast differences in mass transfer in fluidized beds depending on the liquid distribution system or the column entrance effects. Unfortunately, it appears that channeling and other irregularities in liquid-carrier contacting may frequently be more important than backmixing or other effects amenable to accurate analysis.

In what may be mostly of interest as a curiosity, DeVera and Varma[101] have shown that a substrate-inhibited enzyme reaction in a tubular reactor with axial dispersion can give rise to multiple steady states. Under certain conditions, these steady states may be attainable. Evidently the middle conversion level is unstable, while the upper and lower ones are stable. This phenomenon had previously been noted for continuous stirred tank reactors.

D. Electrostatic Effects

In contrast to other influences on mass transfer rates that have been the subject of much quantitative analysis and experimental study, electrostatic effects have usually been involved only as qualitative explanations for phenomena such as pH optima shifts. Early attempts to deal with these effects quantitatively were simplistic and empirical in nature, such as the approach of Hornby et al.,[102] who lumped electrostatic effects together with diffusion-related effects by using a modified Michaelis constant. Shuler et al.[103] developed a quantitative treatment of the effects of electrostatic fields and diffusion on the rates of reactions catalyzed by immobilized enzymes. Their approach was based on the assumption of a distribution of potential for the electrical double layer that is valid only for small surface potentials. By employing the complete Guoy-Chapman solution, Hamilton et al.[104] extended this treatment to the higher surface potential case. However, the results of Shuler et al.[103] are sufficiently accurate for most practical applications. Hamilton et al.[104] also discussed the implications of the theoretical predictions by giving examples of the curvature of Lineweaver-Burk plots that can result from significant diffusional and electrical effects. They also pointed out that for substrate and surface of opposite charge, the attractive forces can enhance mass transfer and increase local concentrations of substrate making possible effectiveness factors exceeding unity. Kobayashi and Laidler[105] have expanded their work with enzymes immobilized in membranes to include the area of electrostatic and diffusive effects on these systems. Karube et al.[106] recently reported modifying immobilized lipase activity by using electric fields.

E. Transient Behavior of Immobilized Enzyme Reactors

Most industrial reactor systems can be described adequately by models assuming steady-state operation. There are, however, some cases in which this approach is not appropriate and analysis of transient behavior is required. In certain analytical applications of immobilized enzymes utilizing small packed columns, the response time and recovery time between samples are important and depend upon transient behavior. A similar situation can be encountered in laboratory or pilot plant studies of immobilized enzymes where operating conditions (flow rate, temperature, etc.) are varied to determine their influence on conversion levels or required residence time. Experiments with immobilized glucose isomerase indicated that after a change in flow rate, several bed volumes (perhaps 2 to 4) of feed must flow through the fixed bed reactor before a new steady state was closely approached. Other systems and large changes in temperature may, in some cases, require an even longer time before reaching steady state. In the absence of continuous conversion level information, several hours should be allowed to insure that steady state has been achieved.

For industrial reactors, cleaning cycles or other operations necessitating frequent start-up and shut-down procedures could result in transient behavior significant to overall operating economics. Frequently, temperature, pH, feed concentration, and even flow-rate may all be changed at once in the course of such procedures. Temporary perturbations in these parameters may also significantly affect reactor operation.

Some theoretical evaluations of transient reactor behavior have been reported in the literature. Gelf et al.[107] numerically solved the case of transient behavior in an ideal plug-flow reactor (in the absence of mass transfer limitations) subject to Michaelis-Menten kinetics. They compared experimental data from a column of immobilized α-chymotrypsin with numerical solutions for cases of time-dependent substrate concentration gradients. Ryu et al.[109] analyzed the transient behavior of a continuous stirred-tank reactor. By considering a multi-stage continuous stirred tank reactor system, they extended this analysis to the plug-flow case. For the case they studied, with both competitive and noncompetitive inhibition, about 1 to 2 hr were necessary to reach steady state in the continuous stirred tank reactor with the higher substrate concentration requiring the longer time.

In their dynamic reactor model, Shyam et al.[108] included mass transfer limitations and transient response. Vieth et al.[4] have also devoted some attention to this subject and reviewed the transient analysis of immobilized enzyme reactors.

V. HEAT TRANSFER AND OTHER TEMPERATURE EFFECTS

Since both enzyme activity and stability vary with temperature, close control of the temperature in the immobilized enzyme reactor is necessary to insure operation under optimal conditions. Generally, the temperature dependence of reaction rate and enzyme activity loss rate can be described by Arrhenius equations. Reaction rates normally increase with temperature as long as enzyme activity loss is not significant over the course of the activity determination. Activation energies for enzyme catalyzed reactions are usually in the 5 to 20 kcal/gmol range. As temperature increases the stability of the enzyme decreases with the resulting deactivation energy ranging from about 10 to 100 kcal/gmol or more. As an example, Reilly[110] found the activation energy for dextransucrase immobilized on alkylamine porous silica to be 6.9 kcal/gmol. His enzyme stability data can be used to estimate a deactivation energy of about 63 kcal/gmol.

In order to achieve the desired temperature control, one might think it necessary to transfer large amounts of heat to or from the reactor. Since the heats of reaction of enzymatically catalyzed reactions are frequently small, this rapid heat transfer is usually not required. However, in cases where substantial heating or cooling is necessary, fluidized-bed and stirred-tank reactors lend themselves more readily to good heat transfer and temperature control. Fixed-bed reactors, unfortunately, present a more difficult problem. Havewala and Pitcher[32] estimated the rate of heat transfer in a bed packed with immobilized glucose isomerase. They concluded that, for a column 15 cm or more in diameter, an adiabatic temperature change resulting from the heat of reaction would occur, at least in the center of the column, in spite of any insulation or even reasonable levels of external heating or cooling. An actual example was described by Marsh and Tsao[111] who monitored temperatures in a packed-bed reactor containing glucoamylase immobilized on porous glass. They used a Plexiglass column of 4.2 cm inside diameter, 1.8 m tall, surrounded by 3.1 cm of insulation. At a flow rate of 44.5 mℓ/min, the feed temperature dropped from 55 to 51°C when they used water and to 50°C when 27% maltose feed was used. A similar 1° difference due to heat of reaction was observed for a 27% solution of 15 DE starch.

VI. IMMOBILIZED ENZYME ACTIVITY LOSS

Enzyme activity loss with time is one of the most important factors in determining or indeed limiting system performance and economics. Three basic types of enzyme activity loss can occur. The enzyme can be lost from the system, it can become inactive, or it can be coated or otherwise blocked from contact with the substrate. Enzyme can be lost from the system because of desorption, severing of chemical bonds, leakage through membranes, or erosion of the support material. Enzyme activity loss can be due to thermal denaturation, inhibition or inactivation by components of the feed, or destruction by protolytic enzymes. The rate of enzyme denaturation can be affected by substrate pH, temperature, or even concentration. Microbial contamination can either destroy the enzyme or block it from access to substrate. Other feed components such as proteins can also cause coating or blocking of porous carriers resulting in apparent activity loss.

Usually enzyme activity loss rate is proportional to the amount of activity remaining resulting in an exponential decay pattern. A plot of the logarithm of activity vs. time gives a straight line. Occasionally, data do not give a straight line for plots of this type. For example, a concave upward semilogarithmic plot of activity vs. time is consistent with two simultaneous decay mechanisms, enzyme species, or enzyme immobilization mechanisms. It is not uncommon for a rapid initial loss in immobilized enzyme activity to be observed as loosely bound enzyme is washed from the system. Havewala and Pitcher[32] reported an exponential decay for immobilized glucose isomerase at 60°C but a concave upward plot for the same system at 50°C as shown in Figures 9 and 10. In this case, severe initial enzyme washout seems unlikely since the effect was noticeable only after more than 3 weeks of operating time. A more likely explanation is that two activity loss mechanisms were involved, each with a different temperature dependence.

Examples of downward concave semilog plots are perhaps even less common and can be more difficult to explain. Pitcher[112] gave an example of this behavior shown in Figure 11. Wallerstein Lactase LP (β-galactosidase from *A. niger*) was adsorbed on Diamond Shamrock Duolite S-30 phenolformaldehyde resin. A reactor containing this immobilized enzyme was operated at 40°C using pH 3.5 acid whey as feed. It was found that the cause of the activity loss was the formation of a protein coating around the resin particles. Lee et al.[23] observed a decay curve of this type for glucose isomerase and attributed it to microbial contamination. Beck and Rase[113] also reported a similarly shaped activity loss curve.

Carberry and Gorring[114] have given plots for the effect of pore mouth poisoning on catalysts. When these graphs are replotted as the semilog of the fractional amount of catalyst remaining as a function of time, the resulting curves are concave downward for the case where the poisoning is not strongly diffusion limited. Viewed more simply, it can readily be seen that if a constant amount of activity is lost to poisoning or any other process during each successive time interval, a downward concave semilog plot will result (see Figure 12).

Other decay models have been proposed. Cardoso and Emery[115] have suggested another model to describe certain cases of enzyme activity decay. They reported that the activity of several immobilized amyloglucosidase preparations could be described as a function of time by the following equation:

$$E = E_o/(1 + K_D t) \tag{27}$$

where E = enzyme activity, E_o = initial enzyme activity, t = elapsed time, and K_D = decay constant. Written in differential form, this equation becomes

$$\frac{dE}{dt} = -\left(\frac{K_D}{1 + K_t}\right) E \tag{28}$$

the same form as the first order decay model, except that the decay constant is no longer invariant with time and is now, $-K_D(1 + K_D t)$. The authors did not give any physical model to explain this behavior other than to suggest that the enzyme molecule is too complex to always be described by a simple exponential decay model. It could also be argued that this decay behavior could be described by a family of exponential decay curves with different decay constants. Even were this the case, the authors' mathematical model may in at least some cases be the easiest way to describe the behavior sometimes observed.

Krishnaswamy and Kittrell[116] discuss another case where more than one model can be used to describe the rate of activity loss. They postulated two models for the deac-

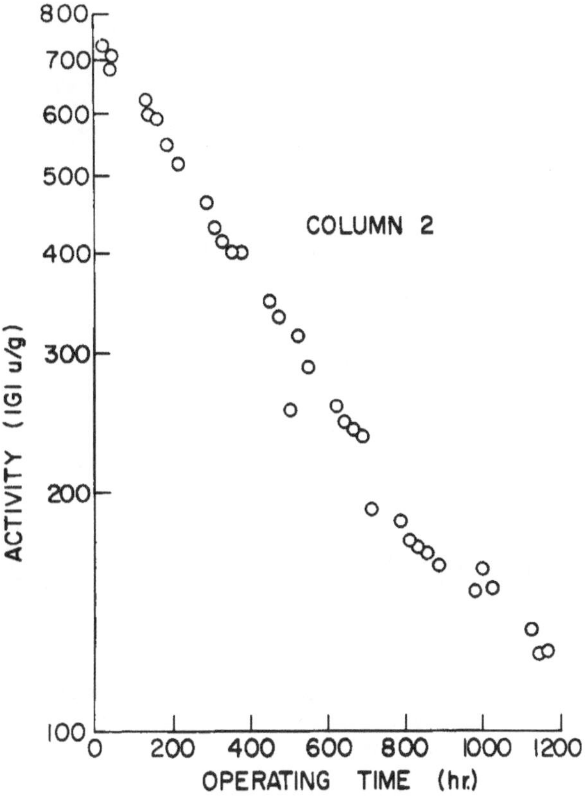

FIGURE 9. Immobilized glucose isomerase activity as a function of time at 60°C. (From Havewala, N. B. and Pitcher, W. H., Jr., in *Enzyme Engineering*, Vol. 2, Pye, E. K. and Wingard, L. B., Jr., Eds., Plenum Press, New York, 1974. With permission.)

tivation of immobilized catalase in a fixed-bed reactor used for decomposing hydrogen peroxide. Both models reduce to the same form and describe the experimental results obtained by Altomare et al.[117] equally well. Altomare et al. represented the conversion of peroxide by first order kinetics and used a first-order, concentration-dependent deactivation model. Krishnaswamy and Kittrell[116] pointed out that a second-order expression for hydrogen peroxide decomposition and a first-order, concentration-independent deactivation model also predicted the behavior equally well. They agree that, in light of known catalase inactivation by high concentrations of peroxide, the first model is preferred.

Carrier attrition has been cited as a cause of enzyme actvity loss by several workers including Regan et al.[118] As referred to in the previous chapter, Weetall and Havewala[119] detected silica eluting from a column of glucoamylase bound to porous glass. The decay rates appeared to depend on the flow rate. When the porous glass carrier was coated with zirconia, the decay rate decreased and showed no flow rate dependence. The speculation that shearing could remove enzymes from carriers must be discounted in light of the calculation by Weetall et al.[120] that disruption of covalent bonds by shearing in fixed-bed operation is practically impossible.

It should also be noted that immobilized enzymes whose apparent activity is limited by diffusion will exhibit a slower apparent loss of activity than if no diffusion limitation were present. As the enzyme activity decreases, the effectiveness factor increases,

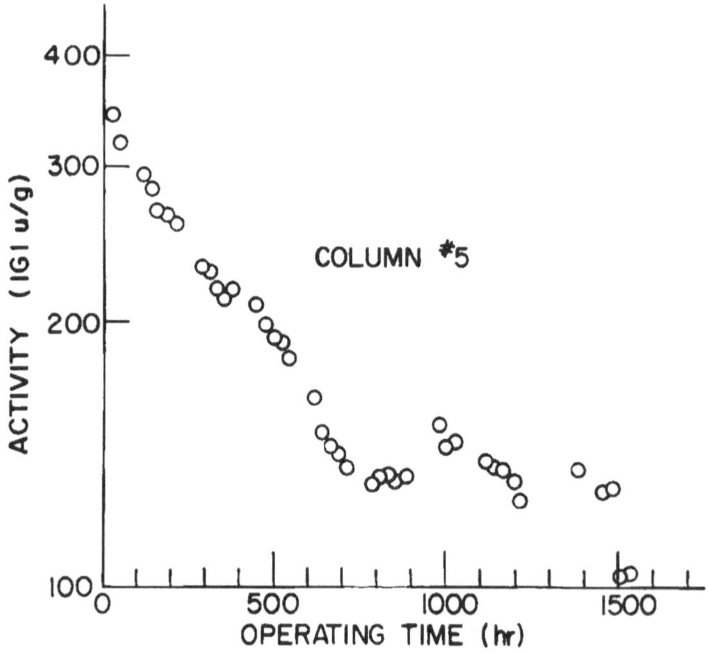

FIGURE 10. Immobilized glucose isomerase activity as a function of time at 50°C. (From Havewala, N. B. and Pitcher, W. H., Jr. in *Enzyme Engineering*, Vol. 2, Pye, E. K. and Wingard, L. B., Jr., Eds., Plenum Press, New York, 1974. With permission.)

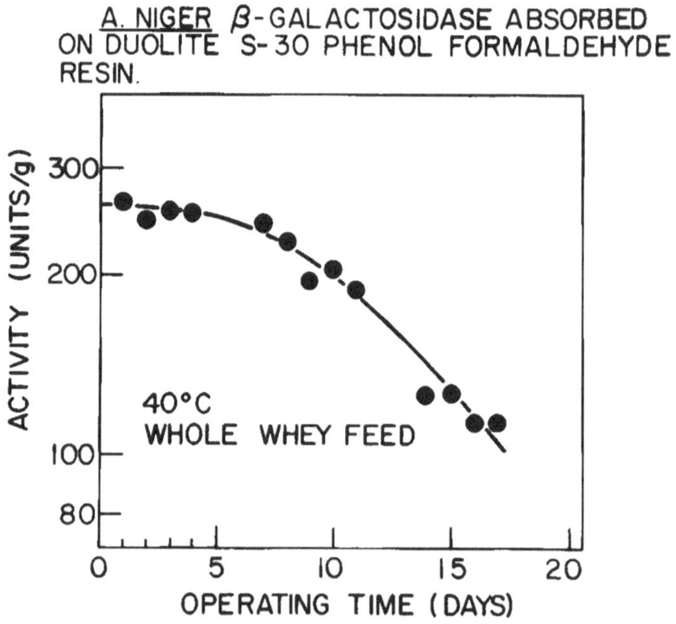

FIGURE 11. The effect of whey feed on immobilized lactase activity.

thus slowing the apparent activity loss rate. Ollis[121] was among the first to comment on this possibility and has discussed the phenomenon in greater detail.

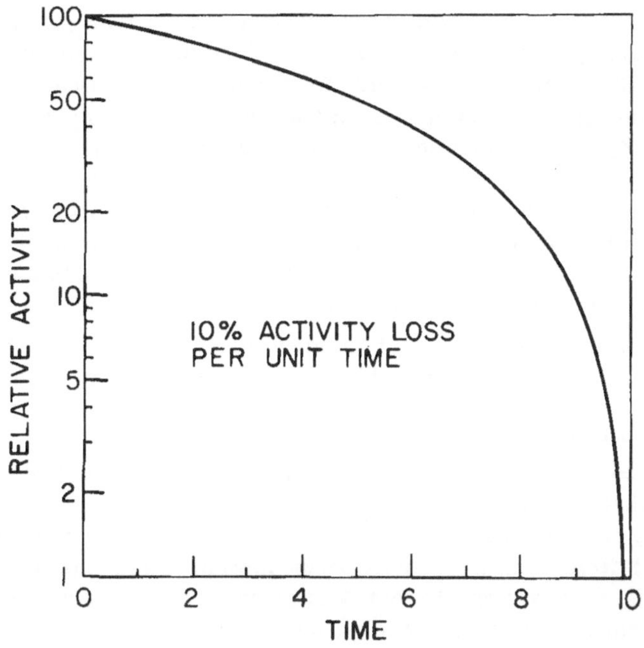

FIGURE 12. Activity vs. time for constant activity loss rate.

Stabilization of the enzymes themselves is a major topic in itself and will only be reviewed briefly here.

In the case of protolytic enzymes, the advantages of immobilization by attachment to a surface are obvious. Even where the enzymes remain in solution, there are steps that can be taken to improve stability. Feder et al.[122] reported stabilization of *Bacillus subtilis* neutral proteases for up to 25 days by the addition of various preparations of proteins including casein, soy protein, and fish meal. For achieving long-term storage stability, they suggested that a reversible neutral protease inhibition such as o-phenanthroline might be used to reduce autolysis.

It is well known that enzymes can be denatured when subjected to shear. However, most pumping or shaking of enzyme solutions during immobilization results in little or no activity loss. Foaming and other excessive exposure to air should be avoided.

Asakura et al.[123] did observe that certain organic solvents in dilute solution had a stabilizing effect on various enzymes and other proteins that normally suffered denaturation when mechanically shaken. Alcohols, ketones, and ether were among the compounds used at concentrations up to 50% (u/v) to retard denaturation. Of the compounds studied, hexamethylphosphoramide was found to be effective at the lowest concentration. The lower the molecular weight of the alcohol the higher the concentration required for the same degree of stabilization. Other compounds such as toluene and chloroform were found to promote denaturation, while yet another group of solvents, including formamide and pentane, had no effect. The products of enzymatically catalyzed reactions can also cause enzyme activity loss (as differentiated from inhibition). Several investigators[124-126] have attempted to improve the stability of glucose oxidase and catalase by controlling hydrogen peroxide levels. Tramper et al.[127] showed that the stability of xanthine oxidase could be increased by coimmobilizing it with superoxide dismutase, catalase or simply protein. The activity loss was proportional to the amount of substrate converted or, perhaps more appropriately, proportional to the amount of peroxide produced.

Johnson[128] reported immobilizing horse liver alcohol dehydrogenase on Enzacryl-TIO (porous titania particles coated with polymerized diazotized m-diaminobenzene). He found that the stability of the derivatives could be increased by increasing the amount of enzyme immobilized on a given quantity of support material. He observed the same effect by adding albumin rather than additional enzyme. Beyond a certain point increasing the amount of protein had no additional effect. Similar effects have been noted before, particularly with regard to the effect of the concentration of soluble enzyme on stability by Johnson and Coughlin.[129] They also reported improved stability for immobilized xanthine oxidase when the substrate contained albumin. Stability was improved by cross-linking the immobilized enzyme with glutaraldehyde.

Along similar lines, Torchilin et al.[130] studied the effect of the length of intramolecular cross-linkages on the thermostability of soluble α-chymotrypsin. Enzyme was first activated with 1-ethyl-3-(3-dimethylaminopropyl) carbodiimide and then treated by diamines of the type, $NH_2(CH_2)_nNH_2$, where n ranged from 0 to 12. The stability of the cross-linked and native enzyme was not affected by enzyme concentration. However, the alkyl chain length of the diamine used in cross-linking did substantially affect enzyme stability. For this particular enzyme best results were obtained with $n = 4$, with a threefold decrease in the activity decay constant versus that for the native enzyme. Additional stabilization, 21-fold, was achieved by succinylating the α-chymotrypsin, which did not affect its stability, before cross-linking with diamine. In this case an alkyl chain length of $n = 2$ proved optimum. The purpose of the succinylation was to increase the number of cross-links. Thus it appears that both the length of the cross-linking agent and the number of cross-links improved the enzyme stability. Likewise, Okada and Urabe[131] improved the stability of α-amylase by acylation.

Immobilization does not always increase enzyme stability. Flynn and Johnson[132] studied the stability of four samples of glucoamylase from different commercial sources. The enzymes were evaluated in the soluble form as well as immobilized by various methods. Substrate or product were found to stabilize both soluble and immobilized glucoamylase. Thermal stability was found to be decreased when the enzymes were immobilized. The silane-glutaraldehyde coupled derivatives were more stable than the metal-link derivatives. Since this difference could not be explained by desorption, the implication remains that the type of immobilization affected the enzyme stability.

Obviously there is much yet to be learned about stabilization of enzymes, but success in this area could change the economics of many potential enzyme-base processes.

VII. PRESSURE DROP

Another specialized consideration, that often affects the design of commercial fixed-bed reactors, is the operational pressure drop. In extreme cases, the build-up of deposits on the carrier particles may cause drastically increased pressure drops or even complete plugging. While small particles are frequently desired to reduce internal diffusion limitations, they also cause higher pressure drops. For large-scale applications, use of particles much below 60 mesh (U.S. standard sieve) is usually impractical.

Leva's correlation as given by Perry[133] affords a means of estimating pressure drop, ΔP, and gives an indication of the effects of various parameters on pressure drop.

$$\Delta P = \frac{2f_m G^2 L_c (1-\epsilon)^{3-n}}{d_p \rho g_c \phi_s^{\,3-n} \epsilon^3} \tag{29}$$

where f_m = $100/N_{Re}$ = $100\ \mu/d_pG$(for $N_{Re} < 10$), n = 1 (for $N_{Re} < 10$), ϕ_s = shape factor (ϕ_s = 1 for spherical particles, less for irregular particles), ε = void fraction, G = fluid superficial mass velocity (lb/ft/sec), L_c = bed depth (ft), μ = viscosity (lb/ft/sec), d_p = particle diameter (ft), g = 32.17 (lb ft/lb$_f$/sec^2), ϱ = fluid density (lb/ft^3), ΔP = pressure drop (lb$_f$/ft^2).

Examination of this expression reveals that pressure drop is extremely sensitive to void fraction, ε. At ε = 0.35 a 10% change in void fraction will change the pressure drop by a factor of 2. The implication, and it holds in actual practice, is that the packing density can significantly affect pressure drop. It is also important to note that a broad particle size distribution may reduce the void fraction and thus increase the pressure drop due to small particles filling the interstices between the larger particles. From Equation 29, it can be seen that the pressure drop is inversely proportional to the square of the particle diameter. It also appears that the pressure drop is directly proportional to the bed height. In practice pressure drop is proportional to the square of the bed height since to retain the same residence time when scaling up, the linear velocity must be increased.

The following example of a pressure drop calculation was given by Havewala and Pitcher[32] for a 3-ft high bed of 20/30 mesh particles in a column operated at 60°C with a 50% glucose feed. The parameter value listed here were substituted into Equation 29.

$$G = 0.356\ \text{lb/ft}^2\text{/sec}$$

$$L = 3\ \text{ft}$$

$$\epsilon = 0.35$$

$$n = 1$$

$$g_c = 32.17\ \text{lb ft/lb}_f\text{/sec}^2$$

$$d_p = 2.32 \times 10^{-3}\ \text{ft (20/30 mesh)}$$

$$\phi_s = 0.65\ \text{(for crushed glass particles)}$$

$$\Delta P = \frac{200\ (3.6\text{cp})\ (0.000672\ \text{cp sec ft/lb})\ (0.356\ \text{lb/ft}^2\text{/sec})\ (3\ \text{ft})\ (0.65)^2}{(144\ \text{in}^2\text{/ft}^2)\ (2.32\times 10^{-3}\ \text{ft})^2\ (32.17\ \text{lb ft/lb}_f\text{/sec}^2)\ (0.65)^2\ (0.35)^3\ (76\ \text{lb/ft})} = 6.3\ \text{psi}$$

This treatment of pressure drop assumes an incompressible bed. Beds of particles that can swell or become compressed are a more difficult problem since the void fraction changes. For those encountering this problem, the paper by Bucholz and Godelmann[134] may be of some interest.

VIII. REACTOR OPERATING STRATEGY

This author[1,2,3] has previously discussed immobilized enzyme reactor operating strategy, an area that has received little attention in spite of its potential importance to immobilized enzyme system economics.

It is usually desirable to maximize the total amount of product produced or feed processed per unit of enzyme, reactor volume, or some other variable with the ultimate goal of minimizing total processing cost.

The total production, designated at P_t, of a reactor during a period of time, t_p, can be related to the feed rate F by the equation

$$P_t = \int_0^{t_p} F\,dt \tag{30}$$

For the most common case, exponential enzyme activity decay and operation at constant conversion,

$$P_t = \int_0^{t_p} F_i e^{-(\ln 2)t/t½} dt \tag{31}$$

where F_i is the initial feed rate and t-½ is the immobilized enzyme half-life (the time required for the activity to fall to one half its original value). This equation can be integrated to obtain

$$P_t = \frac{F_i t-½}{\ln 2} [1 - e^{-(\ln 2)t/t½}] \tag{32}$$

Operating isothermally at constant conversion requires decreasing flow rates or production rates to maintain the conversion level as the enzyme activity drops. Usually such drastic changes in production rate during the operational life of the immobilized enzyme are unacceptable in commercial practice. The most frequently utilized solution to this problem is to use a multiple-reactor system. Start-up or reloading times for individual reactors are then staggered with the reactors operated either in series or parallel. The variation in production rate can be limited to any desired level by using a sufficient number of reactors. The number of reactors required to maintain the production rate within a given range is a function of the number of half-lives for which the reactor is operated before the immobilized enzyme is replaced.[32]

$$R_p = e^{\frac{-H\ln 2}{N}} \tag{33}$$

where N is the number of reactors, R_p is the ratio of the maximum to the minimum production rate, and H is the number of half-lives for which the immobilized enzyme is used. An example, shown in Figure 13, shows the production rate as a function of time from original start-up for a case where $R_p = 0.82$, $N = 7$, and $H = 2$. This is the optimal strategy if there are only constraints of temperature. If the percentage decrease in half-life with temperature is greater than the percentage increase in activity, then the minimum allowable temperature is optimal. If the reverse is true, then the maximum allowable temperature will yield the highest total productivity for a given amount of enzyme. Since the former case is more common, the temperature is usually limited on the low side by such considerations as microbial contamination, capital requirements (lower temperature means larger reactors for the same throughout), or product quality (such as competing side reactions).

Another reactor operation strategy is to raise the operating temperature to compensate for activity losses, thus maintaining the original production rate and conversion level. This strategy may be useful in cases where half-lives are extremely long and the start-up of multiple reactor systems poses problems.

Other constraints placed on operating conditions necessitate other types of strategies to achieve optimal enzyme usage. This author[135] discussed a case where other con-

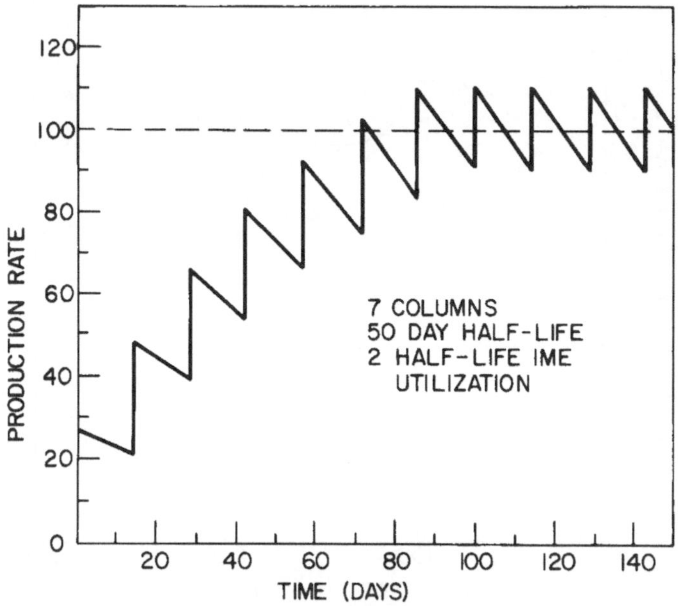

FIGURE 13. Multiple column system production rate.

straints were placed on the strategy selection. These limits included upper and lower operating temperature limits and a fixed length of operating time. The optimal operating strategy was determined for an actual immobilized lactase system where 40 and 50°C temperature limits and a 300-day operating period were assumed. The optimal temperature policy was found to consist of three distinct phases. First the reactor must be operated at the minimum temperature, in this case 40°C. The constant conversion must be maintained by increasing temperature until the upper limit, 50°C, is reached. The third and final phase consists of isothermal, 50°C, operation. Figure 14 shows the flow rate (production rate) as a function of time. This operating strategy can be approximated by using 2°C temperature increment and is also shown in Figure 14. This approximation results in productivity only 0.1% less than the optimal policy. For actual systems, the large drop in production rate during the final phase of operation may be unattractive. A simplified strategy consisting of 40°C isothermal operation followed by a constant conversion temperature policy for 168 and 132 days, respectively, yields a total productivity only 1% less than the optimum and is probably more practical. Crowe[136] has given a more theoretical treatment of this type of reactor operating strategy problem. Care must be exercised that all the conditions are clear for such analysis. For example, the solution to the immobilized lactase case described here is only valid when the increase in reaction rate with temperature is less than the decrease in immobilized enzyme half-life. It should also be obvious that if the operating period had been substantially less than 300 days that a higher starting temperature would be chosen along with other operating strategy changes. Levenspiel and Sadana[137] commented on a similar case but with a slightly different definition of the quantity to be maximized and no lower temperature constraint. Their solution thus consisted simply of a constant conversion policy designed to end at the maximum allowable temperature.

IX. MULTIENZYME SYSTEMS

Although multienzyme systems abound in nature and have been the subject of nu-

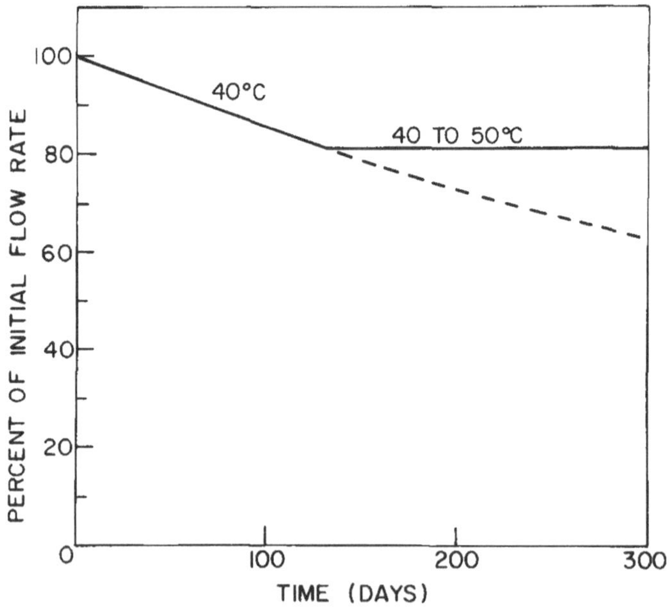

FIGURE 14. Optimal temperature policy.

merous studies where as many as four enzymes have been acting simultaneously, little attention has been paid to optimization. Goldman and Katchalski[138] performed a theoretical analysis of the kinetics of a two-enzyme system. Ford[139] analyzed the oxidation of ethanol to acetic acid by alcohol dehydrogenase and aldehyde dehydrogenase in a hollow fiber reactor. He determined the optimal ratio of the enzymes for the overall conversion of ethanol to acetic acid. The glucoamylase-glucose isomerase system has also received some attention. Kent and Emery[140] combined the two enzymes in varying proportions and studied the conversion of maltose to fructose at various pH levels. The problem is an interesting one since the two enzymes have widely separated pH optima. Reilly[110] also studied the conversion of dextrin to fructose using these two enzymes. He discussed the problem of optimizing catalyst profiles for two-step reactions. Choi and Perlmutter[141] studied a dual enzyme sequential reaction system using arginase and urease to decompose arginine to ammonia. They showed analytically that a bang-bang policy (a bed of one immobilized enzyme followed by a bed of the other immobilized enzyme without any mixed bed) is optimal for the distribution of the immobilized enzymes for an irreversible two-step consecutive reaction system in a plug flow tubular reactor. The authors found that for the case where Michaelis-Menten kinetics apply to both reaction, the optimal switching point location (percentage of the distance down the reactor where the second bed of immobilized enzyme begins) depends on system parameter values. As the second reaction rate increases, the optimal switching point moves toward the exit end of the reactor. They also investigated a suboptimal policy consisting of a uniform mixed bed of the two immobilized enzymes. This policy gave 41% less productivity than did the optimal bang-bang operating strategy. Results were verified experimentally. Chang and Reilly[142] performed a similar theoretical and experimental study of the glucoamylase-glucose isomerase system. They too found the bang-bang policy optimal with the suboptimal uniform mixed bed policy substantially inferior in converting maltose to fructose. They also found an optimal bang-bang-bang policy maximizing the conversion of fructose by a glucose isomerase-glucose oxidase-catalase system.

In multienzyme systems diffusion effects can also complicate the analysis. Ho and Kostin[143] derived expressions for end product concentrations for consecutive reactions where diffusion limitations occur. They gave both steady-state and transient solutions. Koch-Schmidt et al.[144] showed that coimmobilizing malate dehydrogenase and citrate synthase improved efficiency over the enzymes separately immobilized on different carriers and then mixed. Apparently diffusion limitations were reduced. They also found that aggregating the enzymes before immobilization, thus increasing to a degree the distance between the two kinds of enzymes, had no effect on efficiency.

Barker and Somers[145] and Mosbach and Mattiason[146] have provided a useful review of immobilized multienzyme systems.

An allied area of interest is the utilization of enzymes that require cofactors or coenzymes. Systems using these enzymes will probably increase in importance in the future since they allow more complicated synthetic reactions to be carried out. Regeneration of cofactors will undoubtedly be an important aspect of practical systems using cofactor requiring enzymes. This regeneration can be accomplished in a number of ways, but perhaps the most intriguing is electrochemically.[147-149]

X. COST ESTIMATES

Preliminary cost estimates are a useful tool in evaluating immobilized enzyme systems during research and development stages. Such cost estimates can also be valuable in helping to identify problem areas in potential systems. One way of looking at these costs is to break them down into several major components. The first is the cost of the immobilized enzyme itself, which is in turn made up of carrier, enzyme, and immobilization costs including labor, capital, and materials. This immobilized enzyme composite cost can be used with the performance data to calculate the cost per unit of product attributable to the immobilized enzyme. Clearly, to do this it is necessary to have sufficient experimental data to project enzyme life as well as initial productivity.

The second cost component is the labor for operating and maintaining the process. Supervision and other overheads will normally bring this cost to two or three times the hourly wage rate or more.

The third cost component, equipment cost, appears as a depreciation charge. Taxes, insurance, maintenance, and other miscellaneous items are frequently estimated as a percentage of capital. The total of these charges may exceed 20% of the capital annually.

These three cost components comprise the processing cost. Raw material costs must also be added to obtain the product cost. Provision for raw material and processed goods losses must also be made. This is probably most easily done by estimating the amount of raw material and process capacity required for actual rather than ideal output. These figures can then be used in the original estimates.

Examples of such preliminary cost estimates for immobilized enzyme systems are difficult to find in the literature. This author[15] contributed several such estimates for immobilized glucoamylase and immobilized lactase systems. Pitcher[112] has also given an example of the effect of relative enzyme and total immobilization costs on the selection of an immobilization method. Wandrey[152] developed equations to find the economically optimal conversion with regard to raw material costs and enzyme costs for an acylase hollow fiber system. Flaschel and Wandrey[153] later reported on a model for the economics of an ultrafiltration membrane reactor containing acylase for optical resolution of methionine. Energy, enzyme, and raw material costs were considered, but not equipment or labor. Another more comprehensive and detailed cost estimate was given by Giacobbe for the production of 6 APA from penicillin G using fiber-entrapped penicillin amidase. Performance data and manufacturing costs were given for the conventional chemical process, a conventional enzymatic process, and an integrated enzymatic process proposed by the authors. The integrated process appeared less expensive because of lower depreciation charges and more efficient utilization of the penicillin G. Other cost estimates encountered occasionally in the literature are generally of a more superficial nature and frequently of little value.

A number of general references on cost estimating are available.[155,156] If care is taken to use the proper information concerning immobilized enzyme performance and realistic allowances for losses are made, preliminary cost estimation procedures need not be unduly complicated nor difficult.

XI. GENERAL SYSTEM DESIGN CONSIDERATIONS

Even though this author[1,2,3] has previously reviewed some of the general considerations affecting immobilized enzyme reactor design, it is certainly appropriate to do so once again.

While the various immobilized enzyme characteristics and reactor properties have been considered individually, in reality they are not independent and affect each other as well as the ultimate efficiency of the entire system. For example, enzyme kinetics, substrate concentration, carrier geometry (as it affects mass transfer), operating temperature, pH, and activators or contaminants in the feed all may affect the rate at which an immobilized enzyme catalyzes a reaction.

Enzyme loading, the amount of enzyme immobilized per unit volume or weight of carrier (or membrane or other immobilization device), and carrier cost both affect total processing cost. For low-cost enzymes high loading is essential to reduce the impact of carrier cost. If the enzyme cost is sufficiently low immobilization may not be attractive at all except for product quality or other secondary reasons. When the cost of the carrier is low relative to the enzyme cost, then high loading is less critical. Reactor size and thus equipment costs are reduced as enzyme loading increases. The enzyme loading achieved depends upon the immobilization technique and the conditions of coupling, carrier properties (surface area, pore size, composition, etc.), and the purity of the enzyme.

Another way of looking at enzyme loading is to determine the coupling efficiency, the percentage of soluble enzyme activity consumed in coupling that appears as immobilized enzyme activity. Coupling efficiency is influenced by the same variables that affect loading. In general, however, the more enzyme immobilized on a specific amount of carrier, the lower the coupling efficiency. Higher purity enzyme can sometimes increase the coupling efficiency for a given activity level. An example of coupling efficiency as a function of enzyme loading for immobilized lactase was given by Eaton et al.[157] and by Ford and Pitcher.[158] Yamane[159] has also presented discussions on this topic.

Feed composition can be an important factor in system performance with Pitcher[135] reporting apparent half-lives for immobilized lactase ranging from 2 days to over 100 days at the same temperature but using different feeds. During development studies it is essential to duplicate as closely as possible the feed intended for commercial use. Microbial growth or trace contaminants such as heavy metals can severely decrease enzyme activity in the course of long-term continuous operation. Salts, protein, or other contaminants must, in some cases, be removed from the feed before it enters the immobilized enzyme reactor. Improved performance must be weighed against feed treatment costs.

The control or prevention of microbial contamination of vital importance commercially has received little attention. Barndt et al.[160] evaluated the effect of various sanitizing agents including hydrogen peroxide, iodophor, an acid cleaner, and a quaternary amine on a lactase-collagen complex. Pitcher et al.[150] used periodic backflushing with dilute acetic acid to control successfully microbial contamination in immobilized lactase columns. In several reported cases,[161,162] immobilized enzymes themselves have been used as bacteriacidal agents.

Plugging problems and high pressure drop, encountered primarily in fixed-bed reactors, result from a number of factors as discussed in Section VII of this chapter. Since pressure drop is sensitive to particle size, a trade-off between pore diffusion problems and pressure drop must sometimes be made.

The fact that an immobilized enzyme processing step is introduced may necessitate or eliminate additional processing steps. For example, sometimes product discoloration or other time-dependent side reactions can be avoided by the short residence time required by certain immobilized enzyme reactor systems. On the other hand additional feed purification may be required to enhance enzyme life. Electrolyte activators used to increase enzyme activity or agents introduced to enhance enzyme stability must sometimes be removed from the product stream.

It is apparent that optimizing such a complex set of parameters could be a lengthy task. In fact, in order to avoid prohibitive development costs, it is usually necessary to make some decisions based on experience or judgment rather than a complete and rigorous optimization.

REFERENCES

1. **Pitcher, W. H., Jr.,** Design and operation of immobilized enzyme reactors, in *Immobilized Enzymes for Industrial Reactors,* Messing, R. A., Ed., Academic Press, New York, 1975, chap. 9.
2. **Pitcher, W. H., Jr.,** Engineering of immobilized enzyme systems, *Catal. Rev. Sci. Eng.,* 12(1), 37, 1975.
3. **Pitcher, W. H., Jr.,** Design and operation of immobilized enzyme reactors, in *Advances in Biochemical Engineering,* Vol. 10, Ghose, T. K., Fiechter, A., and Blakebrough, N., Eds., Springer-Verlag, Heidelberg, 1978, 1.
4. **Vieth, W. R., Venkatasubramanian, K., Constantinides, A., and Davidson, B.,** Design and analysis of immobilized-enzyme flow reactors, *Appl. Biochem. Bioeng.,* 1, 221, 1976.
5. **Butterworth, T. A., Wang, D. I. C., and Sinskey, A. J.,** Application of ultrafiltration for enzyme retention during continuous enzymatic reaction, *Biotechnol. Bioeng.,* 12, 615, 1970.
6. **O'Neill, S. P., Wykes, J. R., Dunnill, P., and Lilly, M. D.,** An ultrafiltration-reactor system using a soluble immobilized enzyme, *Biotechnol. Bioeng.,* 13, 319, 1971.
7. **Coughlin, R. W., Aizawa, M., and Charles, M.,** Preparation and properties of soluble-insoluble nicotinamide coenzymes, *Biotechnol. Bioeng.,* 18, 199, 1976.
8. **Robinson, P. J., Dunnill, P., and Lilly, M. D.,** The properties of magnetic supports in relation to immobilized enzyme reactors, *Biotechnol. Bioeng.,* 15, 603, 1973.
9. **Havewala, N. B. and Weetall, H. H.,** Catalytic Stirrer, U.S. Patent 3,767,535, 1973.
10. **Dinelli, D.,** Fibre-entrapped enzymes, *Process Biochem.,* 7, 9, 1972.
11. **Morisi, F., Pastore, M., and Viglia, A.,** Reduction of lactose content of milk by entrapped β-galactosidase. I. Characteristics of β-galactosidase from yeast and *Escherichia coli, J. Dairy Sci.,* 56, 1123, 1973.
12. **Marconi, W., Gulinelli, S., and Morisi, F.,** Properties and use of invertase entrapped in fibers, *Biotechnol. Bioeng.,* 16, 501, 1974.
13. **Giacobbe, F., Iasonna, A., and Cecer, F.,** Production of 6APA in the penicillin G fermentation plant by using fiber-entrapped penicillin amidase, in *Enzyme Engineering,* Vol. 4, Broun, G. B., Manecke, G., and Wingard, L. B., Jr., Eds., Plenum Press, New York, 1978.

14. **Closset, G. P., Shah, Y. T., and Cobb, J. T.,** Analysis of membrane reactor performance for hydrolysis of starch by glucoamylase, *Biotechnol. Bioeng.*, 15, 441, 1973.
15. **Closset, G. P., Cobb, J. T., and Shah, Y. T.,** Study of performance of a tubular membrane reactor for an enzyme catalyzed reaction, *Biotechnol. Bioeng.*, 16, 345, 1974.
16. **Emery, A. H.,** Annular column enzyme reactors, in *Enzyme Engineering*, Vol. 2, Pye, E. K. and Wingard, L. B., Jr., Eds., Plenum Press, New York, 1974.
17. **Emery, A., Sorenson, M., Kolarik, M., Swanson, S., and Lim, H.,** An annular bound-enzyme reactor, *Biotechnol. Bioeng.*, 16, 1359, 1974.
18. **Venkatasubramanian, K. and Vieth, W. R.,** Effect of pressure on the hydrolysis of sucrose by invertase immobilized on collagen, *Biotechnol. Bioeng.*, 15, 583, 1973.
19. **Benoit, M. R. and Kohler, J. T.,** An evaluation of a ceramic monolith as an enzyme support material, *Biotechnol. Bioeng.*, 17, 1617, 1975.
20. **Horvath, C., Solomon, B. A., and Engasser, J. M.,** Measurement of radical transport in slug flow using enzyme tubes, *Ind. Eng. Chem. Fundam.*, 12, 431, 1973.
21. **Coughlin, R. W.,** Method of Carrying out Enzyme-Catalyzed Reactions, U.S. Patent 3,928,143, 1975.
22. **Charles, M., Coughlin, R. W., Allen, B. R., Paruchuri, E. K., and Hasselbergen, F. X.,** Increasing economic value of whey waste waters using immobilized lactase, Paper 17b, presented at AIChE 66th Annual Meeting, Philadelphia, 1973.
23. **Lee, Y. Y., Wun, K., and Tsao, G. T.,** Kinetics and mass transfer characteristics of glucose isomerase immobilized on porous glass, Paper 11a, presented at the AIChE 77th National Meeting, Pittsburgh, 1974.
24. **Emery, A. N. and Revel-Chion, L.,** Hydrodynamic effects in the design of fluidized bed immobilized enzyme reactors, Paper 11b, presented at the AIChE 77th National Meeting, Pittsburgh, 1974.
25. **Emery, A. N. and Cardoso, J. P.,** Parameter evaluation and performance studies on a fluidized-bed immobilized enzyme reactor, *Biotechnol. Bioeng.*, 20, 1903, 1978.
26. **Chung, S. F. and Wen, C. Y.,** Longitudinal dispersion of liquid flowing through fixed and fluidized beds, *AIChE J.*, 14, 857, 1968.
27. **Allen, B. R.,** personal communication, 1978.
28. **Scott, C. D., Hancher, C. W., and Shumate, S. E., II,** A tapered fluidized bed as a bioreactor, in *Enzyme Engineering*, Vol. 4, Broun, G. B., Manecke, G., and Wingard, L. B., Jr., Eds., Plenum Press, New York, 1978.
29. **Gelf, G. and Boudrant, J.,** Enzymes immobilized on a magnetic support-preliminary study of a fluidized bed enzyme reactor, *Biochim. Biophys. Acta*, 334, 467, 1974.
30. **Lilly, M. D.,** Comments at enzyme engineering conference, Bad Neuenahr, West Germany, 1977.
31. **Siimer, E.,** Generalized rate equation for one-substrate enzymatic reactions, *Biotechnol. Bioeng.*, 20, 1853, 1978.
32. **Havewala, N. B. and Pitcher, W. H., Jr., Eds.,** *Enzyme Engineering*, Vol. 2, Pye, E. K. and Wingard, L. B., Jr., Eds., Plenum Press, New York, 1974.
33. **Dixon, M. and Webb, E. C.,** *Enzymes*, Academic Press, New York, 1964.
34. **Plowman, K. M.,** *Enzyme Kinetics*, McGraw-Hill, New York, 1972.
35. **Reiner, J. M.,** *Behavior of Enzyme Systems*, Van Nostrand Reinhold, New York, 1969.
36. **Westley, J.,** *Enzymic Catalysis*, Harper & Row, New York, 1969.
37. **Atkins, G. L. and Nimmo, I. A.,** The reliability of Michaelis constants and maximum velocities estimated by using the integrated Michaelis-Menten equation, *Biochem. J.*, 135, 779, 1973.
38. **Bates, D. J. and Friedon, C.,** Treatment of enzyme kinetic data, *J. Biol. Chem.*, 248, 7878, 1973.
39. **Bates, D. J. and Briedon, C.,** Full time course studies on the oxidation of reduced coenzyme by bovine liver glutamate dehydrogenase, *J. Biol. Chem.*, 248, 7885, 1973.
40. **Halwachs, W.,** K_m and Y_{Max} from only one experiment, *Biotechnol. Bioeng.*, 20, 281, 1978.
41. **Eisenthal, R. and Cornish-Bowden, A.,** The direct linear plot, *Biochem. J.*, 139, 715, 1974.
42. **Cornish-Bowden, A. and Eisenthal, R.,** Statistical considerations in the estimation of enzyme kinetic parameters by the direct linear plot and other methods, *Biochem. J.*, 139, 721, 1974.
43. **Cornish-Bowden, A. and Eisenthal, R.,** Estimation of Michaelis constant and maximum velocity from the direct linear plot, *Biochim. Biophys. Acta*, 523, 268, 1978.
44. **Yun, S. L. and Suelter, C. H.,** A simple method for calculating K_m and V from a single enzyme reaction progress curve, *Biochim. Biophys. Acta*, 480, 1, 1977.
45. **Storer, A. C., Darlison, M. G., and Cornish-Bowden, A.,** The nature of experimental error in enzyme kinetic measurements, *Biochem. J.*, 151, 361, 1975.
46. **Fjellstedt, T. A. and Schlesselman, J. J.,** A simple statistical method for use in kinetic analysis based on Lineweaver-Burk plots, *Anal. Biochem.*, 80, 224, 1977.

47. **Duggleby, R. G. and Morrison, J. F.**, The analysis of progress curves for enzyme-catalyzed reactions by non-linear regression, *Biochim. Biophys. Acta*, 481, 297, 1977.
48. **Duggleby, R. G. and Morrison, J. F.**, Progress curve analysis in enzyme kinetics model discrimination and parameter estimation, *Biochim. Biophys. Acta*, 526, 398, 1978.
49. **Wilson, E. J. and Geankoplis, C. J.**, Liquid mass transfer at very low Reynolds numbers in packed beds, *Ind. Eng. Chem. Fundam.*, 5, 9, 1966.
50. **Rovito, B. J. and Kittrel, J. R.**, Film and pore diffusion studies with immobilized glucose oxidase, *Biotechnol. Bioeng.*, 15, 143, 1973.
51. **Raman, S. V., Horbett, T. A., and Hoffman, A. S.**, *J. Mol. Catal.*, 2, 275, 1977.
52. **Tourine, P., Laguerie, C., and Coudrec, J. P.**, Mass transfer in a liquid fluidized bed at low Reynolds numbers, *Chem. Eng. Sci.*, 32, 1259, 1977.
53. **Vanadurongwan, V., Laguerie, C., and Coudrec, J. P.**, Effect of the physical properties on mass transfer in liquid fluidization, *Chem. Eng. J. (Lausanne)*, 12, 29, 1976.
54. **Engasser, J. M.**, A fast evaluation of diffusion effects on bound enzyme activity, *Biochim. Biophys. Acta*, 526, 301, 1978.
55. **Satterfield, C. N.**, *Mass Transfer in Heterogeneous Catalysis*, MIT Press, Cambridge, 1970, chap. 2.
56. **Satterfield, C. N.**, personal communication, 1972.
57. **Brian, P. L. T. and Hales, H. B.**, Effects of transpiration and changing diameter on heat and mass transfer to spheres, *AIChE J.*, 15, 419, 1969.
58. **O'Neill, S. P.**, External diffusional resistance in immobilized enzyme catalysis, *Biotechnol. Bioeng.*, 14, 675, 1972.
59. **Horvath, C. and Solomon, B. A.**, Open tubular heterogeneous enzyme reactors: preparation and kinetic behavior, *Biotechnol. Bioeng.*, 14, 885, 1972.
60. **Bunting, D. S. and Laidler, K. J.**, Flow kinetics of L-aspariginase attached to nylon tubing, *Biotechnol. Bioeng.*, 16, 119, 1974.
61. **Kobayashi, T. and Liidler, K. H.**, Theory of the kinetics of reactions catalyzed by enzymes attached to the interior surfact of tubes, *Biotechnol. Bioeng.*, 16, 99, 1974.
62. **Narinesingh, D., Ngo, T. T., and Laidler, K. J.**, Flow kinetics of β-galactosidase chemically attached to nylon tubing, *Can. J. Biochem.*, 53, 1061, 1975.
63. **Ngo, T. T. and Laidler, K. H.**, Immobilized chemically attached to nylon tubing, *Biochim. Biophys. Acta*, 377, 317, 1975.
64. **Blaedel, W. J., Kissel, T. R., and Boguslaski, R. C.**, Kinetic behavior of enzymes immobilized in artificial membranes, *Anal. Chem.*, 44, 2030, 1972.
65. **Fink, D. H., Na, T. Y., and Schultz, J. S.**, Effectiveness factor calculations for immobilized enzyme catalysts, *Biotechnol. Bioeng.*, 15, 879, 1973.
66. **Gondo, S., Isayama, S., and Kusunoki, K.**, Effects of internal diffusion on the Lineweaver-Burk plots for immobilized enzymes, *Biotechnol. Bioeng.*, 17, 423, 1975.
67. **Gondo, S., Sato, T., and Kusunoki, K.**, Lineweaver-Burk plots for the insolubilized enzyme particle, *Chem. Eng. Sci.*, 28, 1773, 1973.
68. **Thomas, D., Broun, G., and Siligny, E.**, Monoenzymatic model membranes: diffusion-reaction kinetics and phenomena, *Biochimie*, 54, 229, 1972.
69. **Kasche, V., Lundqvist, H., Bergman, R., and Axen, R.**, A theoretical model describing steady-state catalysis by enzymes immobilized in spherical gel particles. Experimental study of α-chymotrypsin-sepharose, *Biochem. Biophys. Res. Commun.*, 45, 615, 1971.
70. **Kobayashi, T. and Laidler, K. J.**, Kinetic analysis for solid-supported enzymes, *Biochim. Biophys. Acta*, 302, 1973.
71. **Kobayashi, T. and Moo-Young, M.**, The kinetic and mass transfer behavior of immobilized invertase on ion-exchange resin beads, *Biotechnol. Bioeng.*, 15, 47,1973.
72. **Lawrence, R. L. and Okay, V.**, Diffusion and reaction in a double enzyme supported catalyst, *Biotechnol. Bioeng.*, 15, 217, 1973.
73. **Mogensen, A. O. and Vieth, W. R.**, Mass transfer and biochemical reaction with semipermeable microcapsules, *Biotechnol. Bioeng.*, 15, 467, 1973.
74. **Sundaram, P. V., Tweedale, A., and Laidler, K. H.**, Kinetic laws for solid-supported enzymes, *Can. J. Chem.*, 48, 1498, 1970.
75. **Vieth, W. R., Mendiratta, A. K., Mogensen, A. O., Saini, R., and Venkatasubramanian, K.**, Mass transfer and biochemical reaction in enzyme membrane reactor systems — I. Single enzyme reactions, *Chem. Eng. Sci.*, 28, 1013, 1973.
76. **Bischoff, K. B.**, Effectiveness factors for general reaction rate forms, *AIChE J.*, 11, 351, 1965.
77. **Moo-Young, M. and Kobayashi, T.**, Effectiveness factors for immobilized-enzyme reactions, *Can. J. Chem. Eng.*, 50, 162, 1972.

78. **Pitcher, W. H., Jr.**, Restricted Diffusion in Liquids within Fine Pores, Sc.D. Thesis, Massachusetts Institute of Technology, Cambridge, 1972.
79. **Satterfield, C. N., Colton, C. K., and Pitcher, W. H., Jr.**, Restricted diffusion in liquids within fine pores, *AIChE J.*, 19, 628, 1973.
80. **Colton, C. K., Satterfield, C. N., and Lai, C. J.**, Diffusion and partitioning of macromolecules within finely porous glass, *AIChE J.*, 21, 289, 1975.
81. **Hamilton, B. K., Gardner, C. R., and Colton, C. K.**, Effect of diffusional limitations on Lineweaver-Burk plots for immobilized enzymes, *AIChE J.*, 20, 503, 1974.
82. **Frouws, M. J., Vellenga, K., and DeWilt, H. G. J.**, Combined external and internal mass transfer effects in heterogeneous (enzyme) catalysis, *Biotechnol. Bioeng.*, 18, 53, 1976.
83. **Kao, H. S-P.**, Effectiveness factors for reversible reactions, *Ind. Eng. Chem. Fundam.*, 7, 664, 1968.
84. **Roberts, G. W. and Satterfield, C. N.**, Effectiveness factor for porous catalysts, *Ind. Eng. Chem. Fundam.*, 4, 288, 1965.
85. **Knudsen, C. W., Roberts, G. W., and Satterfield, C. N.**, Effect of geometry on catalyst effectiveness factor, *Ind. Eng. Chem. Fundam.*, 5, 325, 1966.
86. **Swanson, S. J., Emery, A., and Lim, H. C.**, Pore diffusion in packed-bed reactors containing immobilized glucoanylase, *AIChE J.*, 24, 30, 1978.
87. **Horvath, C., Sherdalman, L. H., and Light, R. T.**, Open tube heterogeneous enzyme reactors: analysis of a theoretical model, *Chem. Eng. Sci.*, 28, 431, 1973.
88. **Rony, P. R.**, Multiphase catalysis. II. Hollow fiber catalysts, *Biotechnol. Bioeng.*, 18, 431, 1971.
89. **Waterland, L. R., Michaels, A. S., and Robertson, C. R.**, A theoretical model for enzymatic catalysis using asymmetric hollow fiber membranes, *AIChE J.*, 20, 50, 1974.
90. **Webster, I. A. and Shuler, M. L.**, Mathematical models for hollow fiber enzyme reactors, *Biotechnol. Bioeng.*, 20, 1541, 1978.
91. **Halwachs, W., Wandrey, C., and Schugerl, K.**, Immobilized α-chymotripsin: pore diffusion control owing to pH gradients in the catalyst particles, *Biotechnol. Bioeng.*, 20, 541, 1978.
92. **Bunting, P. S. and Laidler, K. J.**, Kinetic studies on solid-supported β-galactosidase, *Biochemistry*, 11, 4477, 1972.
93. **Marsh, D. R., Lee, Y. Y., and Tsao, G. T.**, Immobilized glucoamylase on porous glass, *Biotechnol. Bioeng.*, 15, 483, 1973.
94. **Wheeler, A.**, Reaction rates and selectivity in catalyst pores, *Advances in Catalysis*, Vol. 3, Frankenburg, W. G., Komarewsky, V. I., and Rideal, E. K., Eds., Academic Press, New York, 1951, 250.
95. **Havewala, N. B. and Pitcher, W. H., Jr.**, Glucose production from cornstarch: reactor system parameter study, unpublished report.
96. **Buchholz, K. and Ruth, W.**, Temperature dependence of a diffusion-limited immobilized enzyme reaction, *Biotechnol. Bioeng.*, 18, 95, 1976.
97. **Horvath, C.**, Pellicular immobilized enzymes, *Biochim. Biophys. Acta*, 358, 164, 1974.
98. **Horvath, C. and Engasser, J-M.**, Pellicular heterogeneous catalysts. A theoretical study of the advantages of shell structured immobilized enzyme particles, *Ind. Eng. Chem. Fundam.*, 12, 229, 1973.
99. **Levenspiel, O.**, *Chemical Reaction Engineering*, John Wiley & Sons, Inc., New York, 1962, chap. 9.
100. **Kobayashi, T. and Moo-Young, M.**, Backmixing and mass transfer in the design of immobilized-enzyme reactors, *Biotechnol. Bioeng.*, 13, 893, 1971.
101. **DeVera, A. L. and Varma, A.**, Substrate inhibited enzyme reaction in a tubular reactor with axial dispersion, *Chem. Eng. Sci.*, 34, 275, 1979.
102. **Hornby, W. E., Lilly, M. D., and Crook, E. M.**, Some changes in the reactivity of enzymes resulting from their chemical attachment to water-insoluble derivatives of cellulose, *Biochem. J.*, 107, 669, 1968.
103. **Hamilton, B. K., Stockmeyer, L. J., and Colton, C. K.**, Diffusive and electrostatic effects with immobilized enzymes, *J. Theor. Biol.*, 41, 547, 1973.
104. **Shuler, M. L., Aris, R., and Tsuchiya, H. M.**, Diffusive and electrostatic effects with insoluble enzymes, *J. Theor. Biol.*, 35, 67, 1972.
105. **Kobayashi, T. and Laidler, K. J.**, Theory of the kinetics of reactions catalyzed by enzymes attached to membranes, *Biotechnol. Bioeng.*, 16, 77, 1974.
106. **Karube, I., Yugeta, Y., and Suzuki, S.**, Electric field control of lipase membrane activity, *Biotechnol. Bioeng.*, 19, 1493, 1977.
107. **Gelf, D., Thomas, D., and Broun, G.**, Water insoluble enzyme columns: kinetic study on steady state and stranient conditions, *Biotechnol. Bioeng.*, 16, 315, 1974.
108. **Shyam, R., Davidson, B., and Vieth, W. R.**, Mass transfer and biochemcial reaction in enzyme membrane reactor systems. II. Expanded analysis for single enzyme systems: effects of enzyme intermediates, denaturation and elution, *Chem. Eng. Sci.*, 30, 669, 1975.

109. **Ryu, D. Y., Bruno, C. F., Lee, B. K., and Venkatasubramanian, K.**, Microbial penicillin amidohydrolase and the performance of a continuous enzyme reactor system, Proc. IV Int. Ferment. Symp., *Fermentation Technol. Today*, 307, 1972.
110. **Reilly, P. J.**, Scale up study on several enzymatic processes for industrial application, *Iowa State Univ. Eng. Res. Inst. Rep.*, ERI-77176, 136, 1976.
111. **Marsh, D. R. and Tsao, G. T.**, A heat transfer study on packed bed reactors of immobilized glucoamylase, *Biotechnol. Bioeng.*, 18, 349, 1976.
112. **Pitcher, W. H., Jr.**, Process engineering for industrial enzyme reactors, *Ann. N.Y. Acad. Sci.*, 326, 155, 1979.
113. **Beck, S. R. and Rose, H. F.**, Encapsulated enzyme: a glucoamylase copolymer system, *Ind. Eng. Chem. Prod. Res. Dev.*, 12, 260, 1973.
114. **Carberry, J. J. and Gorring, R. L.**, Time-dependent pore-mouth poisoning of catalysts, *J. Catal.*, 5, 529, 1966.
115. **Cardoso, J. P. and Emery, A. N.**, A new model to describe enzyme inactivation, *Biotechnol. Bioeng.*, 20, 1471, 1978.
116. **Krishanswamy, S. and Kittrell, J. R.**, Deactivation studies of immobilized glucose oxidase, *Biotechnol. Bioeng.*, 20, 821, 1978.
117. **Altomare, R. E., Greenfield, P. F., and Kittrell, J. R.**, Inactivation of immobilized fungal catalase by hydrogen peroxide, *Biotechnol. Bioeng.*, 16, 1675, 1974.
118. **Regan, D. L., Dunnill, P., and Lilly, M. D.**, Immobilized enzyme reaction stability: attrition of the support material, *Biotechnol. Bioeng.*, 16, 333, 1974.
119. **Weetall, H. H. and Havewala, N. B.**, Continuous production of dextrose from cornstarch: a study of reactor parameters necessary for commercial application, *Enzyme Engineering*, Wingard, L. B., Jr.,Ed., *Biotechnol.Bioeng.*, Symp. No. 3, John Wiley & Sons, New York, 1972, 241.
120. **Weetall, H. H., Havewala, N. B., Garfinkel, H. M., Buehl, W. M., and Baum, G.**, Covalent bond between the enzyme amyloglucosidase and a porous glass carrier: the effect of shearing, *Biotechnol. Bioeng.*, 16, 169, 1974.
121. **Ollis, D. F.**, Diffusion influences in denaturable insolubilized enzyme catalysts, *Biotechnol. Bioeng.*, 14, 871, 1972.
122. **Feder, J., Kochavi, D., Anderson, R. G., and Wildi, B. S.**, Stabilizing of proteolytic enzymes in solution, *Biotechnol. Bioeng.*, 20, 1865, 1978.
123. **Asakura, T., Adachi, K., and Schwartz, E.**, Stabilizing effect of various organic solvents on protein, *J. Biol. Chem.*, 253, 6423, 1978.
124. **O'Neill, S. P.**, Inactivation of immobilized catalase by hydrogen peroxide in continuous reactors, *Biotechnol. Bioeng.*, 14, 201, 1972.
125. **Cho, Y. K. and Bailey, J. E.**, The influence of peroxide-stabilizing agents on enzyme deactivation by H_2O_2, *Biotechnol. Bioeng.*, 19, 157, 1977.
126. **Cho, Y. K. and Bailey, J. E.**, Enzyme immobilization on activated carbon: alleviation of enzyme deactivation by hydrogen peroxide, *Biotechnol. Bioeng.*, 19, 769, 1977.
127. **Tramper, J., Muller, F., and Van Der Plas, H. C.**, Immobilized xanthine oxidase: kinetics, (in) stability, and stabilization by coimmobilization with superoxide dismutase and catalase, *Biotechnol. Bioeng.*, 20, 1507, 1978.
128. **Johnson, D. B.**, Horse liver alcohol dehydrogenase immobilized on inorganic supports: stabilizing effect of bound protein, *Biotechnol. Bioeng.*, 20, 1117, 1978.
129. **Johnson, D. B. and Coughlan, M. P.**, Studies on the stability of immobilized xanthine oxidase and urate oxidase, *Biotechnol. Bioeng.*, 20, 1085, 1978.
130. **Torchilin, V. P., Maksimenko, A. V., Smirnov, V. N., Berezin, I. V., Klibanov, A. M., and Martinek, K.**, The principles of enzyme stabilization. III. The effect of the length of intra-molecular cross-linkages on thermostability of enzymes, *Biochim. Biophys. Acta*, 522, 277, 1978.
131. **Okada, H. and Urabe, I.**, Alteration of enzymatic characteristics of bacterial alpha-amylase by acylation, in *Immobilized Enzyme Technology-Research and Applications*, Weetall, H. H. and Suzuki, J., Eds., Plenum Press, New York, 1978, 37.
132. **Flynn, A. and Johnson, D. B.**, Some factors affecting the stability of glucoamylase immobilized on hornblende, and on other inorganic supports, *Biotechnol. Bioeng.*, 20, 1445, 1978.
133. **Perry, J. H.**, *Chemical Engineers Handbook*, 4th ed., McGraw-Hill, New York, 1963, 5.
134. **Bucholz, K. and Godelmann, B.**, Pressure drop across compressible beds, in *Enzyme Engineering*, Vol. 4, Broun, G. B., Manecke, G., and Wingard, L. B., Jr., Eds., Plenum Press, New York, 1978.
135. **Pitcher, W. H., Jr.**, Immobilized lactase for whey hydrolysis: stability and operating strategy, in *Enzyme Engineering*, Vol. 4, Broun, G. B., Manecke, G., and Wingard, L. B., Jr., Eds., Plenum Press, New York, 1978.
136. **Crowe, C. M.**, Optimization of reactors with catalyst decay I — single tubular reactor with uniform temperature, *Can. J. Chem. Eng.*, 48, 576, 1970.

137. **Levenspiel, O. and Sadana, A.**, The optimum temperature policy for a deactivating catalytic packed bed reactor, *Chem. Eng. Sci.*, 33, 1393, 1978.
138. **Goldman, R. and Katchalski, E.**, Kinetic behavior of a two-enzyme membrane carrying out a consecutive set of reactions, *J. Theor. Biol.*, 32, 234, 1971.
139. **Ford, J.R.**, Enzyme Engineering. I. Characterization of Immobilized Enzymes. II. Mutli-enzyme Reaction Systems, Ph.D. Thesis, Tulane University, New Orleans, 1972.
140. **Kent, C. A. and Emery, A. N.**, Reactor design for the immobilized enzyme system amyloglucosidase-glucose isomerase. Paper 11e, presented at AIChE 77th Annual Meeting, Pittsburgh, 1974.
141. **Choi, C. Y. and Perlmutter, D. D.**, Optimal catalyst distribution in a dual enzyme sequential system, *AIChE J.*, 23, 310, 1977.
142. **Chang, H. N. and Reilly, P. J.**, Experimental operation of immobilized multienzyme systems in optimal and suboptimal configurations, *Biotechnol. Bioeng.*, 20, 243, 1978.
143. **Ho, S. P. and Kostin, M. D.**, Kinetics of immobilized multienzyme systems, *J. Chem. Phys.*, 61, 918, 1974.
144. **Koch-Schmidt, A-C., Mattiasson, B., and Mosbach, K.**, Aspects on microenvironmental compartmentation. An evaluation of the influence of restricted diffusion, exclusion effects, and enzyme proximity on the overall efficiency of the sequential two-enzyme system malate-dehydrogenase-citrate synthase in its soluble and immobilized form, *Eur. J. Biochem.*, 81, 71, 1977.
145. **Barker, S. A. and Somers, P. J.**, Biotechnology of immobilized multienzyme systems, in *Advances in Biochemical Engineering*, Vol. 10, Ghose, T. K., Fiechter, A., and Blakebrough, N., Eds., Springer-Verlag, Berlin, 1978.
146. **Mosbach, K. and Mattiasson, B.**, Multistep enzyme systems, in *Methods in Enzymology*, Vol. 44, Mosbach, K., Ed., Academic Press, New York, 1976.
147. **Coughlin, R. W., Aizawa, M., Alexander, B. F., and Charles, M.**, Immobilized-enzyme continuous-flow reactor incorporating continuous electrochemical regeneration of NAD, *Biotechnol. Bioeng.*, 17, 515, 1975.
148. **Coughlin, R. W. and Alexander, B. F.**, Simplified flow reactor for electrochemically driven enzymatic reactions involving cofactors, *Biotechnol. Bioeng.*, 17, 1379, 1975.
149. **Kelly, R. M. and Kirwan, D. J.**, Electrochemical regeneration of NAD+ on carbon electrodes, *Biotechnol. Bioeng.*, 19, 1215, 1977.
150. **Pitcher, W. H., Jr., Ford, J. R., and Weetall, H. H.**, The preparation, characterization and scale-up of a lactase system immobilized to inorganic supports for the hydrolysis of acid whey, in *Methods in Enzymology*, Vol. 44, Mosbach, K., Ed., Academic Press, New York, 1976.
151. **Pitcher, W. H., Jr. and Weetall, H. H.**, Cost of saccharification by immobilized glucoamylase, *Enzyme Technol. Dig.*, 3, 127, 1975.
152. **Wandrey, C.**, Macrokinetics and reactor design for the industrial application of enzymes in L-amino production, in *Enzyme Engineering*, Vol. 3, Pye, E. K. and Weetall, H. H., Eds., Plenum Press, New York, 1978.
153. **Flaschel, E. and Wandrey, C.**, Economical aspects of continuous operation with biocatalysis, in *Enzyme Engineering*, Vol. 4, Broun, G. B., Manecke, G., and Wingard, L. B., Jr., Eds., Plenum Press, New York, 1978.
154. **Giacobbe, F., Iasonna, A., and Cecer, F.**, Production of GAPA in the penicillin G fermentation plant by using fiber-entrapped penicillin amidase, in *Enzyme Engineering*, Vol. 4, Broun, G. B., Manecke, G., and Wingard, L. B., Jr., Eds., Plenum Press, New York, 1978.
155. **Popper, H., Ed.**, *Modern Cost-Engineering Techniques*, McGraw-Hill, New York, 1970.
156. **Guthrie, K. M.**, *Process Plant Estimating Evaluation and Control*, Craftsman Book Company of America, Solana Beach, Calif., 1974.
157. **Eaton, D. L., Ford, J. R., and Pitcher, W. H., Jr.**, The use of controlled pore ceramic bodies for enzyme immobilization, Paper 11d, presented at AIChE 77th Annual Meeting, Pittsburgh, 1974.
158. **Ford, J. R. and Pitcher, W. H., Jr.**, Enzyme engineering case study: immobilized lactase, in *Immobilized Enzyme Technology — Research and Applications*, Weetall, H. H. and Suzuki, S., Eds., Plenum Press, New York, 1975.
159. **Yamane, T.**, A proposal for accurate determination of "true" fractional retention of enzyme activity on immobilization, *Biotechnol. Bioeng.*, 19, 749, 1977.
160. **Barndt, R. L., Leeder, J. G., Giacin, J. R., and Kleyn, D. H.**, Sanitation of a biocatalytic reactor used for hydrolysis of acid whey, *J. Food Sci.*, 40, 291, 1975.
161. **Karube, I., Suganuma, T., and Suzuki, S.**, Bacteriolysis by immobilized enzymes, *Biotechnol. Bioeng.*, 19, 301, 1977.
162. **Mattiasson, B.**, The use of coimmobilized lysozyme as a bacteriacide in enzyme columns: a step toward the design of self-sterilizing enzyme columns, *Biotechnol. Bioeng.*, 19, 777, 1977.

Chapter 3

REQUIREMENTS UNIQUE TO THE FOOD AND BEVERAGE INDUSTRY

J. F. Roland

TABLE OF CONTENTS

I. Introduction 56

II. Parameters Affecting the Activities of the Food and Beverage Industry 57
A. Nutritional Mission 57
B. Industry Size 57
C. Energy Requirements 58
D. Productivity-Growth 58

III. Governmental Regulations Affecting Food Processing 59
A. Overview of Regulatory Agencies 59
B. Origins of the FDA 59
1. Delaney Clause 61

IV. Governmental Regulation of Enzymes 61
A. GRAS Food Ingredients 61
B. Enzymes as Food Additives in U.S. 62
1. GRAS Petition for FDA Approval 63
2. Toxicological Format for FDA Approval 63
C. Enzymes as Food Additives in Europe 65
D. Immobilized Enzyme as Food Additives 65

V. Applications of Immobilized Enzymes in Food Processing 66
A. Areas of Food Modification 66
B. Insolubilizing Methods 66
1. Stability of Immobilized Enzymes 67
2. Testing Chemicals and Solvents Used for Insolubilizing Enzymes 67

VI. Economic Constraints on the Application of Immobilized Enzymes Systems in Food and Beverage Processes 69
A. Large-Scale Continuous Process Plants 69
B. Long-Term Capital Investments 70
C. Economic Overview of Immobilized Enzyme Processes 70
D. Innovation in Process Development 70
E. Hydrolyzed Lactose Syrup Production 71
1. World-Wide Impact 71
2. Chemical and Enzymatic Hydrolysis Processes 71
3. Economic Evaluation of Syrup Process 72

VII. Technological Factors in Scaling Up Immobilized Enzyme Systems in the Food Industry 72
A. Problems of Scale-Up 72
B. Complexity of Fluid Food Flavors 73
C. Role of the Pilot Plant in Scale-Up 74
1. Development of the Rate Equation 74
2. Determination of Process Economics 74

VIII. Microbial Control of Immobilized Enzyme Reactors 75
A. Aspects of Quality Assurance 75
B. Establishment of Process Control — HACCP 75
1. Food Poisoning Microorganisms 75
2. Microbial Attachment Mechanism — "glycocalyx" 76
C. Sanitizing Immobilized Enzyme Systems 76
1. Surface Sanitizer 76
2. FDA Approved Antimicrobials 76
3. Experimental Control Chemicals 77

References 77

I. INTRODUCTION

Food processors have had a long and continuing interest in developing the potential of enzyme systems to modify foods and beverages. In recent years, this interest has been further intensified by the development of procedures that allow enzymes to be attached to solid surfaces while still retaining their functionality.

These immobilization processes of enzymes now allow food manufacturers a level of process control previously unavailable. Simply by altering flow rate of the feed stock, the desired degree of process modification can be attained. The added benefit of reuse of the enzyme over a period of many weeks or months provides an obvious economic attraction. Finally, as the enzyme no longer appears as an additive in the final product, an additional processing step to remove it in order to prevent overtreatment during storage due to continuing activity, is not necessary.

Although early reports some 60 years ago described immobilization of enzymes,[1] interest in the application of these observations was not further developed until Mitz[2] reported the successful adsorption of catalase to a cellulose anion exchanger with retention of activity. Subsequently, he used these catalase-cellulose particles in a plug-flow column to decompose 3 to 6% solutions of hydrogen peroxide. The process was operated semicontinuously for several days. Mitz later showed the active enzyme could be displaced from its attached form by the addition of salt solution.

In the 18 years following these early observations, over 1000 papers have been published dealing with immobilized enzymes. A variety of enzymes have now been adsorbed to modified celluloses (DEAE) which, in two instances, have led to the development of major industrial-scale immobilized enzyme processes. In one, case Tosa et al.[3] brought to industrial scale a process in which various amino acylases adsorbed to DEAE cellulose or DEAE Sephadex are employed to convert tonnage quantities of synthetic DL amino acids to the biologically available L-form.

Most recently, a process utilizing glucose isomerase adsorbed to DEAE cellulose has been patented and used in the full-scale conversion of corn syrup sugar to fructose.[4,5] This process received the IFT Industrial Achievement Award in 1975 and, in conjunction with five other more recently developed systems,[6] will permit the annual production of over four billion pounds of isomerized glucose.

Thus, it is evident the food and beverage industry, past and present, is becoming more closely involved in the utilization of this new form of enzyme technology. As research continues, newly developed systems will evolve and establish their economic merits. This will lead to improved processes designed to provide foods of enhanced flavor and nutritional character.

II. PARAMETERS AFFECTING THE ACTIVITIES OF THE FOOD INDUSTRY

Before delineating the various areas of potential application for immobilized enzymes and also some of the problems related to their development into scaled up food processes, it is important to provide an overview of the food and beverage industry. Some definition of the industry's mission is essential to a fuller understanding of the vital role it plays in feeding the nation.

A. Nutritional Mission

Based on its ability to supply the nutritional needs of over 250 million people in the U.S., plus a partial supplementation of 350 million people in Western Europe, our food industry is unquestionably the most important in the world. Recent estimates now indicate that about 70% of all the food consumed is processed in some degree.[7] In order to accomplish this monumental feeding task, and regardless of climatic variations which affect crop yields, highly efficient systems of crop-food production, processing, storage, and distribution are required. Likewise, the interfacing of individual unit processes into integrated food systems is becoming more and more essential to the implementation of new major food processes. The goals of the industry are to achieve higher productivity and reductions in production costs.

B. Industry Size

Some concept of the current size of the present food and beverage industry may be gained from a recent National Restaurant Association study which estimated total food sales in 1977 of $86.9 billion. The bulk of domestic food processing is carried out from a broad base of 22,000 food processors.[9] About 100 of these are major food producers, and within this group 28 are billion-dollar multinational food and beverage companies.[10] These latter organizations, in addition to their domestic production, have food processing capabilities in one or more of the 16 countries in Western Europe.

A domestic work force of about 1.53 million people is presently engaged in food processing, indicating the labor-intensive character of the industry.[8] The results of their productive endeavors are transferred to the rapidly increasing (10% annual growth rate) food service industry, with its even broader based work force of about 8 million people. Most of these workers are directly engaged in serving food to the consumer.

Since nutrition is involved in all aspects of food production, packaging, distribution, and marketing, a broad range of activities must be continuously monitored in order to ensure quality food products. The basic principles of food processing technology are primarily concerned with microbial and chemical safety, preservation of the inherent nutritional value of food products, and maintenance of flavor quality during storage.

C. Energy Requirements

In order to achieve this goal, significant energy expenditures are required. Recent estimates indicate that about 12% of the nation's total energy output is expended in carrying out the mission of the food processing industry; i.e., satisfying the nutritional needs of the consumer.[11] These energy requirements are channeled mainly into a variety of thermal processes. They are primarily designed to eliminate microorganisms and enzymes which, if unchecked, would result in deterioration of food upon storage, or would subsequently present a health hazard to the consumer. However, concomitant with preservation, the thermal process used must be operated to maximize the retention of nutrients. A recent review evaluating the effect of the three major preservation processes — blanching, pasteurization, and commercial sterilization — concluded that the food process industry is, in general, utilizing thermal processes which result in maximum nutrient retention.[12]

Another major area of energy consumption involves the concentration of liquid foods. There are a variety of fluid food products in this category, which include sugar syrups, milk and whey, citrus juice, soluble coffee and tea, and tomato paste. In order to concentrate these products, it is estimated that about 184 billion lb of water are removed annually. Energy requirements to accomplish this particular task are estimated at 59×10^{12} Btu/year. This represents about 4.4% (or roughly one third) of all the energy used in processed foods in the U.S. Specific energy requirements for fruit and vegetable canning plants indicate the use of about 3.5×10^6 Btu/ton of finished product. Similarly milk processing operations utilize about 1.3×10^6 Btu/ton of raw product. The conversion of 100 lb of milk to cheese and dry whey requires about 81,620 Btu.[13] Additional requirements for electrical power used for product cooling and refrigerated storage are inherent parts of dairy-cheese operations under current processing systems.

Programmed efforts to reduce energy requirements are an on-going part of virtually all food plant operations and a variety of mechanisms aimed at water conservation (recycling) and the controlled use of steam and electricity are being utilized. One whey processing plant has recently installed an electronically controlled six-effect evaporator which will further serve to reduce energy expenditures. This highly efficient unit removes 6.6 to 6.7 lb of water per pound of steam.[14]

D. Productivity-Growth

One measure of the economic viability of an industry (food processing) is productivity, a term which, from an economist's point of view, assesses the quantitative relationship between output and one or more inputs.[15] In practical terms, it is frequently expressed as:

$$\text{Labor Productivity} = \frac{\text{Labor Output}}{\text{Labor Input}}$$

This formula, however simple, clearly provides a basis for establishing corporate motivations to transform labor-intensive food processes into more capital-intensive, continuous operations, offering constant quality in production, reduced plant size, and the use of personnel primarily for controlling operations. It is for reasons such as these that insolubilized enzymes offer tangible advantages to industry.

As pointed out recently by Instrati,[16] as population growth slows to zero levels, industrial growth potential based on volume production, may become limited for many food manufacturers in the U.S. In terms of available populations that can sustain growth and productivity, only Western European markets appear viable. Primarily this is due to income limitations of the less developed countries. Obviously, the European market area is only open to U.S. companies with full-scale international operations that can compete on equal terms with the large continental food processors. It is evident, as indicated early by Terleckyj,[17] that increases in industry growth potential or productivity will mainly arise only from technical innovations in methods of production and the development of new products. Gains in productivity from these sources occur because improved production processes reduce real unit costs, and new products provide new marketing targets for sustained growth and output patterns. Developments stemming from these two major sources can only come about through the activities of organized research and development (R&D). Terleckyj's study on the relationship between R&D and productivity concluded there is a strong positive relationship between company financed R&D and productivity growth.[17] He further points out that R&D-intensive inputs from other companies or industries often are transferred or modified to fit particular circumstances within individual companies with R&D capabilities.

III. GOVERNMENTAL REGULATIONS AFFECTING FOOD PROCESSING

A. Overview of Regulatory Agencies

Every food manufacturer actively engaged in marketing products, modifying products, or developing new products recognizes his efforts are overshadowed by a variety of governmental laws and regulations. The agencies which are currently responsible for establishing guidelines and standards for food manufacturing are: OSHA, EPA, USDA, Department of the U.S. Treasury, U.S. Public Health, and FDA. At present, the agencies most concerned with the use of enzymes, either free or immobilized as food or beverage additives, are the FDA, the USDA, and the U.S. Department of the Treasury (Figure 1)

In order to more clearly understand the ever-increasing role the government has come to play in regulating the food industry and its efforts directed toward product development, it is necessary to review both the origins of the agencies and the broad mandate handed down by Congress regarding regulation of foods (Table 1).

B. Origins of the FDA

Historically the first Federal Food and Drug law was established in 1906 and provided a basis for the implementation of several of the first food Standards of Identity and also for the prevention of the sale of foods that were toxic, unsanitary, adulterated, or misbranded. The most important provision of the law was that it required the government to prove the foods under investigation were truly toxic or unsafe for human consumption.

Only minor amendments to the original occurred (1907, 1912) until the tragic deaths of 70 people who had taken the drug "Elixir Sulfanilamide" (1937) brought about new and more comprehensive legislation. In 1938, Congress enacted the Federal Food, Drug, and Cosmetic Act. This legislation, unique to the rest of the world, later came to serve as a guide to the formulation of food safety laws in many other countries. One of the major provisions of the Act was to establish and fund an agency known as the Food and Drug Administration. The primary functions of the agency were mainly administrative and provided only for the prohibition of the use of known toxic substances as food additives. However, programs for establishment of additional stand-

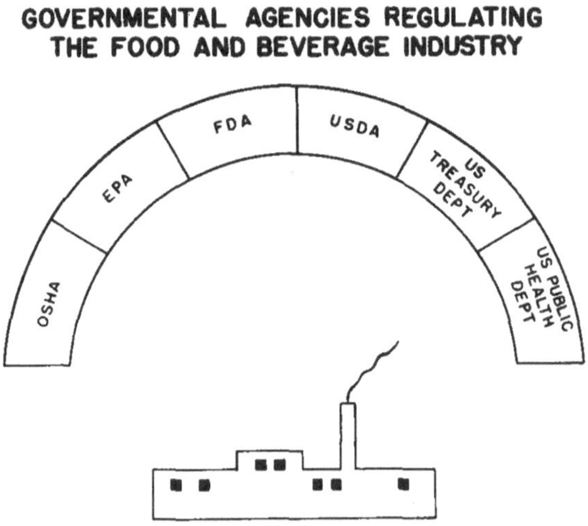

FIGURE 1. Governmental agencies regulating the food and beverage industry.

Table 1
LEGISLATIVE RESPONSE TO CALAMITIES

	Calamity		Response
1.	Ten children died of diptheria antitoxin	1.	Biologics Control Act of 1902
2.	Soldiers died as result of tainted canned beef	2.	Food & Drug Law — 1906
3.	"Elixir sulfanilimide" event — 70 people died	3.	Federal Food, Drug & Cosmetic Act 1938
4.	Physicians meeting in Milan, Italy on cancer	4.	Food Additives Amendment to the Food, Drug & Cosmetic Act 1958
5.	"Thalidomide" incident — Birth defects caused by new drug	5.	New Drug Act — 1962 (Public Law 87-871)
6.	Over 2,000 deaths of children Due to accidental poisoning	6.	Poison Prevention Packaging Act of 1970
7.	Over 350 reports annually of allergic reactions to cosmetics	7.	Cosmetic ingredient labeling-voluntary & mandatory provisions

ards of identity and of food quality were gradually developed, along with mandates to provide honest labeling and safe packaging.

The next major legislative development had its origins stemming from a cancer convention held in Milan, Italy, in 1956. One concensus decision arising from the physicians attending the Milan seminars was the observation that the presence of chemical additives in foods may be an initiating cause of cancer. This finding eventually came to the attention of the Congress, in particular Rep. James J. Delaney (D-N.Y.). He, along with Sen. Estes Kefauver (D-Tenn.), drafted a new amendment to the 1938 Act, known as the Food Additives Amendment of 1958. It was designed to provide enhanced public health protection by controlling all of the chemical additives entering the food supply. The food additive provisions of the law require the FDA to consider "safety factors which, in the opinion of experts qualified by scientific training and experience to evaluate the safety of food additives, are generally recognized as appropriate for the use of animal experimentation data". This change in legislation further served to remove the burden on the government to prove that an added chemical was toxic or deleterious.

1. Delaney Clause

One additional clause added to the legislation has become a matter of major scientific and public controversy. It is known as the Delaney clause, which states: "Provided, that no additive shall be deemed safe if it is found to induce cancer when ingested by man or animal, or if it is found, after tests which are appropriate for the evaluation of the safety of food additives, to induce cancer in man or animal . . ." Proper interpretation of this anticancer clause containing the above stated broad mandate has led to considerable discussion between the scientific community, government regulators, producers, and consumers.[18] Questions regarding the use of zero tolerance levels, dose-response relationships, "tests which are appropriate", and other vagaries in the interpretation of the 49 words in the Delaney clause will undoubtedly provide a basis for continuing discussion for many years in the future.

The problem of interpreting the significance of analytical results has become increasingly difficult and controversial as analytical systems designed to detect carcinogens have increased in sensitivity levels from 10^6 to 10^9 in recent years. Some researchers now claim that almost anything can be detected everywhere.[19] Designing new food systems utilizing immobilized enzyme technology under these restrictive conditions presents a formidable problem to the food industry.

IV. GOVERNMENTAL REGULATION OF ENZYMES

A. GRAS Food Ingredients

As indicated in early paragraphs, with the adoption of the 1958 Amendment, a variety of substances routinely added to foods were required to be cleared prior to their continued use in foods.[20] This requirement, relating to the use of enzymes in food processing in particular, is evident from the definition of food additives, ". . . any substance the intended use of which results or may reasonably be expected to result, directly or indirectly, in its becoming a component or otherwise affecting the characteristics of any food (including any substance intended for use in producing, manufacturing, packing, processing, preparing, treating, packaging, transporting, or holding food; and including any source of radiation intended for any such use; if such substance is not generally recognized among experts qualified by scientific training and experience to evaluate its safety as having been adequately shown through scientific procedures (or, in the case of a substance used in food prior to Jan. 1, 1958, through either scientific procedures or experience based on common use in food) to be safe under the conditions of its intended use . . ."

Thus it can be seen that, in addition to a large number of chemical substances, the use of protein catalysts (enzymes) which are added directly to food with the intention of changing the character or functionality of the food, would fall within the provision of the law.

In order to partially cope with the thousands of substances which require evaluation, the FDA established a category of food additives Generally Recognized as Safe (GRAS). These substances were eliminated from the original definition based on the results of studies conducted by scientists who are qualified to establish the safety of the food additives. Under this provision, many food ingredients and foods of natural biological origin that were widely consumed for the nutrient properties prior to 1958 would ordinarily be recognized as GRAS without specific listing. In the past, enzymes and foods containing enzymic ingredients were included in this category, even though the food and not the enzyme is consumed for its nutrient content.

During the period of 1962 to 1974, considerable confusion arose between the FDA and producers concerning the criteria for establishing GRAS substances and it has been only recently that the criteria for GRAS and Prior Sanction for food ingredients have been finalized by the FDA.[21] These new regulations were recently published in the Federal Register.[22] Under these latest regulations, recognition of GRAS status, based on common use in food (sanctions issued prior to January 1, 1958), will not require the same quality and quantity of scientific evidence required for approval of a food additive regulation. In this instance, the term "common use" has the connotation that there is a substantial history of consumption of the food ingredient by a significant body of consumers.

Three categories of food ingredients which can be defined as GRAS have now been delineated. Under these criteria, a substance may be:

1. GRAS with no limitations other than good manufacturing practice, as long as the conditions of use are not significantly different from those on which the GRAS status was affirmed
2. GRAS with specific limitation, such as category of food, use function, and use level
3. GRAS for a specific use only

B. Enzymes as Food Additives in the U.S.

The status of regulatory concern with particular reference to enzymes has recently been reviewed by Miles.[23] This report indicates that all food processing enzymes, even those commonly in use prior to 1958, are now in the process of being reevaluated for safety. With this basis in mind, the FDA has recently requested that all information relating to the manufacturing and use of these food ingredients be submitted by either producers or processers concerned with enzyme modification of foods. Among the 2100 direct food, flavor, and color additives now under reevaluation by FDA are a number of enzyme preparations used in food processing. These enzymes of microbial, plant, or animal origin are listed in the *First Supplement to the Food Chemical Codex*, 2nd Edition.[24] Most of the enzymes under review have a history of common use in foods in the U.S. prior to 1958 and have previously been classified as GRAS for prior sanction,[25] or GRAS affirmation petitions containing adequate data to support their use have been submitted.

The policy now established is that any new food ingredient (enzyme) petition must show publication of required safety data available to food and toxicological scientists for evaluation before the new ingredient can be recognized as safe. Additionally, accurate information regarding the source of the enzyme, its specific and unspecific function in manufactured foods, and methods of production and recovery must be fully disclosed if the enzyme product under review is seeking a GRAS affirmation. This latter disclosure regarding manufacturing processes serves as a point of distinction between GRAS petitions and food additive petitions. While both forms of petition require a full consideration of the manufacturing process by the FDA, food additive regulations are treated to protect the trade secrets of the manufacturer and do not specify the manufacturing procedure. Recently, after review by FDA, several enzyme preparations and microorganisms were affirmed as regulated food additives.[26]

It is important to note that enzyme petitions are unusual in that the petition data must generally demonstrate safety clearance not only to the enzyme, but also for the final products of its action. Thus, it is necessary to follow petition format requirements for both the final product and the enzyme preparation.

1. GRAS Petition for FDA Approval

1. Clearly state the intent of the petition to demonstrate safety based upon one common use in foods for the enzyme and/or the end product. Alternatively, establish safety through the publication of adequate supporting safety data.

2. Identification of the enzyme
   a. Class
   b. Species
   c. Strain — variety
   d. Enzyme source

3. Method of enzyme preparation
   a. Complete medium composition
   b. Mechanical units used to produce
   c. Fermentation conditions, aeration, agitation, temperature, pH

4. Enzyme specifications
   a. Single enzyme, multiple enzyme, or whole cell
   b. Impurities, i.e., heavy metals, solvents
   c. Analytical methods for compositional and impurities data
   d. Comparative compositional data on a minimum of five process batches of the enzyme

5. Methods of enzyme recovery
   a. Types and levels of solvents employed
   b. Types of processing aids, e.g., filtration
   c. Other

6. Toxicological data on enzyme or microorganism
   a. Organism must be nonpathogenic
   b. Organism must not produce mycotoxins or other toxic chemicals
   c. Organism must not produce antibiotics

2. Toxicological Format for FDA Approval

The format for a typical toxicological evaluation of a new food additive has been outlined by Ebert.[27] The various animal feeding trials involving acute, subacute, chronic, and other specialized tests are illustrated (Table 2).

Recently, Sontag has reviewed the limitations and economics of bioassay testing for carcinogens.[28] In it he stresses the need to know the limitations of various assay systems prior to selecting an appropriate test system. Influences such as sex of the test animal, light stress, density of animals per cage, temperature, humidity, and altitude have been shown to effect tumor incidences in mice. Dietary composition and caloric intake also have marked influence on incidence and organ site of both spontaneous and induced tumors in experimental animals.

A bioassay of a single chemical fed to mice and rats using the standard National

Table 2
TYPICAL TOXICOLOGIC EVALUATION OF FOOD ADDITIVES

Study	Scope
Acute	
Oral LD_{50}	Two species (min) range find for subacute studies
Dermal	
Inhalation LD_{50}	As needed (OSHA requirements, etc.)
Subacute	
90-Day-feeding	Two species: one nonrodent, dog preferred 3 levels (min) plus control Complete biochemical and histopathologic work-up on control, highest level fed. Other levels if needed. Define no-effect level. High level to give effect.
Chronic 2 Years +	Two species — rat/dog. Carry rat to 25% survival of controls. Perhaps combine with 3 generation reproduction study. 3 levels plus control. complete biochemical and histopathologic work-up.
Metabolic fate	Search for animal model for man. Absorption, distribution kinetics, organ/tissue concentration (if any) metabolic products defined.
Special	Carcinogenicity, teratology, cytogenicity (host mediated, dominant lethal, chick embryo)
Interaction	Special dietary restrictions, drug interactions.

Cancer Institute protocol involving both sexes, 50 animals per group, two dose levels and a study period of 24 months, now costs \$200,000 to \$210,000.[29] More complicated experimental testing, such as a gavage study, will cost \$275,000.

Further, it is pointed out that both FDA and USDA may call upon a number of advisory groups to aid in the evaluation of a new food additive. Some of the organizations that have impact on the approval of petitions are illustrated in Table 3. Ebert notes the importance of making sure the data necessary to clear a given substance gets into the hands of the appropriate advisory groups.[27]

The procedure involved if the new food additive (enzyme) is to be approved for international use has also been outlined. Ebert states the only viable international organization active in food safety evaluation is the joint FAO/WHO Food Standard Program, or Codex Alimentarius.[27] This organization is operated by and is open to all U.N. members. The flow pattern for evaluation of a new food additive (enzyme) would involve the Joint Expert Committee on Food Additives who, after establishing appropriate use levels, would transfer the required information to the food additive committee. Toxicology and safety evaluations similar to those required by FDA and USDA would be reviewed by the Joint Expert Committee on Food Additives. They have established the following four criteria for the technological justification for the use of the food additive (enzyme):

1. Maintenance of nutritional quality
2. Reduction of waste by enhancing the keeping quality
3. Making foods attractive to the consumer
4. Essential aids in food processing

The use of food additives may not cover up manufacturing defects, reduce the nutritive value of the food, or deceive the consumer.

Table 3
ORGANIZATIONS IMPACTING ON FOOD ADDITIVES

Government advisory groups	Government	Petitioners
National Academy of Sciences		Food additives producers-users
Federation of American Societies for Experimental Biology	FDA USDA	Consumers Center for Science in the Public Interest
Special advisory committees		Media
Individual consultants		Press-trade and popular radio, television
	Overseas Activities FAO/WHO Individual countries	

C. Enzymes as Food Additives in Europe

A review of the attitudes of foreign countries regarding the use of enzymes as food additives has been presented by Ilany.[30] In Germany, according to the new "Food Law" definition of "Food Additives" (Zusatzstoffe) replacing the concept of "Foreign Subtances" (Fremdstoffe), all food additives must be listed. In neither case is the use of enzymes defined.

In Sweden, food additives are defined as substances not contained in the raw material of the foodstuff but are added and remain in it in final form. Enzymes in this category must be listed before their use in foods is permitted. The Swedish Board of Trade annually publishes a list of enzymes which may be used in foods.

In the U.K., enzymes are included in the definition of food in the Food and Drug Act, which encompasses all ingredients in the preparation of food and drink.

France has published several authorizations on the use of a number of enzymes. Special and temporary authorizations have been granted.

Canada includes some enzymes in a "positive" list of food additives and these are generally subject to food regulation principles of abuse, prohibiting poisonous or deleterious substances in food. No other enzymes may be used.

In Holland, the use of enzymes is regulated by means of food standards for particular products and by a positive list of food additives.

Belgium's Food Law (not mentioning enzymes) defines additives as food substances which are not foods themselves but have a positive effect by their presence within or on foodstuffs (independent of their possible nutritive value or content in vitamins, flavoring, and seasoning components).

Switzerland considers enzymes as foreign substances. According to Swiss Food Law, food additives and foreign substances must be approved before use.

D. Immobilized Enzymes as Food Additives

As pointed out recently by Nelson, the commercial use of immobilized enzymes is still too new to have created any easily identifiable regulatory or labeling problems.[31] At present, six or seven immobilized enzyme processes for the production of high fructose corn syrups are now in full-scale operation in the U.S. GRAS petitions for several of these processes are on file with the FDA. However, no approvals have yet been granted.

Immobilized enzyme systems for the production of high fructose corn syrup (HFCS) routinely include a terminal carbon filtration step which serves to polish the syrup and also to remove any minute attritional losses of particulate support material which might enter the product. Processing systems for HFCS systems call for operational temperatures of 60 to 65°C. Thus, problems associated with microbial growth are sharply reduced.

Presumably, immobilized enzymes will qualify as processing aids. This means they are either not added to the food, or if added, are subsequently removed before the food is packaged in finished form. Immobilized enzymes may reasonably achieve consideration and acceptance as incidental additives because they are (1) present at insignificant levels, and (2) without technical or functional effect in food.

Clearly the use of immobilized enzyme systems provide food processors the opportunity for controlled modification of food systems without the stigma of adulteration by the use of "additives". Additionally, many of the confusing problems concerned in multiple ingredient (enzymes) label declarations would be resolved.

V. APPLICATIONS OF IMMOBILIZED ENZYMES IN FOOD PROCESSING

A. Areas of Food Modification

For the successful industrial application of immobilized enzymes to foods and beverages, it is necessary to satisfy a number of conditions. From a processor's point of view, it is recognized that enzymes are utilized for two major purposes — to improve the production process and reduce its cost, and to improve some quality of the food itself. The area of food modification in which enzymes can be employed may involve the taste, color, texture, aroma, digestibility, or viscosity of the food. Additionally, enzyme modification may enhance shelf life or serve to eliminate certain undesirable characteristics originally present in the food.

In the past, it was necessary to utilize low-cost enzymes that were relatively impure yet readily available. Many commercial enzyme products still contain multiple activities, although customarily the subordinate activities are present at significantly lower levels than the major activity for which the product is sold. These minor enzymic activities in some instances (e.g., lipase) may significantly reduce shelf life of the product due to the liberation of fatty acids. This results in off-flavor development (rancid or soapy), even when the product is stored at refrigerator or freezer temperatures for prolonged periods.

B. Insolubilizing Methods

In recent years, a number of comprehensive reviews have been published regarding the application of immobilized enzymes to foods and beverages. The general methods by which enzymes are fixed to insoluble supports (adsorption, entrapment, copolymerization, ion exchange, or covalent attachment) are well known. They have been presented in detail, both in food oriented reviews[32-39] and in more generalized reviews dealing with the preparation of water insoluble enzyme derivatives.[40-44] The use of a wide variety of organic and inorganic materials such as organic polymers, glass, mineral salts, silica, and metal oxides which may serve as the insoluble support to which the enzyme is adsorbed or attached has also been described.[45]

In many instances a bifunctional alkyl chain or ring chemical may first be attached to the support prior to the enzyme fixation, enhancing the efficiency of the enzyme by reducing diffusional effects. Additionally, a large number of individual reactive organic chemicals and organic polymers have been utilized to cross-link the adsorbed or covalently bound enzyme. This further ensures the integrity of the immobilized enzyme system. To illustrate the diverse nature of the possible reactants in immobilized enzyme systems, many of the carrier supports, reactant chemicals, spacers, and cross-linking chemicals are presented in Tables 4 and 5.

1. Stability of Immobilized Enzymes

Before discussing the varied problems involved in utilizing immobilized enzymes in food processing systems, it would be appropriate to preface this portion of the chapter with the current, environmentally inspired caveat, "What you put in the river may end up in your liver." It is with this concern in mind that food processors must approach the utilization of immobilized enzymes in food systems. For in this context, serious consideration must be given to the possible dissolution or degradation of a "stable" immobilized enzyme into the food system it is modifying. There are a variety of displacement mechanisms potentially operative in food systems which may serve to partially degrade an enzyme reactor, thereby causing transference of minute or larger amounts of chemical reactants into the processed food or beverage. Some of the factors that may cause reactor degradation are listed in Table 6.

In all cases in which potential dissolution of the immobilized enzyme system could occur, evaluation of the process at the R&D level must consider a "worst case" scenario and anticipate the hazard level which might develop under adverse conditions.

2. Testing Chemicals and Solvents Used for Insolubilizing Enzymes

The use of immobilized enzyme systems which utilize cyano- or halo-genated chemical derivatives for enzyme attachment, spacing, or cross-linking would not be regarded as safe for food processing due to the obvious toxic or carcinogenic potentials of these compounds. For similar toxicological reasons most, if not all, cross-linking agents listed in Table 1 (with the exception of glutaraldehyde), would not be acceptable for use in food modifying systems.

The use of glutaraldehyde as an effective cross-linking reagent for proteins has been recognized for many years and appears to warrant continued use in these applications. However, data on carcinogenicity, mutagenicity, and teratogenicity has not been established. In recent times, its use for cross-linking enzymes to carrier supports has become very popular due to the ease with which water solutions of the reagent can be utilized in the attachment or cross-linking process and also due to the low cost of the compound.

As a note of caution, it would appear to be highly desirable to use purified glutaraldehyde solutions for immobilized enzyme attachments as commercial solutions have been shown to contain various impurities including acrolein, glutaric acid, and glutaraldoxine.[46]

The term toxicity is described by the National Research Council as the capacity of a substance to produce injury. On the other hand, the term hazard is defined as the probability that injury will result from the use of the substance in the quantity and manner proposed.

Methods for establishing neoplastic hazard levels have been limited to controversial in vivo animal bioassay systems in which extremely high dose levels are employed to elicit mutagenic responses. More recently, the Ames Salmonella/microsome mutagenicity assay has been adopted by many manufacturers as part of an early warning screening system for the detection of potential carcinogens.[47] Acceptability of the Ames test arises from the observation that 90% of all known carcinogens can be detected as

Table 4
CHEMICAL AGENTS USED FOR IMMOBILIZING ENZYMES

Cross-linkers

Class	Examples
Aldehydes	Formaldehyde, glutaraldehyde′, polyacrolein, glyoxal, oxidated carbohydrates
Isocyanates	Toluene-2,4-diisocyanate, hexamethylene-diisocyanate, diphenylmethane-diisocyanate
Thioisocyantes	Hexamethylene-diisothiocyanate
Anhydrides	Polymethacrylic anhydride, polyethylene-maleic anhydride
Azides	Azides of polyacrylate and of carboxymethycellulose
Carbodiimides	1-Cyclohexyl-3(2-morpholineothyl) carbodiimide metho-p-toluene sulfonate, n,n-dicyclohexylcarbodiimide, dicyclohexylcarbodiimide
Chloro-triazines	Cyanuric chloride, 2-amino-4,6-dichloro-s-triazine
Cyanogen halogenide	Cyanogen bromide
Diazo compounds	Bis-diazobenzidine-3,3′-disulfonate tetraazotized O-dianisidine
Difluorobenzenes	1,3-Difluoro-4,6-dinitrobenzene
Epoxy compounds	Epichlorohydrin, polyepoxides
Phosgene derivatives	Ethylchloroformate
Halogenoalkyl-derivatives	Bromoacetylbromide, polyiodobutylacrylate
Metal salts	$CrCl_3$, $TiCl_4$, Co^{+++}, Al^{+++}, Ca^{++}, Zn^{++}
Miscellaneous	Tannic acid, Woodward's reagent K, tetrakis-(hydroxymethyl)-phosphonium chloride

Table 5
CHEMICAL SOLVENTS USED FOR IMMOBILIZING ENZYMES

Class	Examples
Alcohols	Methanol, isopropanol, butanol, amyl alcohol, ethylene glycol, glycol cyclohexanol
Ketones	Acetone, methyl ethyl ketone, methyl isobutyl ketone, cyclohexanone
Ethers	Dioxan, diethyl ether, tetrahydrofuran
Ether alcohols	Ethoxy-ethanol
Amides	Formamide, dimethylformamide
Amines	Ethanol amine, triethyl amine, pyrrolidine, 2-pyrrolidone
Aromatics	Benzene, toluene, pyridine, collidine, pyridazine
Esters	Methyl acetate, hydroxypropyl methacrylate, methyl sulphate
Acetals	Methylal
Miscellaneous	Dimethyl sulfoxide, methylene chloride

Table 6
FACTORS THAT MAY CAUSE ENZYME REACTOR DEGRADATION

1. Degradative attack of enzyme or cross-linker by bacteria or mold
2. Protein displacement of adsorbed or cross-linked enzyme
3. Ionic displacement of enzyme
4. Adverse pH environment
5. Degradation due to enzymes normally present in the food or beverage
6. Adverse cleaning or sanitation procedures by plant personnel
7. Particle attrition due to recycle cleaning in plug flow reactors, or in continuously stirred reactor processes

mutagens. The relative simplicity of the test and the rapidity (49 to 72 hr) by which semiquantitative data can be obtained using standard microbial plate counting techniques are obvious advantages. This sensitive procedure can detect many potentially hazardous chemicals which can be eliminated from consideration as possible food additives at an early stage in R&D programs.

In addition to the Ames test, there are 80 other short-term tests undergoing evaluation in mutagenicity testing. Descriptive characterizations of these tests and their sensitivity to potential chemical carcinogens has been reviewed recently.[48] It is important to note that chemicals may vary over a million fold range in carcinogenic potency.[49]

Tests such as these have applications in the development of safe immobilized enzyme systems intended for use in food modification. In these instances, preliminary testing of the carrier, and all derivatized forms of the enzyme which could fragment into food systems, should be evaluated. Additionally, testing of the modified food product resulting from the enzyme action may also be important.

VI. ECONOMIC CONSTRAINTS ON THE APPLICATION OF IMMOBILIZED ENZYME SYSTEMS IN FOOD AND BEVERAGE PROCESSES

A. Large-Scale Continuous Process Plants

During the past three decades, there has been an increasing trend on the part of food processors to build bigger plants, incorporating the latest technological innovations. Competition in the market is intensive and can only be met by building larger processing units which permit economies of scale operation. One most striking example of this trend is the newly completed corn syrup processing plant of A. E. Staley. This plant is considered to be the most highly computerized controlled food plant in the world. It is now "on stream" (utilizing immobilized enzyme technology) with a current HFCS (42%) capacity of over 1 billion lb/year.[50]

This move toward bigger, more sophisticated plants has in large part been motivated by observation that proportional increases in productive capacity can be obtained from less than proportional increases in capital expenditures. These findings in part stem from the earlier publications of Chilton,[51] who reported the general applicability of the relationships listed below for cost estimations of equipment or plant capacity to process plants.

$$\frac{C_1}{C_2} = \left(\frac{V_1}{V_2}\right)^{\delta}$$

Where $0 \leqslant \delta \leqslant 1$ and δ = an exponential value of 0.6 to 0.7 (the two thirds rule); and C_1 = installed plant cost — size 1, C_2 = installed plant cost — size 2, V_1 = production capacity, plant size 1, and V_2 = production capacity, plant size 2.

B. Long-Term Capital Investments

In all cases concerned with establishing large-scale continuous process plants, in order to achieve economic growth, significant increases in capital investment are required from the corporate structure.[52] Considerable effort must, therefore, be made both at marketing and R&D levels to provide management with in-depth information accurately identifying cost/benefit ratios and alternative process pathways which may influence the economic life of the potential product. Decisions regarding the investment of capital for large-scale continuous process plants are extremely difficult. Many factors may influence the ultimate merits of the process. Among these are the levels of process flexibility (1) permitting product change or improvement, and (2) allowances for the introduction of new innovative technology. Changes in the market due to competitive processes or in the availability and price of raw material may also have strong influence on the ultimate fate of the capital-funded process plant. Capital investments of this magnitude are based on estimations of corporate marketing plans for a decade or more in the future. The consequences of a decision to commit capital to a new process plant often are not fully realized for several years after the plant has become operational.

C. Economic Overview of Immobilized Enzyme Processes

Sweigart recently presented an overview of the current stage-of-the-art of industrial applications of immobilized enzymes.[53] His report stresses the importance of economics to the commercial success of any immobilized enzyme system. Some of the primary factors involved in these economic considerations are listed. Among these are cost, enzyme availability, immobilization costs, system performance, capital investment, cleanup costs (substrate, reactor, final product), and marketing costs.

D. Innovation in Process Development

In addition to these primary considerations, there are other factors which relate to process developments utilizing enzyme technology. One is concerned with the innovative character of the process. The term "innovation", defined by Bradbury relates to the whole process of R&D leading to a new product.[54] The process, in this sense being subject to several inputs or inventions which require considerable research effort, along with the interaction of ideas related to the total development.

One important aspect of the innovative concept is concerned with the relationship between the introduction of a new technology into an already existent older technology. In this case, the innovation must compete with established plants where capital costs have already been expended. For most producers it is economic wisdom to continue producing, utilizing the older technology as long as the operation provides a profitability pattern which allows the recovery of more than variable recurrent costs. This point of view can perhaps be exemplified in one case by the failure of immobilized glucoamylase technology to become established at industrial-scale levels. For this particular innovation, the benefit/cost ratio for newer technology does not appear economically attractive enough to warrant capital expenditures required to install the process.[55] In instances such as this, where enzymatic modification of the product can be carried out batch style using low-cost readily available enzyme in existing facilities, there is little motivation for producers to consider change. This picture will change when modifications of the newer technology permits significantly greater productivity and total costs of the new get under the variable costs of the old.

Another important factor concerning utilization of a new enzyme technology is related to its position or fit in the systems development of the manufacturer's total production operation. In the food industry, the innovation is usually viewed as part of a larger context and must be integrated into the overall company objectives and plans. An illustration of this integrating concept may be found in the development of a process for the production of a sugar syrup derived from enzyme (lactase) treated whey or whey by-product (permeate).

E. Hydrolyzed Lactose Syrup Production

In recent years, innovations in membrane technology (ultrafiltration-reverse osmosis) have led to economical processes for the recovery of the nutritionally important proteins (PER 3.1) present in whey.[56,57] Utilization of these whey protein concentrates in dairy based foods presents little problem to manufacturers. However, salvaging the lactose-mineral permeate fluid portion of the process has initiated considerable R&D effort for the dairy product manufacturers, the commercial organizations with strong enzyme technology bases, and the governmentally funded university (NSF) groups.

As a result of these efforts, the use of immobilized enzyme (lactase) technology to produce a concentrated syrup from whey permeate, with a sweetness profile comparable to commercial corn syrups, has been proposed as a unit process.[58] This process presents one possible salvage solution which could be integrated into a broad whey production-recovery plan. Even if syrup costs are marginal or slightly higher, incorporation of the unit process may allow optimization of the overall system.

1. World-Wide Impact

Some understanding of the potential for a hydrolyzed sugar syrup is indicated by the following considerations. At present, the world production of cheese is 10 million tons, which provides about 60 million tons of liquid whey, including the whey from casein factories, for processing.[59] If 50% of the available world lactose supply could be converted into a hydrolyzed sugar syrup, it would provide about 1.5 million tons of product. When we recall that world consumption of sucrose is over 81 million metric tons/annum, it can be seen that a whey syrup would represent less than 1% of the world's sugar production.[60] It would, therefore, have virtually no impact on world sugar economics.

2. Chemical and Enzymatic Hydrolysis Processes

Lactose can be hydrolyzed by various chemical and enzymatic methods. Among the potentially competitive options which have been proposed are the following:

1. Batch process using soluble yeast enzyme[61]
2. Continuous hydrolysis using a yeast enzyme — membrane reactor process[62,63]
3. Direct (batch) hydrolysis using heat and mineral acids[64,65]
4. Continuous H^+-induced catalysis using heat and acidified ion exchange resin[66]

Each of these process proposals has its particular advantages and disadvantages which must be fully evaluated prior to the selection and development of a full-scale syrup production process. In all cases, some method of utilization or disposal of by-product mineral salt generated by sugar purification (deashing) or acid base neutralization of ion exchangers must be anticipated.

3. Economic Evaluation of Syrup Process

Factors which must also be considered in the economic evaluation of the syrup process are related to the plant site where the whey permeate is produced. Some of these are summarized below:

1. Production volume of whey permeate
2. Overall plant efficiency
3. Size and efficiency of evaporation equipment
4. Location of the plant in relation to water disposal of whey salts
5. Location of the plant in relation to transportation of syrup to food processing plants

Immobilized enzyme systems are intended for continuous operation and must have large volumes of raw materials available for processing. Syrup production would not be feasible or attractive to many low-volume whey plants.

For cheese manufacturers who have large volumes of whey available, the scale factor of size is an advantage. Significant improvement in process economics can be found between plants producing only 10 million lb of whey by-product a year and one which produces 3 to 4 times this volume.

Some preliminary data regarding production costs of an immobilized enzyme syrup process indicate the enzyme cost is only one of six major cost items Figure 2. If depreciation and repair are taken together, capital, fuel, ion exchange regeneration, labor, and enzyme replacement comprise the other five factors. If processing costs and return on capital expenditure are plotted as functions of enzyme half-life, extrapolation of the data indicates that little reduction in cost occurs if enzyme activity can be maintained beyond 100 days.[67]

VII. TECHNOLOGICAL FACTORS IN SCALING UP IMMOBILIZED ENZYME SYSTEMS IN THE FOOD INDUSTRY

A. Problems of Scale-Up

Scale-up of production operations in the food industry may often be a more complex process than is usually encountered in the chemical and pharmaceutical industries. There are several basic factors involved in scaling up food and beverage processes which cause special problems.[68] They include:

1. Variability in composition of the raw material to be processed
2. Difficulty in monitoring in-line process modifications in continuous operations
3. Difficulty in maintaining safe and sanitary conditions during processing operations
4. Lack of highly trained personnel to monitor each unit operation
5. Lack of basic scientific knowledge concerning food technology

As we are concerned with introducing immobilized enzyme technology into food processing, we should first examine the available fluid foods which could potentially be modified. In Table 7 are listed most of the fluid products produced by the food

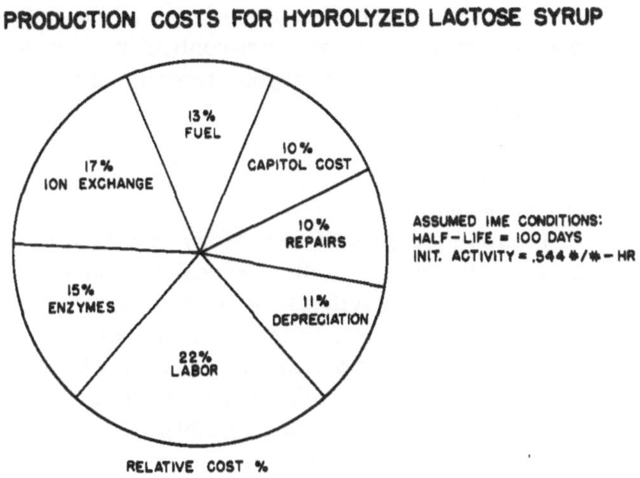

FIGURE 2. Production costs for hydrolyzed lactose syrup.

Table 7
LIQUID CONSUMPTION TRENDS IN U.S. FROM 1967 TO 1976[a]

	1976[P]	1975	1974	1973	1972	1971	1967
Soft drinks	34.2	31.4	31.6	31.9	30.3	28.7	23.6
Coffee	32.5	33.1[R]	33.8[R]	35.1[R]	35.2[R]	35.3[R]	37.0
Milk	24.5	24.4	24.3	25.1[R]	25.3	24.9	25.3
Beer	21.8	21.6	21.1	20.2	19.4	19.0	16.8
Tea	7.8	7.5	7.5	7.6	7.2	7.1	6.4
Juices	6.2	6.2	6.0	5.7	5.4	5.3	4.7
Powdered drinks	5.8	5.2	NA	NA	NA	NA	NA
Distilled spirits	2.0	2.0	2.0	1.9	1.9	1.9	1.6
Wine	1.8	1.7	1.7	1.7	1.6	1.5	1.0
Subtotal	136.6	133.1	128.0	129.2	126.3	123.7	116.4
Imputed water consumption*	45.9	49.4	54.5	53.3	56.2	58.5	66.1
Total	182.5	182.5	182.5	182.5	182.5	182.5	182.5

Note: * includes all others, R — revised, P — preliminary, NA — not available.

[a] Gallons per capita per year.

and beverage industry and their consumption trends. One additional fluid beverage not listed in the table is vinegar, a flavor and and preservative commodity with an annual production volume of 120 million gal/year.

B. Complexity of Fluid Food Flavors

It is apparent that all of these fluid foods are subject to great variations in acidity, sugar, fat, protein, and the more complex lactones and phenolic constitutents that determine acceptable flavor profiles. One recent review shows the presence of 352 volatile substances in tea, 438 in coffee, and 284 in cocoa.[69]

At the present time, most food production operations have no in-line sensors available which can provide the complete subjective information required to establish continuous control of food flavor profiles. However, some in-line levels of information have recently been achieved using GC and HPLC analyses for monitoring grape must in the wine industry. It is virtually certain that in the future all the fluid food processes in Table 3 will become fully automated computer-controlled operations. In Europe, the first completely computer-controlled dairy plant recently went on stream, producing over 400 individual products.[70]

C. Role of the Pilot Plant in Scale-Up

1. Development of the Rate Equation

The introduction of immobilized enzyme technology into food processing requires process conditions more closely related to the chemical and pharmaceutical industry than to conventional food engineering systems. In most cases, these principles are well understood. As described earlier by Allen,[71] the theoretical basis for plant design usually evolves from the development of an overall rate equation which expresses the rate at which the physical or chemical process takes place. A number of parameters such as size, time, a numerical component, operating variables, and physical and chemical data are all coordinated to develop the rate equation.

A second engineering precept utilizing the principle of similarity (if two differently sized pieces of equipment are of similar configuration and similar processes are carried out in them, similar results will be obtained) is usually also employed to establish the basic information required for the design of a full scale process.

Immobilized enzymes behave essentially the same as other heterogenous catalysts employed in chemical processes. In each instance, the enzyme catalyzes a specific chemical reaction. As outlined earlier, establishment of the rate equation for the process involves the determination of factors such as temperature, pH, feedstock and product concentration, flow properties of the immobilized enzyme, and critical bed height of the enzyme reactor. Application of data derived from studies of the above factors for the characterization of a newly established glucose isomerase process has recently been outlined by Hupkes.[75]

Most often this information is generated from smaller-scale pilot plant operations. The importance of using pilot scale operations to establish the data required for scale-up design has been strongly stressed by process development engineers.[72-74] Data derived from laboratory experiments frequently has little relationship to plant design operations.

2. Determination of Process Economics

By using these procedures, it is possible to determine process economics and initiate designs for production plants. In processes employing immobilized enzymes, the best experimental procedures for establishing the reliability of the enzyme system performance requires the continuous operation of the reactors under the conditions which will be used in the large-scale plant production process. This can be accomplished in the pilot plant which permits the following developments to occur:

1. Translation of laboratory scale studies into the initial stages of a plant design embodying today's high technology computer-based control systems
2. Evaluation of the proposed process to establish its long-term reliability, and to test other competitive systems using different attachment mechanisms and support materials under similar conditions
3. To permit the production of sufficient quantities of product for market development
4. Analyze the data obtained from the above functions to establish the process economics

VIII. MICROBIAL CONTROL OF IMMOBILIZED ENZYME REACTORS

A. Aspects of Quality Assurance

The one factor of major importance to food processors is the safety of their products during all processing and packaging stages. Food becomes a health hazard if undesirable bacteria, chemicals, or other foreign substances enter into the food system. In the case of immobilized enzyme systems, any of these three hazards may be source related to the biocatalytic process unless adequate control points are analyzed and established. In the food industry, control and responsibility for food safety of all products are integral functions of quality assurance. Most often this responsibility is assigned to a Quality Assurance Department established within the corporate company structure.[78] Major objectives of the operation are to monitor quality control at all levels of manufacturing and to assure that regulatory standards of compliance are met for each food product. These standards are now recodified by the FDA as the Good Manufacturing Practice Regulations (Part 113 and Part 114) and are designed to support the manufacture of high-quality wholesome foods.[77,78]

Therefore, one of the major design problems associated with the commercial development of immobilized enzyme systems (IMES) is concerned with providing a unitized system capable of controlling microbial contamination. Other researchers involved in studying immobilized enzymes have observed the close relationship between increases in bacterial populations and declines in enzymatic activity of the system.[79,80] The control of microbial contamination becomes an essential part of maintaining the operational half-life of the bioreactor and in preventing the development of food-borne infections.

B. Establishment of Process Control — HACCP

Control of microbial contamination in IMES used for food processing can best be achieved by utilizing a Hazard Analysis and Critical Control Point (HACCP) evaluation.[81] This type of hazard analysis identifies all points in the process which may represent potential sources of hazards to the safety of the final product.

Obviously the type of hazard and the critical control points will vary considerably with each type of product. Several factors such as operational temperatures, pH, salt concentration, and buffer capacity of the food system are important parameters in the evaluation of the system. Also, the composition of the food or beverage being processed is important. The natural flora of many foods are frequently distinctive and may be used to anticipate subsequent contamination problems. Shotwell et al.[83] have shown that contamination of soybeans with *Aspergillis flavus* may approach 50% in commercial samples. Likewise, *Geotrichum candidum* is a frequent contaminant of dairy and tomato products. Molds that occur most frequently in sugar syrups are *Stemphylium, Sterigmatocystis, Cladosporium, Monila,* and *Penicillium*. Yeasts (osmophilic) frequently contaminate sugar products, dried and frozen eggs, and dried fruits.

1. Food Poisoning Microorganisms

The types of bacteria of most concern to food processors are those associated with food poisoning. Among these are *Salmonella, Shigella, Bacillus cereus,* coagulase-positive Staphlococci, and the anaerobic *Clostridia*. The latter two genera produce toxins in the temperature range of 10 to 40°C. None of the above organisms grow (nor do spores germinate) at pH levels below 4.5, but growth may occur from pH 4.5 to 8.5. Although *Clostridia* are anaerobic, growth and toxin production can occur under aerobic conditions.[84]

The significance of molds and yeasts in foods has been reviewed recently by Anderson.[85] It is now recognized that molds and yeasts are also able to grow in foods and elaborate mycotoxins or cause disease. Among the toxigenic genera of molds are *Aspergillus, Alternaria, Cladosporium, Fusarium, Penicillum,* and *Mucor.* Toxic substances produced by molds include ergot, alkaloids, lactones, dicoumarol, ochratoxins, aflatoxins, rubratoxins, zearalenone sterigmatocystin, kojic acid, and aspergillic acid. Additionally, patulin, a toxic metabolite of *Aspergillus* and *Penicillium* species, is a known carcinogen for mice. Many *Penicillium* species grow at refrigerator temperatures and may produce toxins under these conditions.

It is, therefore, apparent that coincident with the frequently destructive effects of microbial growth on the activity of the biocatalytic reactor, the potential for microbial toxins to enter the food system does exist. Improper or inadequate sanitation procedures further enhance this possibility.

2. Microbial Attachment Process — "Glycocalyx"

Recognition that a unique biological process exists which permits bacteria growing in natural environments to adhere to solid surfaces has only occurred within the last decade. These findings, reviewed recently by Costerton et al.,[86] now have shown that, unlike pure laboratory cultures, wild-type bacteria growing in natural environs produce a complex mass of polysaccharide fibers "glycocalyx" which surrounds each cell surface. This felt-like mass permits the bacteria to bond to solid surfaces and initiate colonization despite strong shear forces.

The mechanism by which the bacterial glycocalyx forms attachments to solid surfaces is still not known. However, the polysaccharide fibers are mostly negatively charged and can form bonds with charged nutrient ions or other soluble molecules in the flow stream such as proteins or polysaccharides. These cross-linking effects lead to the accumulation of thick films which may, in the case of immobilized enzyme systems, explain the rapid loss of functionality which occurs when bacterial populations increase. Research is in progress to find inhibitors which would prevent the formation of the polysaccharide glycocalyx and thus reduce adhesion.

C. Sanitizing Immobilized Enzyme Systems

1. Surface Sanitizers

The conventional compounds most frequently employed in the food and beverage industry for surface sanitizing are chlorine-based compounds. Other choices are quaternary ammonium chloride, iodophor, hydrogen peroxide, and acid-anionic germicidal compounds. In most cases, the chlorine-based and quaternary ammonium chloride germicidal reagents are employed at levels of 200 ppm.

Application of several of these sanitizing chemicals at periodic cleaning intervals to an immobilized lactase-whey system has been reported.[87] Results show that loss of activity occurred rapidly when chlorine sanitizers, iodophor, or acid cleaner were employed as control agents. Only quaternary ammonium chloride gave effective control and did not cause loss of enzyme activity.

2. FDA Approved Antimicrobials

There is also a small number of FDA approved antimicrobial preservatives which may be utilized in food systems.[88] These are mainly salts (sodium, potassium, or calcium) of benzoic acid, sorbic acid, and propionic acid. The antimicrobial effect of these compounds is due primarily to the undissociated acid molecule. Their effectiveness is pH dependent and is most effective on the acid side of neutrality. In actual practice, the benzoates are poor inhibitors of bacteria, yeasts, and molds at pH levels above 4.0. In addition, the use level for benzoates is restricted to 0.1% in foods and beverages. There are no absolute tolerance levels for the propionates and sorbate other than as specified for individual foods. However, propionates are only effective against molds and have no antibacterial or inhibitory activity against yeast. Sorbic acid salts (0.03 to 0.3%) show a wide spectrum of activity against molds, yeasts, and bacteria and would appear to be the most effective acid compound for use in maintaining microbial control of immobilized enzyme reactors.

3. Experimental Microbial Control Agents

Several workers have reported on the use of phenolic antioxidants as possible antimicrobial agents in food systems.[89,90] Among the compounds investigated are butylated hydroxyanisole (BHA), propyl gallate, and nordihydroguararetic acid (NDGA). Shih and Ayres have reported that propylgallate and NDGA at 400 ppm have a strong lethal effect on *E. coli*, and NDGA alone at 50 ppm is highly inhibitory to *S. aureus*.[91] There are no literature reports regarding the use of these compounds as control agents for immobilized enzyme reactors.

Glutaraldehyde (1,5 pentanedial) is important both as a disinfectant and as an effective cross-linking agent for immobilizing enzymes. Its effectiveness against bacteria, spores, and fungi on the alkaline side (pH 7.9) has been studied by several workers.[92,93] A study showing the effectiveness of glutaraldehyde against salmonella at levels above 0.05% has been published by Thomson et al.[94] In this possible connection, the use of glutaraldehyde as a sterilant for an immobilized lactase reactor was proposed by Pitcher.[95] Recently a comprehensive overview of problems encountered in the microbial control of an immobilized enzyme reactor (lactase) has been provided by Harju.[96] Additionally, studies on the control of bacteria in a model immobilized enzyme system (sulfhydryl oxidase-GSH) using 10% ethanol or 0.01% hydrogen peroxide as preservatives during storage was reported by Swaisgood et al.[97]

REFERENCES

1. **Nelson, J. M. and Griffin, E. G.**, Adsorption of invertase, *J. Am. Chem. Soc.*, 38, 1109, 1916.
2. **Mitz, M. A.**, New insoluble active derivative of an enzyme as a model for study of cellular metabolism, *Science*, 123, 1076, 1956.
3. **Tosa, T.**, Studies on continuous enzyme reactions I. Screening of carriers for preparation of water-insoluble aminoacylase, *Enzymologia*, 31, 214, 1966.
4. **Thompson, K. N., Johnson, R. A., and Lloyd, N. E.**, U.S. Patent 3,788,945, 1974.
5. **Coher, W. P., Lloyd, N. E., and Hinman, C. W.**, U. S. Patent 3,623,953, 1971.
6. **Sweigart, R. D.**, What's next for immobilized enzymes? *Food Eng.*, 50(5), 80, 1978.
7. **Gortner, W. A.**, The impact of food technology on nutrient supplies, *Food Technol. Aust.*, 24, 504, 1972.
8. **Andres, C.**, Foodservice market growing at 10% per year, *Food Process. (Chicago)*, 38(9) 28, 1977.
9. **Fawcett, J. E.**, Competition in the Food Industry, Grocery Manufacturers of America, Washington, D.C.
10. **Research Department**, FPI trends, *Food Process*, 36(12), 21, 1975.
11. **Hirst, E.**, Food-related energy requirements, *Science*, 184, 134, 1974.
12. **Lund, D. B.**, Design of thermal processes for maximizing nutrient retention, *Food Technol. (Chicago)*, 31(2), 71, 1977.
13. **Rippen, A. L.**, Energy conservation in the food processing industry, *J. Milk Food Technol.*, 38, 715, 1975.
14. **Lemaire, W. H.**, Six-effect evaporation at Wisconsin dairy, *Food Eng.*, 50(2), 67, 1978.
15. **Cocks, D. L.**, Drug-firm productivity, R&D, and public policy, *Pharm. Technol.*, p. 21, 1977.

16. **Instrati, G.**, International growth prospects for major food companies, *Food Process.*, 38(10), 31, 1977.
17. **Terleckyj, N. E.**, Effects of R&D on the Productivity Growth of Industries: an Exploratory Study, National Planning Association, Washington, D.C., 1974.
18. **Miller, S. A.**, Risk/benefit, no-effect levels, and Delaney: is the message getting through? *Food Technol.*, 32(2), 43, 1978.
19. **Whelan, E. M. and Stare, F. J.**, Panic in the pantry, in *Food Facts and Fallacies*, Athenum, New York, 1977, 149.
20. **Wood, L. I. and Weitzman, S. A.**, Regulation of enzymes, *Food Drug Cosmet. Law J.*, 28, 286, 1973.
21. **Hickman, D. H.**, The Role of Government in Regulating the Commercial Use of Enzymes, Conf. on Enzyme Economics, Chicago, Ill., June 1978.
22. Federal Register, GRAS Substances, U.S. Department of Health, Education and Welfare, Washington, D.C., December 1976.
23. **Miles, C. I.**, Regulatory Aspects in the Use of Lactase and Other Enzymes in Food, Inst. Food Tech. Conf., Philadelphia, Penn., June 1977.
24. *Food Chemical Codex*, 2nd ed., National Academy of Science, Washington, D.C., 1972.
25. **Beckhorn, E. J., Labee, D., and Underkofler, L. A.**, Production and use of microbial enzymes for food processing, *J. Agric. Food Chem.*, 13, 30, 1965.
26. **Anon.**, Part 173 — Secondary direct food additives permitted in food for human consumption, *Fed. Regist.*, 42, 14526, 1977.
27. **Ebert, A. G.**, Advice to the technologist developing food additives, *Food Prod. Dev.*, 10(5), 40, 1976.
28. **Sontag, J. M.**, Bioassay limitations and economics, *Food Prod. Dev.*, 10(5), 105, 1977.
29. **Sontag, J. M.**, Guidelines for Carcinogen Bioassay in Small Rodents, DHEW Pub. No. (NIH) 76, U.S. Government Printing Office, Washington, D.C., 1976.
30. **Ilany, J.**, Industrial enzymes as food additives, *Gordian*, 75, 4, 1975.
31. **Nelson, J. H.**, Enzyme Producers, Food Processors and the Regulators, Conf. on Enzyme Economics, Chicago, Ill., June 1978.
32. **Weetall, H. H.**, Immobilized enzymes: some applications to foods and beverages. I. Immobilization methods, *Food Prod. Dev.*, 7(2), 46, 1973.
33. **Weetall, H. H.**, Immobilized enzymes: some applications to foods and beverages. II. Applications, *Food Prod. Dev.*, 7(4), 94, 1973.
34. **Richardson, T.**, Immobilized enzymes in food systems, *J. Food Sci.*, 39, 645, 1974.
35. **Hultin, H. O.**, Symposium: immobilized enzymes in food systems. Characteristics of immobilized multi-enzyme systems, *J. Food Sci.*, 39, 647, 1974.
36. **Olson, N. F. and Richardson, T.**, Symposium: immobilized enzymes in food systems. Immobilized enzymes in food processing and analysis, *J. Food Sci.*, 39, 653, 1974.
37. **Stanley, W. L. and Olson, A. C.**, Symposium: immobilized enzyme in food systems. The chemistry of immobilizing enzymes, *J. Food Sci.*, 39, 660, 1974.
38. **Lee, R. R. and Tsao, G. T.**, Symposium: immobilized enzymes in food systems. Mass transfer characteristics of immobilized enzymes, *J. Food Sci.*, 39, 667, 1974.
39. **Olson, N. F.**, Potential uses for immobilized enzymes in the dairy and food industries, *Sci. Tec. Latt.-Casearia*, 27, 365, 1976.
40. **Melrose, G. J. H.**, Insolubilized enzymes; biochemical applications of synthetic polymers, *Rev. Pure Appl. Chem.*, 21, 83, 1971.
41. **Dinelli, D.**, Fibre-entrapped enzymes, *Process Biochem.*, 7, 9, 1972.
42. **Aiba, S., Humphrey, A. E., and Millis, N. F.**, Immobilized enzymes. An alternative to whole cells or enzymes, in *Biochemical Engineering*, 2nd ed., Academic Press, New York, 1973, chap. 14, p. 393.
43. **Vieth, W. R. and Venkatasubramanian, K.**, *Enzyme Engineering*, American Chemical Society, Washington, D.C., 1973.
44. **Manecke, G.**, Immobilized enzymes, *Chimia*, 28, 467, 1974.
45. **Messing, R. A.**, *Immobilized Enzyme for Industrial Reactors*, Academic Press, New York, 1975, chap. 1.
46. **Anderson, P. J.**, Purification and quantitation of gluteraldehyde and its effects on several enzyme activities in skeletal muscle, *J. Histochem. Cytochem.*, 15, 652, 1967.
47. **McCann, J., Choi, E., Yamasaki, E., and Ames, B. N.**, Detection of carcinogens as mutagens in the Salmonella/microsome test: assay of 300 chemicals, *Proc. Natl. Acad. Sci., U.S.A.*, 72, 5135, 1975.
48. **Maugh, T. H.**, II, Chemical carcinogens: the scientific basis for regulation, *Science*, 201, 1200, 1978.

49. **Maugh, T. H.**, II, Chemical carcinogens: How dangerous are low doses? *Science*, 202, 36, 1978.
50. **Homan, J., Brennecke, O., and Forwalter, J.**, Computer control, innovative process/energy design maximizes quality, reduces energy needs 30%, *Fed. Proc.*, 37(8), 54, 1978.
51. **Chilton, C. H.**, Cost data correlated, *Chem. Eng.*, 6, 97, 1949.
52. **de la Mare, R.**, The economic implications of plant reliability, *Chem. Ind. (N.Y.)*, p. 366, 1975.
53. **Sweigart, R. D.**, State of the Art — Immobilized Enzymes, 4th Enzyme Engineering Conf., Bad Neuenahr, Germany, September 1977.
54. **Bradbury, F. R.**, Innovation in the chemical and allied industries, *Chem. Ind. (N.Y.)*, p. 852, 1975.
55. **Weetall, H. H.**, Immobilized enzyme technology, *Cereal Foods World*, 21, 581, 1976.
56. **Sharples, A.**, An introduction to reverse osmosis, *Chem. Ind. (N.Y.)*, p. 322, 1970.
57. **Nielsen, I. K., Bundgard, A. G., Olsen, O. J., and Madsen, R. F.**, Reverse osmosis for milk and whey, *Process Biochem.*, 7(9), 17, 1972.
58. **Pitcher, W. H., Jr.**, Design and operation of immobilized enzyme reactors, in *Immobilized Enzymes for Industrial Reactors*, Messing, R. A., Ed., Academic Press, New York, 1975.
59. **Aries, R.**, New profitable products from whey, *North Eur. Dairy J.*, p. 354, 1977.
60. **Gramera, R. E.**, The interrelationship between agricultural food crops for the production of sweeteners, *Staerke*, 30, 20, 1978.
61. Technical Bulletin — MaxiLact, GB Fermentation Industries, Inc., Des Plaines, Ill.
62. **Kowalewska, J., Poznanski, S., Bendarski, W., and Sulima, K.**, The application of membrane techniques in enzymatic hydrolysis of lactose, and repeated use of beta-galactosidase, *North Eur. Dairy J.*, p. 20, 1978.
63. **Norman, B. E., Severinsen, S. G., Nielsen, T., and Wagner, J.**, Enzymatic treatment of whey permeate with recovery of enzyme by ultrafiltration, *The World Galaxy for the Dairy World*, p. 7, 1978.
64. **Coughlin, J. R. and Nickerson, T. A.**, Acid-catalyzed hydrolysis of lactose in whey and aqueous solutions, *J. Dairy Sci.*, 58, 109, 1974.
65. **Lin, A. Y. and Nickerson, T. A.**, Acid hydrolysis of lactose in whey versus aqueous solutions, *J. Dairy Sci.*, 60, 34, 1976.
66. **Demaimay M., LeMenaff, Y., and Printemps, P.**, Hydrolysis of lactose on catalytic resin, *Process Biochem.*, 13(4), 3, 1978.
67. **Frook, D. H.**, unpublished data, 1978.
68. **Taylor, A. W.**, Scaling-up of process operations in the food industry, *Chem. Ind. (N.Y.)*, p. 102, 1977.
69. **Reymond, D.**, Flavor chemistry — coffee, cocoa, and tea, *Chem. Technol.*, 7, 664, 1977.
70. **Attiyate, Y.**, Europe's newest computerized dairy, *Food Eng.*, 50(2), 84, 1978.
71. **Allen, D. H.**, Technological and economic factors of scaling up in the food industry. Some general principles and problems of scale-up, *Chem. Ind. (N.Y.)*, p. 98, 1977.
72. **Hockenhull, D. J. D.**, Why a pilot plant? Fermentation pilot plants and their aims, *Chem. Ind. (N.Y.)*, p. 461, 1973.
73. **Wandrey, C. and Flaschel, E.**, Optimal Particle size of Biocatalysts I — Reaction Engineering Aspects for Continuous Operation, 4th Enzyme Engineering Conf. Session 2, paper 3, Bad Neuenahr, Germany, September, 1977.
74. **Flaschel, E. and Wandrey, C.**, Optimal Particle Size of Biocatalysts, II — Economic Aspects of Continuous Operation, 4th Enzyme Engineering Conf., Session 2, paper 4, Bad Neuenahr, Germany, September 1977.
75. **Hupkes, J. V.**, Practical process conditions for the use of immobilized glucose isomerase, *Staerke*, 30, 24, 1978.
76. **Newton, S. B.**, Kraft's approach to quality assurance — interactions with quality control, *Food Prod. Dev.*, 11(10), 12, 1977.
77. **Anon.**, Code Federal Regulations, Title 21, Parts 10 to 129, U.S. Government Printing Office, Washington, D.C., 1975.
78. **Anon.**, Part 128g — Current good manufacturing practices regulation — Pickled, fermented foods. Fed. Reg. 41, 30458. Part 90., *Fed. Regist.*, 41, 30442, 1976.
79. **Pastore, M., Morisi, F., and Viglia, A.**, Reduction of lactose content of milk by entrapped beta-galactosidase, II. Conditions for an industrial continuous process, *J. Dairy Sci.*, 56, 269, 1973.
80. **Okos, E. S. and Harper, W. J.**, Activity and stability of beta-galactosidase immobilized on porous glass, *J. Food Sci.*, 39, 88, 1974.
81. **Curtis, J. E. and Huskey, G. E.**, HACCP analysis in quality assurance, *Food Prod. Dev.*, 8(3), 19, 1974.
82. **Somers, I. I.**, FDA — HACCP — Inspection Suggestion for the Canner, Bulletin 35-L National Canners Association, Washington, D.C.

83. **Shotwell, O. L., Hesseltine, C. W., Burmeister, H. R., Kwolek, W. F., Sharmon, G. M., and Hall, H. H.**, Survey of cereal grains and soybeans for the presence of aflatoxin, II, *Corn and Soybean Cereal Chem.*, 46, 454, 1969.
84. **Morisetti, M. D.**, Public health aspects of food processing, *Process Biochem.*, 6(6), 21, 1971.
85. **Anderson, A. W.**, The significance of yeasts and molds in foods, *Food Technol.*, 31(2), 47, 1977.
86. **Costerton, J. W., Geesey, G. G., and Cheng, K. J.**, How bacteria stick, *Sci. Am.*, 238(1), 86, 1978.
87. **Barndt, R. L., Leeder, J. G., Giacin, J. R., and Kleyn, D. H.**, Sanitation of a biocatalytic reactor used for hydrolysis of whey, *J. Food Sci.*, 40, 291, 1975.
88. **Sauer, F.**, Control of yeasts and molds with preservatives, *Food Technol.*, 31(2), 66, 1977.
89. **Chang, H. C. and Branen, A. L.**, Antimicrobial effects of butylated hydroxyanisole (BHA), *J. Food Sci.*, 40, 439, 1975.
90. **Kaufmann, H. P. and Ahmad, A. K. S.**, Pro and anti-oxidants in the field of fats. 22. Effect of antioxidants on the growth and lipid metabolism of *Saccharomyces cerevisiae*, *Fette Seifen Anstrichm.*, 69, 837, 1967.
91. **SH. H, A. L. and Harris, N. D.**, Antimicrobial activity of selected antioxidants, *J. Food Prot.*, 40, 520, 1977.
92. **Gorman, S. P. and Scott, E. M.**, A quantitative evaluation of the antifungal properties of glutaraldehyde, *J. Appl. Bacteriol.*, 43, 83, 1977.
93. **Munton, T. J. and Russell, A. D.**, Aspects of the action of glutaraldehyde on *Eschericia coli*, *J. Appl. Bacteriol.*, 33, 410, 1970.
94. **Thomson, J. E., Cox, N. A., and Bailey, J. S.**, Control of Salmonella and extension of shelf-life of broiler carcasses with a gluteraldehyde product, *J. Food Sci.*, 42, 353, 1977.
95. **Pitcher, W. H., Jr.**, Hydrolysis of whey by immobilized lactase, *North Eur. Dairy J.*, 42, 219, 1976.
96. **Harju, M.**, Microbiological control of an immobilized enzyme reactor, *North Eur. Dairy J.*, 43, 155, 1977.
97. **Swaisgood, H. E., Janolino, V. G., and Horton, R. H.**, Immobilized sulfhydryl oxidase, *AIChE Symp. Ser.*, 172, (74), 25, 1978.

Chapter 4

MANUFACTURE OF HIGH FRUCTOSE CORN SYRUP USING IMMOBILIZED GLUCOSE ISOMERASE

R. V. MacAllister

TABLE OF CONTENTS

I. Introduction 82
 A. Scope 82
 B. Definitions 82
  1. HFCS 82
  2. Immobilized Enzymes 82
  3. Nutritive Sweetener 82
  4. Starch 83
 C. Historical Background 83

II. General Principles 87
 A. Overall Process 87
 B. Manufacture of Starch 89
 C. Production of Immobilized Glucose Isomerase 89
 D. Hydrolysis of Starch 92
 E. Isomerization of Glucose to Fructose 93
 F. Refining and Concentration 96
 G. Fractionation 97

III. Specific Processes 98
 A. Production of Glucose Isomerase 99
  1. Organisms that Require Xylose (or Xylan) for Production of Glucose Isomerase 99
  2. Organisms that do not Require Xylose for Production of Glucose Isomerase 100
  3. Improved Methods for Conducting Fermentations to Produce Glucose Isomerase 100
 B. Production of Immobilized Glucose Isomerase 100
  1. Glucose Isomerase Immobilized with Bacterial Cells 101
  2. Cell-Free Glucose Isomerase Immobilized on Insoluble Carriers 102
  3. Cell-Free Glucose Isomerase Immobilized by Covalent Bond Formation between the Enzyme and an Insoluble Carrier 103
 C. The Isomerization Reaction Glucose ⇌ Fructose 105
 D. Isomerization Reactions to Attain High Ratios of Fructose to Glucose 106
 E. Fractionation of Standard HFCS Into Fructose-Rich and Glucose-Rich Portions 107
 F. Crystallization 107

References 108

I. INTRODUCTION

A. Scope

In this chapter the events which led up to the development of processes for the manufacture from corn starch of nutritive carbohydrate sweeteners having relatively high fructose levels will be described briefly. The major emphasis will be placed on the technical aspects and underlying scientific principles of those manufacturing processes.

B. Definitions

1. High Fructose Corn Syrup (HFCS)

Currently the most widely produced HFCS, hereinafter called the standards HFCS, contains about 42% D-fructose, 52% D-glucose, and 6% other saccharides on a dry basis. They are usually marketed as 29% water content syrups. These products are made by partial isomerization of the glucose in a starch hydrolysate which contains about 94% D-glucose on a dry basis. The other 6% are predominantly maltose, isomaltose, and trisaccharides. The HFCS products are practically all carbohydrate and water. The nitrogen (Kjeldahl) and sulfated ash contents are of the order of 7 ppm and 200 ppm, respectively.

Products of 55, 60, and 90% fructose are also being manufactured from corn starch,[1] though at present in much lower quantities than the 42% fructose products.

2. Immobilized Isomerase

The isomerase discussed in this chapter is classified as intramolecular oxidoreductase. The particular isomerase of this class which is now used most extensively in making HFCS is D-xylose isomerase, CBN code number 5.3.1.5, that catalyzes the reversible isomerization of D-xylose to D-xylulose and also catalyzes the reversible isomerization of D-glucose to D-fructose.[2] This enzyme is commonly called "glucose isomerase" in the HFCS industry. D-glucose isomerase, CBN code number 5.3.1.18,[2] catalyzes the reversible isomerization of D-glucose to D-fructose but does not catalyze the isomerization of D-xylose.[3]

The term "immobilized D-glucose isomerase" as used in this chapter refers to any composition that is insoluble in water solutions of D-glucose and/or D-fructose of low ionic strength and of pHs of about 4 to 9 and that can catalyze the reversible isomerization of D-glucose to D-fructose when placed in contact with a solution of D-glucose through the action of the enzyme which makes up a part of that composition.

3. Nutritive Sweetener

As this term is used in this chapter, "nutritive" is applied to substances that, in the quantities usually consumed by humans, contribute significantly to growth and to sustenance of life through the normal digestive and metabolic processes.

By "sweetener" is meant materials that are recognized by most humans as having a sweet taste. An operational definition of a "sweet taste" is "a taste sensed by most humans as having substantially the same quality as that noted when sucrose and sucrose-water solutions are tasted." Examples of nutritive sweeteners are such compounds as D-fructose, D-galactose, D-glucose, lactose, maltose, and sucrose. Mixtures of water soluble carbohydrates, such as occur in D-glucose syrups (commonly called corn syrup in the U.S. and Canada when made from corn starch) made by partial hydrolysis of starches and in honey and maple syrup are also examples of "nutritive sweeteners."

4. Starch

Starch, the raw material from which the HFCSs are made, is usually a mixture of two polymers, one of which, amylose, is a linear (1 → 4)-α-D-glucan of DP ∼ 400. The other polymer is a branched D-glucan having mostly α-D-(1 → 4) linkages, but with ∼ 4% of linkages of the α-D-(1 → 6) type. This branched polymer designated amylopectin, consists of short amylose-like chains of DP 12 to 50 with an average of ∼ 20 linked into a branched structure. The component amylose chains in amylopectin have been classified as types A, B, and C by French.[4] Type A is unsubstituted except at the reducing end (which is involved in a α-D-(1 → 6) linkage). The B chains are substituted on one or more 6-hydroxyl groups by A chains or other B chains, and are involved in a α-D-(1 → 6) linkage at the reducing end. The C chains are like the B chains except they are not substituted at the reducing end. There is only one C chain per molecule of amylopectin. Structures of amylose and amylopectin are shown in Figure 1.

Starches of various plants differ with respect to the proportions of amylose and amylopectin. The starch in corn (maize) of the variety mostly grown in the U.S. contains ∼26% amylose and 74% amylopectin. Some other varieties of corn contain starches which are substantially 100% amylopectin, and there are others having amylose contents as high as 80%.

Starch occurs in the plants as discrete granules. In corn these granules are irregular polyhedra of a width of 10 to 20 μm. The granules can be isolated from plants as concentrated, (up to ∼ 46% d.s.) free flowing suspensions in water because the granules retain their discrete character and do not swell, dissolve, or adhere to one another so long as the temperature of the suspension is not allowed to exceed about 130°F (∼ 54.5°C). Such suspensions can be filtered readily, producing solid, thick, dense filter cakes of ∼65% d.s.

C. Historical Background

In 1811 Kirchoff discovered that a sugar-like substance could be made by treating a water-starch mixture with acid at a high temperature. Later, Sausaure identified the substance as D-glucose. In the late 1800s large factories were built in the U.S. for isolating starch from corn and then hydrolyzing the starch by acid catalysis to make corn syrups. Through the efforts of W. B. Newkirk,[5] the manufacture from acid hydrolyzed starch of pure crystalline D-glucose as either the monohydrate or anhydrous form was established and became a major new food product ingredient.

By 1965 there were ten major corn wet milling companies in the U.S., all of which could produce corn starch of high purity. Nine of those companies produced a variety of corn syrups and three of them produced crystalline glucose.

By 1965 the total production in the U.S. of corn syrup and crystalline glucose was 5.10^9 and 1.10^9 lb d.s. per year.

Before about 1940, practically all corn syrup was made by acid hydrolysis of starch followed by refining and concentrating. At about that time Dale and Langlois[6] introduced enzyme conversion systems for greatly extending the range of compositions and related useful properties that could be obtained by hydrolysis of starch. These syrups were usually made by a preliminary acid hydrolysis, followed by the enzyme conversion. Some idea of the range of compositions attainable by acid conversions and by combined acid and enzyme conversions of starch can be obtained from Table 1.

Prior to about 1960, the manufacture of crystalline glucose involved crystallization of the product from water solutions prepared by acid catalyzed hydrolysis of starch followed by refining and concentrating. Subsequently, the manufacturers of crystalline glucose introduced into their operations the enzymatic hydrolysis of starch. This proc-

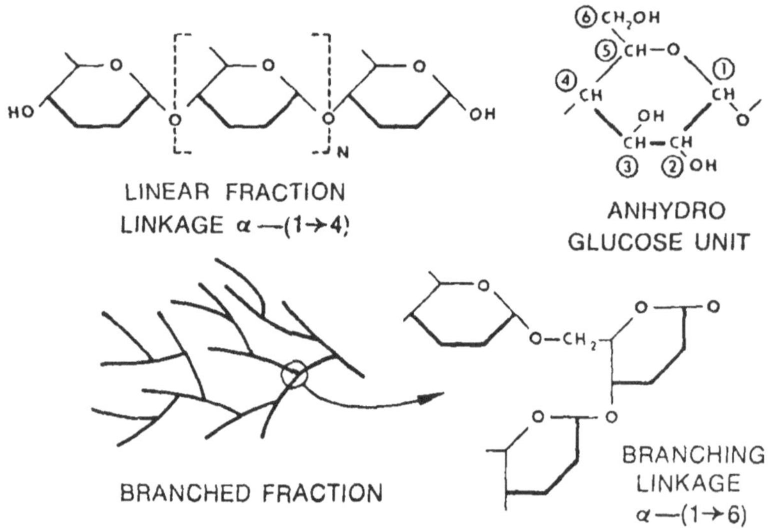

FIGURE 1. (Reproduced with permission from *Nutritive Sweeteners from Corn*, 1976. Copyright by Corn Refiners Association, Washington, D.C.)

ess made use of glucoamylase, CBN 3.2.1.3,[7] to effect the major glucose producing reaction in a solution of starch which had been prehydrolyzed to about 15 to 20 DE with an acid catalyst or with a starch liquefying enzyme such as alpha amylase, CBN 3.2.1.1.[7] The enzymatic conversion of starch for glucose production was a major development because, with it, glucose concentrations of the hydrolysate could attain 94 to 95% glucose (on a dry basis) in a reaction mixture of 30% d.s. (and even higher conversions at lower d.s. values in the reaction mixture). There were relatively small amounts of carbohydrate degradation products in the hydrolysate, and the residual small quantities of protein, fat, and ash could be easily removed by conventional refining operations to yield sweet, colorless solutions of almost pure carbohydrate having no off flavor. High yields of pure-crystalline glucose could be obtained from such liquors very efficiently. In contrast, the acid hydrolysis systems designed to attain high glucose concentrations in the hydrolyzates could attain only about 88% glucose (dry basis) in a reaction mixture of about 20% d.s. The resulting liquor, even after extensive refining had residual impurities and a very bitter flavor and the yields of crystalline glucose attainable were much lower than those obtainable from the enzymatically hydrolyzed starch liquors.

In 1965, the total market in the U.S. for nutritive sweeteners was about 24.10^9 lb/year, with sucrose and the 50-50 mixture of glucose and fructose (invert sugar) obtained by hydrolysis of sucrose having the major share. That share amounted to about 18.10^9 lb, as compared to the total of 6.10^9 lb of nutritive sweeteners derived from corn.

The nutritive sweeteners derived from corn starch were generally lower in price per pound d.s. than sucrose or invert sugar. Because of the lower price, the products from starch could compete with sucrose in those areas where nutritive value per pound product was the key issue. Also, corn syrups had great value in some uses because of the body or viscosity they could contribute to foods and also because of their great resistance to crystallization.

However, the nutritive sweeteners made from starch could not invade most of the traditional sucrose market because they were less sweet than sucrose or invert when

Table 1
COMPOSITIONAL DATA[a]

Sample	Dextrose equivalent	% Saccharides, carbohydrate basis						
		DP_1[b]	DP_2	DP_3	DP_4	DP_5	DP_6	DP_7+
Maltodextrin	12	1	3	4	3	3	6	80
Corn Syrup AC[c]	27	9	9	8	7	7	6	54
Corn Syrup AC	36	14	12	10	9	8	7	40
Corn Syrup AC	42	20	14	12	9	8	7	30
Corn Syrup AC	55	31	18	12	10	7	5	17
Corn Syrup HM, DC	43	8	40	15	7	2	2	26
Corn Syrup HM, DC	49	9	52	15	1	2	2	19
Corn Syrup DC	65	39	31	7	5	4	3	11
Corn Syrup DC	70	47	27	5	5	4	3	9
Corn Syrup DC	95	92	4	1	1	sum	of DP	2
						5,	6,	7+

[a] Data supplied by cooperating Member Companies of the CRA.
[b] DP = degree of polymerization.
[c] AC = Acid Conversion, DC = Dual Conversion (Acid-Enzyme), HM = High Maltose.

Reproduced with permission from "Nutritive Sweeteners From Corn", 1976. Copyright by Corn Refiners Association, Washington, D.C.

compared on equal weight basis. The sweeteners made from starch were generally more expensive than sucrose or invert sugar when compared on an equal sweetness basis. Furthermore, even if the cost of the starch based sweeteners was competitive with sucrose or invert sugar on an equal sweetness basis, the foods made up to desired level of sweetness by using large amounts of the starch base sweeteners would, in many cases, be less acceptable than those made with sucrose or invert sugar because of the undesirable textural characteristics of the food that would be imparted by the relatively large amounts of the starch based sweetener that would have to be used.

It was, of course, recognized for many years that the market for nutritive sweeteners made from starch could be greatly increased by an efficient method for converting glucose to fructose, because, in principle, a glucose-fructose mixture of the same composition as invert sugar could be achieved.

With the advent of high glucose yields and good flavor characteristics of the liquors which could be made efficiently by the enzyme catalyzed hydrolysis of starch, it became obvious that products similar in composition and taste to invert sugar might be achievable by enzymatic starch hydrolysis and subsequent or concurrent isomerization of glucose to fructose without the necessity of crystallizing the glucose from the liquor prior to isomerizing it to fructose. Thus, substantially all of the carbohydrate derived from the starch would go into the primary product. No by-product such as the mother liquor which would necessarily accompany a precrystallization of glucose would have to be processed and sold at a relatively low price.

Efficient isomerization reactions were still not available at the time that the opportunities opened up by enzymatic hydrolysis of starch to high glucose yields were layed open.

Many efforts had been made to exploit the Lobry de Bruyn reaction to convert glucose to fructose, principally by base catalysis.[8-12]

Invariably though, the best systems developed were defective to some degree because of low fructose yields, or because of excessive color, acid, and off-flavor formation through the degradation of glucose and fructose during exposure to the alkaline conditions of the reaction.

The formation of glucose-1-phosphate from starch and subsequent transformations to glucose-6-phosphate, fructose-6-phosphate, and finally to fructose are reactions well known from studies of carbohydrate intermediary metabolism.[13] The enzymes by which the phosphate systems could effect these transformations were well known, so in principle the conversion of starch to fructose could be carried out. Commercial processes for making fructose from starch through the phosphate intermediates have not been developed, probably because of the complexity of the multiple enzyme systems required and the difficulty of obtaining the enzymes for large-scale operations.

A landmark discovery was that of an enzyme in an organism which could catalyze the direct isomerization of an aldose (D-erythrose) to a ketose (D-glycero tetrulose)[14] without the need to go through a series of phosphate ester intermediates.

Cohen[15] then found an *Escherichia coli* which produced an enzyme that catalyzed the isomerization of D-arabinose to D-ribulose, and also the isomerization of L-fucose to the corresponding ketose. Cohen expressed the thought that since D-arabinose and L-fucose have the same configurations of substituents on carbon atoms 2 through 4 those configurations define specificity of the enzyme. Subsequently, other pentose and then also hexose isomerases produced by a variety of organisms were discovered.[16,17] The specificities of these latter enzymes were generally consistent with Cohen's suggestion regarding the significance of the configuration on carbon atoms 2 through 4 of the aldose sugar to the specificity of isomerase action. Marshall and Kooi[18] and Marshall[19] showed the isomerization of D-glucose to D-fructose by an enzyme which also catalyzes the conversion D-xylose to D-xylulose which substantiate further the specificity control of the configurations of substituents on carbon atoms 2 through 4.

Other bacterial sources of xylose isomerase were identified by Mitsuhashi and Lampen,[20] Hockster and Watson,[21] Slein,[22] Littauer et al.,[23] and Cohen and Barner.[24] As far as is known, all enzymes which catalyze the isomerization of D-xylose to D-xylulose also catalyze the isomerization of D-glucose to D-fructose.

Major improvements in the efficiency of production and use of xylose isomerase for conversion of glucose to fructose were made by discovery of an organism and fermentation process for producing xylose isomerase as disclosed by Takasaki and Tanabe,[25] and by a unique, efficient system for immobilizing and using xylose isomerase to isomerize D-glucose described by Takasaki.[26] These discoveries provided the basis for an efficient, enzymatic conversion of glucose to fructose.

Thus, in brief, the emergency of a major new industrial product, the high fructose corn syrup resulted from:

1. The gradual development, particularly in the U.S., of improved types of corn seed, cheap abundant fertilizers, and efficient agricultural machinery and practices by which corn can be grown, harvested, stored, and distributed very efficiently and economically and in enormous quantities, $\sim 300 \cdot 10^9$ lb of corn d.s. per year (containing about $200 \cdot 10^9$ lb of starch)
2. The development over a number of years by the corn wet milling industry of technology by which starch of high purity could be produced efficiently in very large quantities from corn (1965 about 5% of the U.S. corn crop was so processed), while at the same time valuable by-products such as corn oil, and nutritious animal feeds could be obtained
3. The development over the years, particularly in the U.S., of large markets for invert sugar, a product which in principle could be duplicated by the isomerization of D-glucose to D-fructose
4. The development of efficient enzymatic systems for converting starch to solutions of substantially pure carbohydrate, of 94% or higher glucose, and having no off flavor or color, and recognition that isomerization reactions carried out on this type of solution could lead to a product which for some uses could replace invert sugar
5. The development of efficient methods for production of D-xylose isomerase and for its use in isomerizing glucose to fructose

II. GENERAL PRINCIPLES

A. Overall Process

In essence, the production of high fructose corn syrups is relatively simple and involves just a few steps. In a typical sequence of operations, a suspension of starch in water of about 30% d.s. is heated either with a small amount of acid (~ suspension pH about 1.8) for a few minutes at about 45 lb in.$^{-2}$ steam pressure, or else it is heated to about 180°F in presence of an alpha-amylase at around pH 6.0 for 10 to 20 min. In either case the starch is partially hydrolyzed to a D.E. of 15 to 25. The starch treated with alpha-amylase may optionally be autoclaved for a short period of time (to improve filtration after the subsequent treatment with glucoamylase). After the conversion to starch to about 15 to 25 D.E., the liquid is adjusted to about pH 4.3, the temperature is lowered to 140°F, and after adding glucoamylase, held at 140°F for time required to attain the desired degree of conversion to glucose, usually about 94 to 95% glucose on a dry basis. The time required to attain 94 to 95% glucose typically is 72 hr, but less time is needed if more glucoamylase is added.

After completion of the treatment with glucoamylase, the so-called saccharification step, the solution contains in addition to glucose, about 5% to 6% other carbohydrates, some small amounts of nitrogen (Kjeldahl), fatty material, fiber, inorganic ions, and a small amount of residual glucoamylase. The fatty material, fiber, and the insoluble nitrogenous material are removed by simple filtration, usually on a rotary vacuum drum filter provided with a precoat of diatomaceous earth.

The liquor is then concentrated under vacuum to about 60% dry substance, and then passed through columns of granular carbon which adsorbs practically all of the noncarbohydrate materials in the solution except the small amount of ionic substances. Alternatively the 60% d.s. solution can be purified by mixing it with powdered carbon, which after adsorbing impurities is removed by filtration. Optionally, the solution having been treated with carbon may then be passed through successive columns of cation exchange resins and anion exchange resins to remove the small amount of ionic material.

Small proportions of magnesium and bisulfite salts are added to the solution, the pH is adjusted to about 7.5, and the temperature is fixed at about 150°F. Then the solution is flowed continuously through a vessel where it makes good contact with immobilized glucose isomerase which stays in the reactor. The residence time in the reactor is typically 20 to 30 min, depending upon the amount of active glucose isomerase present in the reactor.

During its passage through the reactor, the glucose contained in the solution is partially isomerized to fructose. Usually the reaction is carried to a composition of about 42% fructose, 52% glucose, and 6% other saccharides (on a dry basis) when the feed to the reactor contains 94% glucose and 6% other saccharides, on the same basis.

During the reaction a very small amount of acid formation occurs due to decomposition of some of the carbohydrate, and the pH decreases a few tenths of a pH unit. The bisulfite salts in the feed liquor, in addition to decreasing the rate of inactivation of glucose isomerase and the rate of color formation, acts as a buffer and moderates the pH change. Some color also tends to form, and if the pH and temperatures were allowed to become too high during the reaction a hexose-ketose known as psicose tends to be formed.

Since the valuable attributes of invert sugar and of the HFCS which competes with invert sugar are, in addition to a high level of sweetness per unit weight, the absence of any other than that sweet taste, the isomerized dextrose solution must be carefully refined in order to remove any traces of off flavor or off flavor precursors. It is important also to remove substantially all the color and color precursors so that the syrup is water-white when made and so that it remains in that condition while in the manufacturer's, distributor's, and consumer's storage or transportation system.

The refining system used to produce the very high purity glucose containing solution that is fed to the isomerase reactors is applied again to the isomerized liquors. The effluent from the isomerization reactors is usually lowered to pH 4 immediately, in order to minimize color formation. Also, that is the best pH for the next operation, that is passage of the liquor through beds of granular carbon, where most of the traces of color bodies and off flavor materials which may have appeared in the solution during the isomerization reaction are removed by adsorption. Then the liquor is passed successively through cation exchange and anion exchange liquors which remove almost all of the inorganic ions and residual color materials. The resultant liquor is substantially colorless, odorless, and has no taste effect other than that of intense sweetness, similar to that of a high grade invert sugar syrup.

The highly refined high fructose corn syrup liquor is then concentrated to about 71% dry substance, and this constitutes the major commercial HFCS product at this time. We will refer to such a product as the standard HFCS in the rest of this chapter.

The isomerized liquor may be treated in various ways in order to attain a product of higher fructose content.

One way to do this is to concentrate the liquor under low pressure to a relatively high dry substance level, then crystallize glucose from the concentrate by carefully controlled seeding, gradual cooling, and agitation. After removing the crystalline glucose by centrifuging, the mother liquor is recovered as a fructose-enriched product. Syrups with up to 70% fructose (dry basis) have been made in this manner from syrups of 42% fructose, 52% glucose, and 6% other saccharides. The crystalline glucose can then be dissolved and recycled to a glucose isomerization reaction.

Enriched fructose products can also be made by large-scale elution chromatography using, for example, columns of certain types of strong acid ion exchange resins in the calcium salt form as the stationary phase. When a refined 42% fructose, 52% glucose, 6% other saccharides liquor is placed on such a column and eluted with water, the other saccharides, being mostly di and higher saccharides, tend to come off of the column first (due to a molecular exclusion effect it is believed), followed by glucose and finally fructose. The efficiency of such a fractionation is dependent upon interaction of the variables of feed liquor concentration, column loading, fluid flow rates, and pattern of blending and recycling column effluents.

The combination of those variables which result in the maximum efficiency is difficult to establish, and is an especially complex system to optimize because the optimum values of the variables changes with the degree of fractionation desired, that is, the composition of the enriched fructose product to be produced.

So, other than for the complexity of the fractionation system mentioned above, the principles involved in the manufacture of HFCS are relatively simple.

Some of the principles involved in the operations described briefly above will now be discussed in more detail.

B. Manufacture of Starch

The HFCS industry exists only because starch can be produced and then converted to HFCS at a low enough cost so that it can compete with that other excellent nutritive sweetener, invert sugar, which can also be produced efficiently and at a low cost.

Corn serves as an excellent source of starch in the U.S. because it is produced in such enormous quantities, far exceeding the amount used to make corn syrup and HFCS, that its price is relatively stable. Furthermore, it has excellent storage equal characteristics, so that it can be used as raw material in year-round manufacturing.

Starch can be recovered from the corn kernels in high yields and high purity, and at the same time, valuable by-products obtained by the corn wet milling process. The term "wet milling" signifies processes in which the corn kernels are steeped in dilute aqueous solutions of sulfurous acid, followed by separation of the steepwater containing the corn solubles, and then separating the insoluble materials (germ, hulls, gluten, and starch) from each other by a series of grinding, screening, and settling operations, all conducted with those insoluble parts of the corn suspended in relatively large volumes of water. There is also a "dry milling" operation in which the various constituents of corn kernels are separated by grinding, screening, and air classification operations on corn that is essentially dry.

The system by which corn is separated into various fractions in a wet milling process is indicated in Figure 2. The starch which is obtained as a suspension in water by this process consists of the small, discrete granules mentioned before. Although the suspension contains 46% dry substance, its viscosity is relatively low and it can be pumped through process lines efficiently as long as the suspension moves rapidly enough and is kept agitated enough to prevent the settling of the granules. The starch at this stage is very pure, containing about 0.05% nitrogen (Kjeldahl), about 0.6% of fatty material, and 0.2% ash (sulfated). This suspension of starch provides the raw material for the hydrolysis and subsequent operations involved in the manufacture of high fructose corn syrups.

C. Production of Immobilized Glucose Isomerase

Glucose isomerase used on a commercial scale is generally produced in submerged aerated fermentations. Many organisms have been found which produce glucose isomerase efficiently.

Many of the organisms for producing glucose isomerase as described in the literature require D-xylose as an inducer of the isomerase formation. Some of those organisms

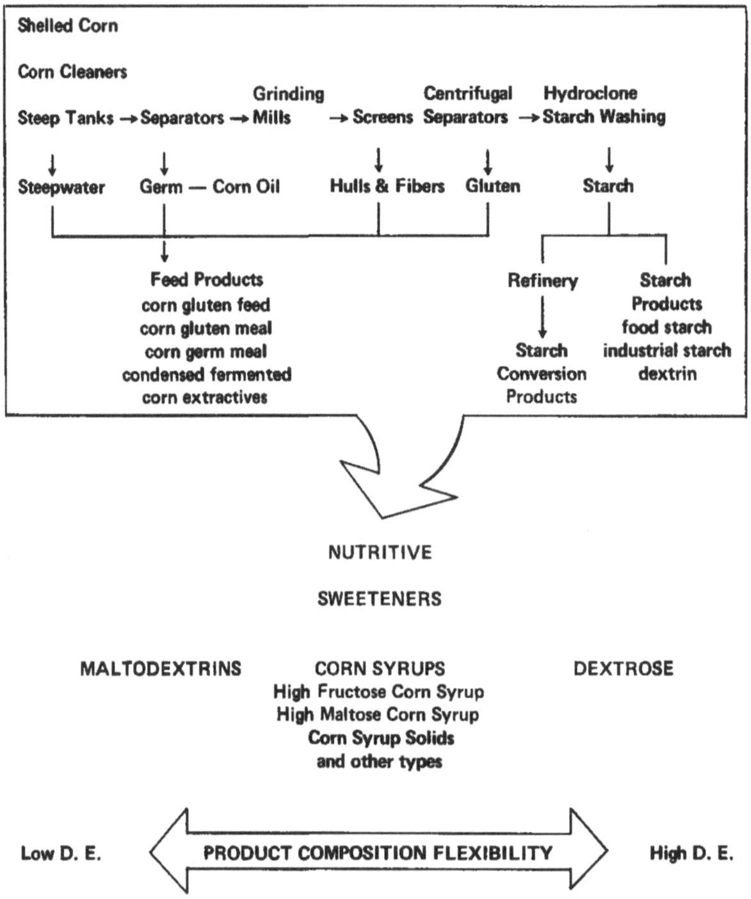

FIGURE 2. Corn wet milling process. (Reproduced with permission from *Nutritive Sweeteners From Corn*, 1976. Copyright by Corn Refiners Association, Washington, D.C.)

can grow on a nutrient containing D-glucose free of D-xylose, but produce very little glucose isomerase. Certain strains of organisms such as *Arthrobacter* and *Streptomyces* produce good yields of the enzyme with D-glucose in the nutrient but no D-xylose.[27] Often good yields of the enzyme are obtained with just corn steep liquor and ammonium phosphate to supplement D-glucose in the fermentation medium.

The fermenters generally must be provided with good agitation, aeration, and pH control in order to obtain good yields of enzyme. It is most important, though, to have cultures capable of producing good yields and to have the equipment, procedures, and operating personnel who can maintain pure culture conditions during the process. If the nutrient becomes contaminated with foreign organisms during fermentation or subsequent processing, the results can be disastrous.

The principle involved in the fermentation and immobilization operations involved in the manufacture of immobilized glucose isomerase is illustrated by the following description of typical processes. After the maximum concentration of enzyme is reached in the fermentation, about 48 to 72 hr, the broth (the whole contents of the fermentor) may be filtered and the mass of organism washed with water on the filter. In this typical process, substantially all of the glucose isomerase is held by the cell mass and the soluble residual sugars, salts, proteins, etc. are washed away. The wet cell mass can then be processed in several alternative ways to make immobilized glucose isomerase prior to using the enzyme in the conversion of glucose to fructose.

1. The cell mass can be stored, usually at low temperatures, and when needed, simply taken to the reactor site. The enzyme in some organisms is immobilized as it naturally occurs on or in the cell mass.[28]
2. The suspension of cells containing the glucose isomerase may contain lytic enzymes which would tend to cause the solubilization of the isomerase when exposed for the long periods of time in the isomerization reactor. The lytic enzymes can be inactivated without inactivating the isomerase by carefully controlled heating of the cell mass suspension.[29] As a result, the heat treated cells can be used in isomerizations with little solubilization and loss of the enzyme from the reactor.
3. The cell mass with or without heat treatment to stabilize the enzyme against solubilization can be dried. Hot air drying is very effective if the air which contacts the cell mass, particularly while the cell mass is still wet, has a wet bulb temperature well below that which corresponds to the onset of rapid heat denaturation of the enzyme. It is important, too, to avoid formation of large masses of the dry enzyme cell complex which cannot be penetrated by the solution to be isomerized. Mixing filter aids of various kinds with the wet cell mass before drying is sometimes effective in developing dry, immobilized enzymes which are of an open structure readily contacted throughout its mass by the solution of reactants.[30]
4. The aqueous suspension of cells containing the glucose isomerase can be treated to solubilize the enzyme. After removing insoluble cell material by filtration or centrifuging, the solution containing soluble glucose isomerase is contacted with an insoluble material referred to as a carrier to which the enzyme is attracted and held. The insoluble carrier-enzyme complex can be removed from the suspension by filtration and washing. The insoluble carrier-glucose isomerase combination can then be dried and adjusted to particle size distribution as needed for a given reactor configuration and reactor flow pattern.

   The agents used for the solubilization of the enzyme are usually surfactants of the quaternary ammonium salt type.[27]

   Many insoluble materials have been advocated as the carrier for forming the insoluble carrier-glucose isomerase complex. The principles involved are illustrated by the use of diethylamino-ethyl cellulose (DEAE) for this purpose.[31] The DEAE-cellulose adsorbs the glucose isomerase effectively, but the solution of glucose isomerase to be treated sometimes contains some proteins other than the isomerase which are also adsorbed. So, in order to make the most efficient use of the DEAE-cellulose and to achieve a high enzyme concentration on the DEAE-cellulose, it is desirable to keep the amount of nonisomerase protein in the solution as low as is practically possible. Toward that same end of economy and efficiency, it is desirable to use no more DEAE-cellulose than is necessary. Adsorbtion isotherm tests carried out on samples of each batch of soluble glucose isomerase to be treated provide a reliable indication of the amount of DEAE-cellulose to be used on the large scale immobilization.
5. The cell containing glucose isomerase can be suspended or solutions of glucose isomerase can be dissolved in solutions of monomers which can be polymerized to entrap the cell-enzyme complex or soluble enzyme within a polymer gel matrix.[32] This material can then be further processed to form a dry solid structure containing the active glucose isomerase within it, and having a pore structure which allows penetration of the solution and flow of reactants and products to

and from the reactive sites of the enzyme, but prevents the enzyme-cell complex or the enzyme itself from leaking out into the reactant solution.

In a variation of this way of making immobilized glucose isomerase, the cells and associated enzyme or the free enzyme are mixed with a solution of a polymer. The polymer is precipitated from the solution, entrapping the cells and/or the enzyme within the polymer structure. This process can be carried out in a way that the reactants and products solution can move to and from the reactive sites while the enzyme cannot dissolve in the reactant solution.

Generally in making immobilized glucose isomerase it is desirable to attain a high specific activity because that tends to decrease the net cost of the carrier per unit substrate processed (other things being equal) and for a given reactor volume, tends to decrease the residence time in the reactor. The latter is especially important because the conditions of isomerization, high pH, and temperature tend to cause formation of undesirable color, taste, acids, and psicose, so the less time under those conditions the better.

D. Hydrolysis of Starch

Most of the HFCS products of current commercial interest are made by isomerization of hydrolysates of starch which have, on a dry basis, 94% or more glucose. The typical HFCS as described before made by partial isomerization of the glucose in such hydrolysates can replace invert sugar in numerous food products. If the hydrolysate contained substantially less than 94% glucose, then the isomerized product would not be as effective as a substitute for invert sugar because it would be less sweet and would result in a food product with somewhat higher concentration of carbohydrate needed to match the taste of a given food product made with invert sugar. On the other hand, the quality of products made by isomerizing to a comparable degree starch hydrolyzates having more than 94% glucose are not, for most food products, much better than the one based on 94% glucose.

In the enzyme catalyzed hydrolysis of starch to make the 94% glucose hydrolyzate, α-amylase may be used to partially hydrolyze the starch so that the solution to be subjected to action of glucoamylase will be of low enough viscosity at the high concentrations (about 30% dry substance) and relatively low temperatures (140°F) required for glucoamylase.

Bacillus subtilis var. *amyloliquefaciens* is a common source of α-amylase. This enzyme is active even at 82.2°C, a temperature at which corn starch granules suspended in water become highly hydrated and readily accessible to the enzyme. α-Amylase catalyzes the hydrolysis of interior α-D-(1 → 4)-glucosidic bonds of amylose and amylopectin. It is called an endo-hydrolase. The bond rupture can occur wherever there are at least six D-glucosyl residues on one side and at least three on the other side of the bond to be broken. Thus extensive hydrolysis by α-amylase results in mixtures of saccharides composed mostly of maltohexaoses, maltoheptoses, and maltotrioses. Branched saccharides such as 6^2-α-maltosyl-maltotriose arise from action of α-amylase on amylopectin.

α-Amylase action on starch is believed to involve protonation of the glycoside bond by an imidazolium ion of the enzyme, followed by action of a carboxyl group of the enzyme on the anomeric carbon resulting in a β-glycosyl ester which is then split by action of water at the anomeric carbon.[33]

After the preliminary hydrolysis by α-amylase, which is usually conducted at a high temperature, the liquid is cooled and subjected to the action of glucoamylase at pH 4.3.

Glucoamylase is classified as an exohydrolase because it attacks starch or the oligosaccharides produced by α-amylase action from the nonreducing end of the molecule and releases D-glucose as the β-anomer from either α-D-(1→4) bonds or α-D-(1→6) bonds. The latter bonds are hydrolyzed much more slowly than are the former. After the (1→6) bonds are broken, the adjacent (1→4) are open to rapid hydrolysis. Maltose is hydrolyzed more slowly than are the bonds in longer α-D-(1→4) linked chains of D-glucose. Isomaltose is very resistant to hydrolytic action of glucoamylase.[34]

Glucoamylase isolated from *Aspergillus niger* consists of a mixture of isozymes, one of which is much more effective in splitting α-D-(1→6) bonds than the other.[35]

Some organisms produce a transglucosidase in addition to the glucoamylases. The transglucosidase can catalyze the formation of isomaltose from the maltose which appears as an intermediate in the glucoamylase catalyzed hydrolysis of starch or oligosaccharides. Since the isomaltose is very resistant to hydrolysis, the result is a decrease in glucose yield. This difficulty can be overcome by removal or selective inactivation of the transglucosidase.[36]

The rate of hydrolysis of corn starch on oligosaccharides becomes limited by the rate of hydrolysis of the α-D-(1→6) bonds in amylopectin. There is an enzyme *Pullulanase* (EC 3.2.1.41) which catalyzes the hydrolysis of α-D-(1→6) linkages in amylopectin, but does not affect the (1→4) bonds. There must be at least two D-glucosyl groups in the group attached to the rest of the molecule through an α-D-(1→6) bond for Pullulanase action.[37] So, if the starch or α-amylase treated starch is treated with Pullulanase prior to or concomitant with the action of glucoamylase, the rate and extent of glucoamylase action to produce glucose can be enhanced.

E. Isomerization of Glucose to Fructose

Glucose isomerase formed by a *Streptomyces* species has four subunits, each having a molecular weight of ~ 41,500.[38] The specific activity of the pure enzyme is about 20,000 International Glucose Isomerase Units (IGIU) per gram. The IGIU unit is defined as the number of micromoles of D-fructose formed from D-glucose per minute under specified conditions as described by Lloyd and co-workers.[39]

The catalytic effect of glucose isomerase is enhanced by divalent cations such as Mn^{2+}, Co^{2+}, or Mg^{2+}, and is inhibited by Cu^{2+}, Hg^{2+}, Zn^{2+} and to some extent Ca^{2+}. Xylitol and D-glucitol inhibit glucose isomerase activity.[40]

The pH range of 6.5 to 8.5 and temperatures from 40°C to 70°C are generally suitable for action of glucose isomerase.

The isomerization of glucose to fructose is reversible. The equilibrium constant is about 1.0 at 60°C, and it increases by ~0.08 for each rise of 10°C.[41]

The enzyme obtained from a *Streptomyces* species has a half-life of several hundreds of hours when used under commercial conditions in which the enzyme is sused in immobilized form, and at about 50% glucose concentration in the substrate which is controlled to a pH of about 7.8 at about 65°C. The presence of Mg^{2+} and HSO_3^- ions in the substrate solution enhances the stability of the enzyme.

It has been proposed that the activity enhancing metal ions act as a bridge between active sites of the enzyme and the substrate molecule. The metal ion coordinates with one or two of the oxygen atoms on carbon atom number 1 (C-1) of the substrate and thus promotes the removal of protons from C-2 of the substrate by basic groups of the enzyme. The strain of the four membered ring that is formed leads to elimination of the ring-oxygen atom, formation of an endiol intermediate, and subsequent transfer of the proton to C-1. The keto furanose formed is then released from the enzyme.[42]

The rates of the enzyme catalyzed isomerizations of glucose to fructose can be rationalized in simple terms G, F, e, q, and E (= e + q) representing concentrations of glucose, fructose, free enzyme, enzyme-substrate complex, and total enzyme, respectively, that are involved in interactions:

$$G + e \underset{k_{-1}}{\overset{k_1}{\rightleftarrows}} q \underset{k_{-2}}{\overset{k_2}{\rightleftarrows}} e + F \tag{1}$$

At the steady state, the following, where t is time, would hold:

$$\frac{-dG}{dt} = \frac{dF}{dt} = k_1 eG - k_{-1} q = k_2 q - k_{-2} eF \tag{2}$$

$$\frac{dF}{dt} = [(k_1 k_2 G - k_{-1} k_{-2} F)E] \,/\, [(k_{-1} + k_2 + k_1 G + k_{-2} F)] \tag{3}$$

assuming that at all times after attaining the steady state $G + F + q = C$ a constant, $C \gg q$ and that at equilibrium, $dF/dt = -dG/dt = 0$, $F = F_e$, and $G = G_e$, it follows that:

$$\frac{dF}{dt} = [E(k_1 k_2 + k_{-1} k_{-2})(F_e - F)] \,/\, [k_{-1} + k_2 + k_1 C + (k_{-2} - k_1)F] \tag{4}$$

At a given concentration of glucose plus fructose ($\simeq$C) the constants may be grouped and the steady rate equation approximated as:

$$\frac{dF}{dt} = kE(F_e - F) \tag{5}$$

$$\ln [(F_e - F) / (F_e - F_0)] = -kEt \tag{6}$$

where t′ is the age of the enzyme in service, E_o is the amount of glucose isomerase present at $t' = 0$, and τ is the half-life of the enzyme.
solutions between the α anomer of glucose (which is the one that reacts with the glucose isomerase) and the β anomer. Fructose exists in four forms in water solutions, but which one interacts with the enzyme is not known yet. Nevertheless, data from experiments carried out with soluble glucose isomerase expressed as change in total glucose and total fructose with time conform well with Equation 6.

The time-concentration relationship observed with a given quantity of substrate containing a given amount of soluble glucose isomerase in the immobilized form on DEAE-cellulose is suspended in the same amount and type of substrate and all other conditions are the same.

When glucose isomerase immobilized on DEAE-cellulose (referred to as DCI) is exposed to glucose solutions under conditions used in isomerization reactions, it loses activity at rate which is first order with respect to concentration of the enzyme. Equation 5 can be changed to take this into account as:

$$\frac{dF}{dt} = kE_0 (F_e - F)\exp(-0.693t' / \tau) \tag{7}$$

Where t′ is the age of the enzyme in service, E_o is the amount of glucose isomerase present at $t' = 0$, and τ is the half-life of the enzyme.

Equations 5, 6, and 7 can be adapted to the continuous flow of substrate through a reactor where it makes contact with an immobilized glucose isomerase (which stays in the reactor). In this case the time, t, value corresponds to the residence time of the substrate in the reactor and inversely to the rate of flow through the reactor. The reciprocal of residence time is referred to as space velocity.

At a given concentration of F + G (≃C) in the solution passed through a reactor containing a given quantity of glucose isomerase (E_o), the residence time t_p required to attain an effluent of fructose content F_p is given by Equation 6. The rate of flow of the solution through the reactor is then proportional to $kE_o/\ln[(F_e - F_o)/(F_e - F_p)]$. The rate of flow (R) at which the value of $F = F_p$ could be maintained will decrease with the age of the enzyme in the reactor as it loses activity. Then $R = kE_o\exp(-0.693t'/\tau)/\ln[(F_e - F_o/F_e - F_p)]$.

The total amount of product Q_p having fructose content F_p which can be made by continuously passing a solution of glucose at concentration C through a reactor containing E_o units of enzyme in it at time $t' = o$, would, in the enzyme service time $t' = t's$, amount to:

$$Q_p = [kE_o/|\ln[(F_e - F_o)/(F_e - F_p)]|] \int_o^{t'_s} \exp(-0.693t'/\tau)dt'$$

$$Q_p = [1.44\ kE_o\tau][1 - \exp(-0.693t'_s/\tau)] / |\ln[(F_e - F_o)/(F_e - F_p)]|$$

Note: Absolute value of the ln function is employed because the in function is negative when $f_o < f_p < F_e$.

Thus, the productivity of a system — the total amount of product (having a fructose to C ratio of F_p) which can be made per unit glucose isomerase introduced into a reactor during a given service life of the enzyme — is proportional to k and to τ, so the product k τ is a useful figure of merit for the glucose isomerase enzyme. Since the k value is a composite of various other constants, including C ≃ G + F, comparisons of various systems should be made at the same value of C.

Productivity approaches a maximum as service life is extended indefinitely of:

$$Q_p\max = 1.44\ kE_o\tau / |\ln[(F_e - F_o)/(F_e - F_p)]|$$

The factor $[1 - \exp(-0.693t'/\tau)]$ is the proportion of the enzyme activity originally introduced which has been lost in the time period t'.

The above analysis of productivity based on models of enzyme catalyzed reactions and enzyme activity loss has served well in developing the structure and operating procedures for isomerization reactors and in analyzing their performance.

Large-scale and laboratory-scale reactor performance generally conform to the productivity analysis based on the models, and analysis of marked deviations from expected performance has often provided clues leading to the solution of production problems.

Reactor productivity can, of course, also be evaluated by mechanical integration of actual production rate vs. enzyme service life plots.

In the continuous flow systems, the isomerization reaction can be continuously monitored by in-line polarimeters. The conversion of glucose to fructose results in a decrease of specific rotation $[\alpha]_D^{25}$ of −140° at concentration of 0.1 *M*, so this provides a sensitive measure of the progress of reaction.

The reactor configuration and fluid flow pattern through reactors containing immobilized glucose isomerase that are best vary with the physical structure of the immobilized enzyme.

For those immobilized enzymes which tend to develop high resistance to flow, the use of relatively thin layers of the enzyme material such as achieved in the cake on a filter press may be desirable. In this case since the reactors have to be able to withstand some considerable pressure, they are generally quite expensive. Therefore, it is especially important to have a relatively high concentration of enzyme on the carrier in order to attain high rates of production per unit immobilized glucose isomerase reactor. It is very important to form the layers of the immobilized enzyme that have high flow resistance in such a way that the resistance is uniform over the whole cross-section of the layer. Otherwise, channeling will develop and the result will be inefficient use of the enzyme.

Immobilized enzymes which have little resistance to flow can be used in deep beds in a simple inexpensive column which can be operated in the downflow mode, but channeling and consequent loss in efficiency of use of the enzyme can become a problem with columns of this type.

The use of continuous upflow of substrate in column type reactors is feasible if the immobilized glucose isomerase has a density sufficient to keep it in the reactor at the desired rate of flow. By keeping the solid enzyme containing particles in a "fluidized" condition, the velocity pressure drop across the reactor becomes very low, and there is good contact between substrate and the immobilized enzyme particles.

The loss of activity which occurs during the service life of the enzyme naturally results in loss of production rate unless more enzyme is added to make up for the loss. The addition of enzyme can be achieved effectively by having a number of separate reactors, either in parallel or series.

When some proportion of the activity of the enzyme remaining in a reactor reaches some particular value, say 12.5% of the original (which corresponds to 3 half-lives), that reactor can be emptied, recharged with fresh enzyme, and returned to service. Since the immobilized glucose isomerase has hundreds of hours half-life values, changes need be made only infrequently, and an average rate of flow can be maintained with a number of reactors in service with relatively little difference between maximum rate (just after a new reactor is put into service) and minimum rate (just before the oldest reactor is taken out of service).

F. Refining and Concentration

The isomerization reactions as typically carried out in the large-scale production of HFCS use substrates of high purity and immobilized enzymes which add very little, if any, substances to the solution phase. During the reaction, practically no change, other than the partial conversion of glucose to fructose, occurs in the composition of the solution phase. However, the HFCS are used in the manufacture of high quality foods to which the syrup must add nothing but the pure carbohydrate material and must impart no color or flavor other than that of sweetness. Therefore, the isomerized syrup is refined to remove the traces of salts, color, and off flavor that are added or formed during isomerization.[43] After refining the solution, it has practically no color and has a clean, sweet taste. Conditions of pH and temperature in subsequent handling of the liquor must be controlled carefully to avoid deterioration in taste and color. The fructose and glucose liquors are most stable at about pH 4.[44]

Too low a pH can cause color and condensation products to form and too high a pH leads to formation of psicose, acidic materials, and color.

The refined isomerized liquor is adjusted to about pH 4 and evaporated at low temperature and pressure in vacuum evaporators. Whatever specific type of evaporator is used in this operation, it is most important to keep the time-temperature pattern to which the liquor is exposed to one which will avoid color and off-flavor formation.

Generally the HFCS are practically without buffer capacity, so it is difficult to keep the pH under control for the evaporation process. Yet it is important to do so in order to avoid changes in the color, flavor, and composition of the product.

As mentioned before, the HFCSs are usually of about 71% dry substance. This is about the highest concentration that can be tolerated without encountering excessive danger of crystallization of glucose from the syrup under the usual storage and transportation conditions. High temperature (say above 100°F) storage and shipment would make higher concentrations possible, but would increase the danger of color and off-flavor development.

Maintenance of high purity in the product and good storage conditions are important because it is then possible to build up inventories which will keep good quality product for sale in later high-demand seasons by keeping production rates at full capacity during periods of lower demand, thus attaining close to peak production year round with its attendant efficiencies.

G. Fractionation

The degree of isomerization of glucose to fructose in water solutions is inherently limited in the direct isomerization reactions to the equilibrium value equivalent to about equal molar proportions of the two sugars. It is possible to exceed that limit by imposing special conditions on the reaction. As far as is known, none of these extended isomerization reactions have been exploited commercially, but some of them will be discussed later in this chapter under specific processes.

There is a need for products having a higher fructose content than can be made by the methods which have been described up to this point in this chapter.

The conventional HFCSs containing about 43% fructose, 51% glucose, and 6% oligosaccharides can be fractionated into high and low fructose content products by a number of different systems which will be described in the next section of this chapter.

Generally the fractionation systems developed to the greatest extent so far are those based upon elution chromatography. The principles involved in commercial scale chromatographic separations are generally the same as those which underlie chromatographic systems designed for analytical procedures. In large-scale chromatography, however, in contrast to chromatography for analytical work, the attainment of the desired degree of resolution of components (which is usually far from a complete resolution) at minimum cost is the important issue. Toward that end, the major problems in the large-scale chromatographic fractionations are: to attain maximum rate of production of the desired fraction per unit volume of fractionating medium, to minimize the amount of elution water needed per unit of product made, and to attain highest possible yield of desired product. The best overall fractionation system will represent a balance in the attainment of those goals which result in the greatest net return. The cost of the material fed to the reactor and the sales volume-price relationship of the fractionated product also enter into the profitability equation. Consequently, the attainment of optimum design in a fractionating system can be a difficult problem.

Cycling zone adsorption is a technique recently invented for separation of multicomponent mixtures. The separations (as quoted directly from Wankat,[45] are achieved by cyclic variable feed to the adsorption column in a series of steps. The amount fed at each step is such that some components will move faster than the wave velocity of the cyclic input and others will move more slowly. The next step causes another component to move faster than the wave velocity. As a result, each component concentrates at the step where its wave velocity first becomes larger than that of the cyclic input.

For instance, consider adsorption with temperature as the cyclic variable with two components (A and B in a nonadsorbed carrier) to be separated. Assume component A is less strongly adsorbed than B at all temperatures. As temperature increases both components are adsorbed less strongly and both concentration waves move through the column at a faster speed. At some low temperature, T_c, both components will move slower than the thermal wave. Since thermal wave velocity is relatively insensitive to the temperature, the concentration wave velocities will eventually become greater than thermal wave velocity but B still moves slower. Component A will now tend to concentrate at temperature T_1 since the thermal wave overtakes all A input at T_c and is over-taken by A input at T_1. Another temperature T_2 is chosen so that both components move faster than the thermal wave velocity. Component B will now concentrate near step at T_2. Component A does not concentrate here, but moves past this thermal wave to temperature T_1. The inlet temperature profile for this two-component separation is shown in Figure 3. The inlet temperature profile is repeated cyclically. If the column is long enough and the times for each temperature are set properly, very large separations can occur.

If more than two components are present the feed can still be separated if additional temperature plateaus are added. The separation can also be obtained if instead of temperature plateaus a continuous change in temperature (either linear or nonlinear) is utilized. Now each component will tend to concentrate at that temperature where it moves at the same velocity as the thermal wave.*

The fractionation of mixtures of glucose and fructose by the cycling zone adsorption technique[46] has been reported. As far as is known, there are no commercial-scale processes based on this technique for the fractionation of glucose-fructose mixtures.

III. SPECIFIC PROCESSES

Since the beginning of manufacture of HFCS in the U.S. by one company in 1967, the rate of production has increased rapidly, and more manufacturers were attracted to production of HFCS as tabulated below.[47]

Year	HFCS produced in U.S. lb dry substance 10^{-6}	Number of manufacturers in U.S.
1972	347	2
1973	625	2
1974	834	2
1975	1469	4
1976	2120	5
1977	2875	7

Forecast for production in U.S. in 1980 and 1985 are $5700 \cdot 10^6$ lb and $8600 \cdot 10^6$ lb, respectively.[48]

In 1976, production of HFCS was $230 \cdot 10^6$ lb in Europe and $400 \cdot 10^6$ lb in Japan.[49]

Among the reasons for the growth of the HFCS industry are the high quality of the HFCS which enabled it to compete with invert sugar, the fact that HFCS created new markets for products from corn (it was not a case of replacement of existing products made from corn), and the development of improved, highly efficient technologies for making HFCS.

The intensity of the work on HFCS technology developments during the past 10 years is indicated by over 200 patents related to HFCS production which were issued during that period.

New technologies developed since 1967 are described in technical and patent literature related to the specific processes used in making standard HFCS that may be classified as:

* Reprinted with permission from Wankat, P. C., *Ind. Eng. Chem. Fundam.*, 14(2), 96, 1975. Copyright by the American Chemical Society, Washington, D.C.

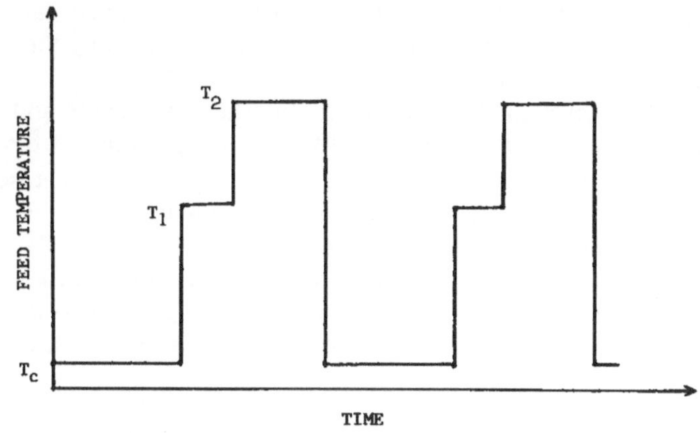

FIGURE 3. Feed temperature profile for two solutes. (Reprinted with permission from Wankat, P. C., *Ind. Eng. Chem. Fundam.*, 14, 96, 1975. Copyright by the American Chemical Society.)

A. Production of glucose isomerase
B. Production of immobilized glucose isomerase
C. The isomerization reaction glucose $\rightleftharpoons$ fructose

In addition there are described in the technical literature and patents processes by which syrups and dry products having higher proportions of fructose than does the standard HFCS. Those processes can be classified as:

D. Isomerization reactions to attain high ratios of fructose to glucose
E. Fractionation of standard HFCS into fructose-rich and glucose-rich portions

Some of the specific processes and developments classified above will now be described.

A. Production of Glucose Isomerase

Since about 1971, more than 20 patents have issued in the U.S. on this subject. All of them deal with various aspects of fermentation processes for producing the enzyme, and some of them also disclose methods for using the enzyme in isomerization reactions.

The various fermentation process patents can be considered in subclasses which describe: systems which require xylose or xylan for production of glucose isomerase, organisms which do not require xylose for production of glucose isomerase, and improved methods for conducting fermentations.

1. Organisms which Require Xylose (Or Xylan) for Production of Glucose Isomerase

Some of the organisms require xylose to form glucose isomerase. As far as is known, all the organisms which can use glucose to produce glucose isomerase in good yields can also produce in good yields with xylose.

The first U.S. patent in this field was that of Marshall,[19] which described the formation of xylose isomerase using xylose as the principal carbohydrate, and disclosed the conditions for its use for isomerizing glucose to fructose.

Another historically important patent was that of Takasaki[25] which issued in the U.S. in 1971, though it was issued in Japan several years earlier. This Takasaki patent was centered on types of *Streptomyces* which were capable of propagation on xylan, a polymer of xylose, and use of the glucose isomerase elaborated by the organism for isomerization of glucose to fructose.

In the fall of 1965, through the Ministry of International Trade and Industry of Japan a license was obtained by the Standard Brands Corporation of the U.S. to develop processes based on the Takasaki patent for production of fructose-containing syrups in the U.S. HFCS was first produced on a commercial scale in the U.S. in late 1967 by processes evolved from the basic Takasaki patent.

Selection of cells which survived action of toxic agents on a population of certain strains of *Streptomyces* and propagation of those cells to produce glucose isomerase resulted in higher yields of the enzyme than could be obtained from the original population.[50]

2. Organisms which do not Require Xylose for Production of Glucose Isomerase.

Xylose has been a relatively expensive ingredient, so identification of organisms which could produce glucose isomerase by use of relatively inexpensive glucose was an attractive goal.

Some organisms were found having the capability to produce glucose isomerase from glucose, and others having that capability were created by subjecting xylose requiring organisms to action of mutagenic agents.

Examples of organisms which produce glucose isomerase in good yields with glucose as the principal carbohydrate are those of genus *Arthrobacter*[51] and genus *Actinoplanes.*[52]

Mutations of organisms which normally would require xylose in order to produce glucose isomerase, did in some cases result in mutations which produced the enzyme even when the organisms were grown on glucose as the principal carbohydrate.[53,54,55]

3. Improved Methods for Conducting Fermentations to Produce Glucose Isomerase

During fermentations of some species of *Streptomyces*, the filamentous cells tend to form spherical aggregates. This tends to decrease the rate of organism growth and of enzyme production. This defect can be overcome by addition of certain materials such as agar to the fermentation broth.[56]

Specific agents recommended for addition to the nutrient medium for fermentations with species of *Streptomyces* in order to increase yields of glucose isomerase are sorbitol[57] and glycine.[58]

Fermentation of a facultative aerobic, glucose isomerase producing organism with a supply of oxygen maintained at growth-limiting levels and with a source of carbon which represses glucose isomerase synthesis, yet is readily converted anaerobically to nonrepressing degradation products added in a small excess, and all other nutrients added in sufficient amounts has been proposed as an efficient system for production of glucose isomerase.[59]

B. Production of Immobilized Glucose Isomerase

The advantages of immobilized enzymes generally over the conventional soluble forms lie in more efficient use of the enzyme (more product per unit enzyme purchased or produced), less contamination of the reaction mixture with the enzyme and impurities usually associated with the enzyme, and lower residence times in reactor, and consequently less by-product formation. All of these issues are of particular importance in the manufacture of HFCS because, especially in the first years of operation, the cost of the enzyme was high. Furthermore, the cost of refining HFCS to a purity demanded by food companies could be greatly decreased by keeping the enzyme and associated impurities out of the reaction mixture solution phase and by preventing (through short residence times) the off flavors and colors which would otherwise develop.

Consequently, since initial production of HFCS in 1967, considerable effort has been exerted to develop immobilized glucose isomerase systems. Since about 1971 over 130 patents have been issued on this subject.

The methods for immobilizing glucose isomerase which have been described in the literature can be classified as:

1. Glucose isomerase immobilized within bacterial cells
2. Cell-free glucose isomerase immobilized by sorption onto an insoluble carrier
3. Cell-free glucose isomerase immobilized by covalent bond formation between enzyme and an insoluble carrier
4. Entrapment of glucose isomerase, with or without bacterial cells within an insoluble matrix which is permeable to glucose and fructose, but from which the enzyme cannot be extracted into the reaction solution phase

1. Glucose Isomerase Immobilized with Bacterial Cells.

In many of the organisms which can produce glucose isomerase efficiently, the enzyme is held within or onto the cells of the organism. The cells can be filtered and washed without losing much enzyme. However, in some instances, while the cells are in contact with a solution of glucose and fructose under conditions which favor isomerization the enzyme gradually becomes solubilized. This severely limits the value of the cells for use in a commercial operation where, if the enzyme were not solubilized during use, it could be kept in service for thousands of hours.

Takasaki, in 1968, discovered a method for immobilizing glucose isomerase within or on the cells of certain species of *Streptomyces* which were good producers of glucose isomerase.[26] This was achieved simply by heating an aqueous suspension of the enzyme-containing cells to about 60° to 80°C for a short period of time. The resultant cell-glucose isomerase could be kept in contact with glucose-fructose solutions for long periods of time without loss of enzyme by solubilization. The effect of the heat treatment was ascribed to selective inactivation of lysing enzymes which would otherwise tend to break up the cell wall structure, liberating the enzyme. The heat treatment method has been used successfully in continuous fixed-bed reactors for commercial scale production of HFCS.

A modification of the basic heat treatment of cells which was introduced involved the further treatment of the treated cells in aqueous suspension with a protease. The effect of the protease treatment was to decrease the pressure drop when the cells were used in a continuous fixed-bed reactor.

A number of processes were developed that involved addition of polyelectrolytes to the fermentor broth after glucose isomerase production was completed. The cells containing the enzyme formed aggregates which could then be filtered and then frozen or dried.[60] The addition of oxides, sulfates, or phosphates of Mg, Ca, Mn, or Fe during formation of the floc resulted in hard aggregate.[61]

Glucose isomerase was fixed within the cells which produced it by adding glutaraldehyde to the suspension of cells which could then be recovered by filtration, washed, and used in continuous flow reactors to catalyze the isomerization reactions.[62] In another process, the suspension of glucose isomerase containing cells was homogenized, and then treated with glutaraldehyde to immobilize the enzyme within the solid phase formed. This material was then dried and shaped into the size and form needed for continuous fixed catalyst bed isomerization reactors.[63]

A further modification involved the recovery of cells containing glucose isomerase from a fermentor broth by filtration followed by suspension of the wet cells in a solution of salts of citric acid of about pH 6. After the citrate treatment the cells could be recovered by filtration and dried to yield an immobilized glucose isomerase preparation.[64]

2. Cell-Free Glucose Isomerase Immobilized on Insoluble Carriers

Glucose isomerase has been immobilized through sorption from water solutions by anion exchange celluloses such as diethylaminoethyl (DEAE) cellulose.[31,65] The enzyme resists elution from the carrier contact with the solution of glucose as long as the pH of the solution is above the isoelectric point of the enzyme and the ionic strength is low enough. Methods for making a DEAE-cellulose which is particularly good for sorption of glucose isomerase have been described.[66] Macroporous anion exchange resins of the quaternary N type have been used for immobilization of glucose isomerase. The use of Fe and Mg salts in the substrate are recommended for these resin-enzyme combinations.[67]

Cation exchange resin saturated with Mg has been used to immobilize glucose isomerase. The immobilized enzyme had a long life under conditions favorable for isomerization. The pH of the effluent from the reactor is 5.7 which is much lower than that usually associated with glucose isomerase action.[68]

A reactor designed for continuous flow through operation featured a macroporous carrier the core of which consisted of a polymer provided with nitrilo, acid imido, and ureido groups onto which the enzyme was sorbed. The enzyme could then be treated with cross-linking agents to hold it in place.[69]

A somewhat different approach involved the dispersion of immobilized enzymes within the interstices of webs made of cellulosic or other fibrous materials.[70]

Swollen membranes or films of proteins such as collagen or zein have been used to immobilize glucose isomerase by adsorption.[71]

Many methods for immobilization of glucose isomerase on insoluble inorganic materials have been developed.

Basic $MgCO_3$ has been used to sorb glucose isomerase and the combination used in a fixed-bed column reactor through which a glucose solution was flowed through continuously to effect isomerization.[72]

Porous alumina of MgO-Al_2O_3 blends of controlled pore diameter usually in the 100 to 1000 Å range has been developed,[73,74] and methods for their regeneration by heating[75] or by use of hypochlorite solutions have been described.[76] The hypochlorite system can be used in regeneration of the carrier without removing it from the reactor, and can be reactivated then by passing a solution of enzyme into the reactor.

Glass has also been used as a carrier for glucose isomerase. In one system the enzyme is first allowed to diffuse into the pores of a porous glass, then locked into place by using cross-linking agents dissolved in an organic solvent.[77]

Freshly fractured fragments of glass, the new surfaces of which must be protected from contact with air before mixing with a solution of enzyme, have been used as the carrier for immobilized glucose isomerase systems.[78]

Porous ceramic materials having pore diameters greater than the equivalent diameter of the enzyme but less than 1000 Å provide good carriers for sorption of glucose isomerase.[79]

Metal oxide powders have been sintered and compacted to produce structures of 10 to 80% porosity and 0.01 to 10 μ pore diameter. When a solution of an enzyme is passed through this metal structure, the enzyme is adsorbed, and can function as an immobilized enzyme system.[80]

Hydrophilic polymers to which hydrophobic groups have been added have been proposed for sorption of enzymes.[81]

Porous titania of controlled pore size, activated with Sn^{++} salts have been advocated as an effective system for immobilization of urease, and may be adapted to immobilization of glucose isomerase.[82]

Immobilized glucose isomerase can also be made from cell-free solutions of the enzyme by coagulation of the enzyme and other proteins which occur with it in a filtered fermentor broth. In effect, the enzyme is sorbed onto proteins associated with it.

One such process uses a polyelectrolyte to effect the coagulation and formation of solid insoluble aggregates containing the glucose isomerase. The aggregate can be used in continuous isomerization reactor.[83]

In another modification of this idea, tannic acid and glutaraldehyde are used to form solid insoluble aggregates containing glucose isomerase in the immobilized form.[84]

3. Cell-Free Glucose Isomerase Immobilized by Covalent Bond Formation between the Enzyme and an Insoluble Carrier

Most of the methods cited above depend on sorption for formation of the insoluble immobilized glucose isomerase structure. Many systems have been developed which depend upon the formation of covalent bonds between the enzyme and the carrier to create an immobilized glucose isomerase. Some of the references to be cited specifically describe immobilization of glucose isomerase, other references cited are not that specific, but could in principle be applied to the immobilization of glucose isomerase.

The processes which are dependent upon the formation of covalent bonds can be classified on the basis of the type of reaction which results in a bond between the enzyme and the insoluble carrier.

In a recent publication,[85] reactions used in making immobilized enzymes by covalent bond formation were classified as:

1. Acylation of an NH_2 group on enzyme by pendent groups of the carrier such as azide, acid anhydride, carbodiimide, sulfonyl chloride, and hydroxysuccinimide esters
2. Arylation or alkylation, in which the function group on the carrier which combines with an NH_2 group of the enzyme is exemplified by the 3-fluoro-4,6-dinitrophenyl group or 2-4 dichloro-s-triazine
3. Cyanogen bromide activation in which method C:N-Br is typically reacted with an insoluble polysaccharide to form an imido carbonate group which then reacts with an NH_2 group of the enzyme to form an N-substituted isourea
4. Carbamylation or thiocarbamylation in which pendent isocyanate or isothiocyantes form substituted ureas or isoureas by interaction with NH_2 groups of the enzyme
5. Aminidenation in which an NH_2 group of the enzyme reacts with a pendent imidoester to form amidines
6. Aldehyde reactions in which aldehyde groups on the carrier can be reacted with NH_2 groups of the enzyme to produce Shiff's base components
7. Glutaraldehyde reactions in which glutaraldehyde reacts with various types of polymers, and enzymes are readily bound to the treated polymer — the nature of the reactions involved are not well understood.
8. Diazo coupling in which aryldiazonium salts existing as pendent groups on the carrier can form azo derivatives with tyrosyl, histidyl, and amino groups of the enzyme

9. Thiol-dissulfide interchange in which, as an example, the treatment of a thiol containing polymer with 2,2′-dipyridyl disulfide to form a pyridyl dissulfide which can react with a sulfhydryl group of an enzyme to form a disulfide bond between the polymer and the enzyme
10. Four component reactions are illustrated by the interaction of acid, amine, isocyanide, and aldehyde groups. The acid and amine groups form an amide, and the combined aldehyde and isocyanide groups become attached to the amide group. The carrier polymer can provide the carboxyl or the amine portion of the reaction, the amine or carboxyl function, respectively, being provided by the enzyme.

In the same reference,[86] the point was made that although sulfhydryl, imidazole, and indole groups of an enzyme can serve as the reactive center for an enzyme immobilizing reaction, most of the methods studied have involved the amino, carboxyl, or phenolic groups of enzymes.

Many of the processes described in the literature for immobilization of glucose isomerase or processes which could in principle be adapted to that purpose may be classified on the basis of the coupling reactions involved.

Acylation reactions — Copolymers of acrylamide and maleic anhydride have been disclosed for enzyme immobilization through the acid anhydride reaction with the enzyme.[87] A system which involves the use of a reagent which contains an acylating or alkylating agent and a group which can form a covalent link with a carrier polymer is another illustration of this class of enzyme immobilization method.[88] This reference also suggests the formation of the polymer simultaneously with the attachment of the acylating or alkylating functional group to the polymer as it forms.

Arylation or alkylation reactions — An example of this method is a process which describes a chloro-S-triazine coupled to an arylazide. The azide is converted to a nitrene which upon activation by light reacts with a polymer carrier to form a covalent bond. The chlorotriazine then couples with the enzyme.[89]

Cyanogen bromide method — Dextran pretreated with epichlorohydrin, mercerized cellulose, agarose, and cross-linked polyacrylamide polymer substituted with p-amino phenol groups, activated with cyanogen halide have been used to bind a variety of proteins such as chymotrypsin, insulin, and glucose oxidase.[90]

Carbamylation and thiocarbamylation reactions — Organic cyanates to couple enzymes to insoluble polymers,[91] and the reaction of isocyanate capped polyurethane with an enzyme under foam forming conditions,[92] illustrate this class of immobilization technique.

Amidination reactions — Methods for formation of imido ester groups on polyacrylonitrile and for reacting them with enzymes to form immobilized enzymes have been described.[93] It is claimed that enzymes in the immobilized form are more stable than the parent enzyme.

Polymeric aldehydes — Numerous patents have described adaptation of this kind of reaction, especially for those designed for industrial application. Dicarbonyl starch[94] and dialdehyde cellulose in the fibrous form,[95] both of which can be made by oxidation with periodate, have been advocated as carriers for immobilization of enzymes. The description of the dicarbonyl cellulose features the idea of *in situ* addition of the enzyme after the carrier has been placed in a reactor. Processes involving polymers containing carbonyl groups have been described in which the carbonyl groups are sulfited[96] or exist as pendent dimethyl acetals[97] are described as having advantages over free carbonyl groups for formation of immobilized enzymes. Sulfited aldehyde or ketone groups on a polymer have been proposed for covalent bonding of enzymes, with the further treatment by dialdehyde cross-linking agents to stabilize the structure.[98]

Glutaraldehyde — Phenol formaldehyde resins have been used to immobilize enzymes through the action of glutaraldehyde.[99] Insoluble supports such as agarose or polyacrylamide which tend to swell in water form immobilized enzymes when mixed with a solution and treated with glutaraldehyde.[100] In a similar process, enzyme is suspended in a solution of a gel-forming protein, which is then treated with glutaraldehyde, yielding an immobilized enzyme[101] chitin granules suspended in a solution of enzyme treated with glutaraldehyde to produce an immobilized enzyme.[102]

Diazonium salt reactions — The process based on the diazonium salt reaction involves the formation of the anthranilates of starch in the granular insoluble form by treatment with isotoic anhydride. The anthranilate is converted to the diazonium salt, which then can couple with enzymes in solution to form covalently linked immobilized products.[103] Surfaces of glass, silica, or alumina can be reacted with 4-4^1-bi(methoxydibenzene diazonium chloride), forming diazo compounds covalently linked to those surfaces. The diazonium salt can then react with enzymes in solution to immobilize them.[104] The treatment of a solution of enzyme and substrate with a glass containing diazo groups is another example of this concept.[105] Monomers containing aromatic nitro groups grafted onto polymers have been reduced, diazotized, and used to immobilize enzymes.[106]

Thiol-disulfide interchange — Polymers having thiol or thiolactone groups reacted with solutions of enzymes to immobilize them.[107] No specific reference to group 10, the four component reaction system, has been found in the patent surveys available. Technical literature references to this system may be found in the general reference cited.[108]

There are a number of methods described for preparation of covalently bonded immobilized enzyme systems which do not fit into any one of the ten categories cited above.

Among those other systems which may be cited are immobilization systems based on silanized clay,[109] electro deposition on porous ceramic carriers,[110] activation of insoluble polysaccharides, nylon, and glass, with salts of Ti, Zr, Zn, Fe,[111] high molecular weight glycidyl ethers of bisphenol A,[112] and chloromethyl polystyrene in which part of the chlorine is substituted with 2,5-dioxo-4-oxaxolidine.[113] In a method referred to as "mechanochemical enzyme immobilization", a partially hydrolyzed nylon fiber, known to have many microcracks on the surface is stretched reversibly and immersed in a solution of enzyme. The fiber is relaxed, then removed from the solution. The fiber retains enzyme even after washing.[114]

C. The Isomerization Reaction Glucose ⇌ Fructose

Many specific procedures and apparatus have been developed for improving the enzymatic isomerization of glucose to fructose. Most of them have to do with the use of immobilized glucose isomerase, though the principles involved could in many cases be applied to the use of soluble glucose isomerase. Some of the methods to be cited were not developed specifically for the glucose to fructose reaction, but the principles involved could be applied to that purpose.

The various methods can be classified as those related to general reactor design, reactor structure operations designed to improve efficiency of use of the enzyme, additives or procedures aimed at improving enzyme stability, and systems for attaining fructose yields beyond the equilibrium value limitations. Beyond this classification are a number of basically different approaches to production of high fructose corn syrups.

Reactors having beds of immobilized glucose isomerase through which the solution of glucose is passed continuously are described with the isomerase bound to cells of organism by heat treatment,[30] and with the enzyme sorbed onto DEAE-cellulose or anionic resins. Similar systems make use of beds of glucose isomerase fixed in a naturally bound form within or on the cells of the organism which produced the enzyme. No special materials or operations were needed to attain the immobilized condition.[28]

Methods in which immobilized glucose isomerase is suspended in the solution of glucose during the reaction have also been described.[115] Related to the idea of this method are the fluidized bed reactors,[116,117] the design and operation of which could, in principle, be used with properly prepared immobilized glucose isomerase.

Methods for improving efficiency of the isomerization reaction, that is the amount of isomerized product that can be produced by a given amount of glucose isomerase in the immobilized form, have been developed on the basis of several principles.

Improved efficiency has been achieved by variation of the temperature of the reaction in accordance with the age of the enzyme in service.[72,118] Reactor performance improvements have been achieved by addition of Fe salts in addition to the conventional use of Mg salts to increase enzyme activity.[67] Improvements in the activity of glucose isomerase bound to cells of organisms is attained by pretreating the cell-enzyme complex with reducing agents such as bisulfite.[119] By controlling the calcium and magnesium quantities and ratios within certain limits, improved reactor efficiency is obtained, and the effluent from the reactor is so low in these salts that ion exchange resin refining is not necessary.[120] *Streptomyces* cells containing glucose isomerase have raised to higher levels of activity by treatment with thioglycolates.[121] It was observed that solutions of HFCS subjected to fractionation on certain types of ion exchange columns produced a glucose-rich fraction which, in order to attain good overall efficiency, had to be reisomerized to the glucose-fructose ratio of the original HFCS. However, in going through the fractionation columns, small amounts of glucose isomerase inhibitors were formed. It was shown that the inhibitors could be removed by passing the glucose-rich fraction through a cation exchange resin in the acid form.[122]

If channeling occurs as the solution being isomerized passes through a column or bed of immobilized glucose isomerase, the efficiency of the reaction is impaired. Flow characteristics have been improved by a uniform distribution of small inert beads throughout the mass of immobilized enzyme[123] or by having the enzyme bound to a self-supporting reticulated cellular material, which also tends to reduce compaction of the enzyme material.[124] Efficiency in the use of the reactor may be improved by procedures in which the immobilized glucose isomerase is regenerated without removing it from the reactor by periodically adding soluble glucose isomerase as required to the solution being fed to the reactor where it is absorbed onto the carrier.[125]

During the isomerization reaction, formation of acids, color, and loss of enzyme activity occur slowly. Methods for decreasing the change in pH caused by acid formation which have been recommended include addition of HSO_3^- salts to the substrate, which not only functions well as a buffer in the 7 to 8 pH range but also tends to stabilize the enzyme and decreases color formation.[126] Other methods for pH control involve the presence of solid phase $CaCO_3$ and $MgCO_3$ or anion exchange resins in the reactor.[127] Cation exchange resins in the form of their Mg and Co salts are used to immobilize glucose isomerase. This system is characterized by good pH control and a long catalytic service life.[68] Long service life of glucose isomerase is achieved by preventing contact between O_2 (air) and the enzyme.

D. Isomerization Reactions to Attain High Ratios of Fructose to Glucose

A method recommended for attaining fructose to glucose ratios larger than that corresponding to the usual solution equilibrium value is based upon the presence of borates during isomerization.[129,130] The removal of borates after the isomerization has been a problem, but purification of sugar solutions containing boric acid or borates can be achieved efficiently by treatment with anion exchange resins in the sulfite or bisulfite form.[131]

Some new approaches to making HFCS through enzymatic isomerization involve the simultaneous action of α-amylase, glucoamylase, and glucose isomerase on slurries of starch granules which are swollen to some extent but not extensively dispersed as in the conventional systems of isomerization.[132,133] Another novel approach for making HFCS recommends the use of glucose isomerase on the mother liquor obtained by crystallization of glucose from solutions of starch hydrolysates of high glucose content. The isomerization is followed by treatment with isomaltase to attain high monosaccharide levels in the HFCS.[134]

E. Fractionation of Standard HFCS Into Fructose-Rich and Glucose Rich-Portions

Systems developed for conversion of the HFCS' of about 42% fructose into products of higher fructose content have been mostly based on elution chromatography.

The method based on use of strong acid cation exchange resins in the Ca^{++} salt form, referred to in a preceding part of this chapter, has been improved and adapted to HFCS fractionation by carefully controlled programs of rate and sequencing of feed liquor, elution water, and feedback of intermediate eluates.[136] Another system defines those feed, elution, feedback systems on the basis of distribution coefficients of the compounds to be separated.[137] Separation on the CA^{++} cation exchange resin columns has been effected by a program of operation calling for the feed liquor and first elutions to be upflow as long as no fructose appeared in the eluate. Subsequent downflow wash resulted in high purity fructose in the eluate coming from the bottom of the column.[138] Feed and elution flow rates and recycle patterns are also featured in other methods described as means of achieving efficient separations of the components of mixtures of carbohydrates.[139]

Anion exchange resins in the HSO_3^- form have also been advocated for separation of fructose-rich and -poor fractions from their mixtures.[140] The elution pattern in another description of this system is designed toward using a volume of external water equal to the feed volume on each pass or cycle.[141] Other column materials which have been used are $AlSiO_3$[142] and artificial zeolites.[143]

Simulated moving bed systems have been described for separation of fructose from mixtures of glucose and fructose. Greater rates of production of high fructose products per unit volume of adsorbent and relatively low elution water requirements characterize this kind of separation scheme.[144]

An entirely different approach to production of products of high fructose content from conventional HFCS or invert sugar depends upon addition of sufficient $CaCl_2$ to solutions of HFCS or invert to cause the fructose $CaCl_2$ compound to crystallize. After separation of the pure crystals from the mother liquor by centrifuging, the fructose $CaCl_2$ compound is dissolved in water, then subjected to electrodialysis, yielding a solution of high purity fructose and $CaCl_2$ solution which is recycled to the HFCS or invert being prepared for the fructose-$CaCl_2$ crystallization operation.[145]

F. Crystallization

Substantially pure solid fructose can be crystallized from solutions containing fructose if the ratio of fructose to nonfructose material in solution is high enough. Solid products (the so-called total sugars) can be made simply by removing most of the water from a HFCS product. In the latter case, the solid is generally of a mixture of crystalline and amorphous material. In general, both the production of pure crystalline fructose and of the so-called total sugar is more readily achieved as the fructose content (on a dry basis) of the HFCS increases.

Manufacture of pure crystalline fructose from water solutions of HFCS requires that the solution have a dry basis purity of about 90% fructose or greater, in order to achieve sufficient super saturation of fructose in a solution which is not so viscous that rate of crystallization and subsequent separation of crystals from mother liquor is impeded. By careful control of degree of supersaturation through temperature control during crystallization and by use of sufficient quantities of pure fructose seed crystals of the correct size, the crystalline fructose can be made.[146] The solid fructose is the β-fructo pyranose.

The dihydrate of fructose crystallizes from water solutions at low temperatures.[147] The dihydrate crystals require careful handling because they will dissolve in their own water of crystallization if the temperature gets much higher than the crystallization temperature.

The crystallization of fructose can be improved by adding polyalcohols to the solution to be crystallized. The polyalcohol decreases the solubility of fructose so that supersaturations sufficient to drive the crystallization can be attained in a relatively fluid solution.[148]

Continuous crystallization of fructose from solutions of high fructose purity (dry basis) and of high methanol concentration has been described.[149]

Among the techniques developed for making dry total sugars of high fructose content is that which in effect precipitates the dry material from very concentrated water solutions of fructose by addition of alcohols.[150]

Mechanical kneading of a mixture of recycled solid phase and a highly concentrated water solution of high purity fructose results in a dry total sugar.[151]

The production of products of this type by use of hot air driers fed with the highly concentrated solution of HFCS of high fructose purity (dry basis) intimately mingled with recycled pulverized product has been described.[152]

Upon concentrating a water solution containing at least 95% fructose (dry basis) to a water content of 2 to 5%, and stirring it vigorously with crystalline fructose at about 60°C, a plastic mass is formed, which upon cooling solidifies and can then be pulverized to size distribution of particles desired.[153]

In summary it is seen that the development of technologies for the production and use of immobilized glucose isomerase has been a dominant factor in the creation of the new HFCS industry and the growth of that industry within just a few years to a position of major importance in the production of food for human consumption.

REFERENCES

1. **Anon.**, Food processing awards, *Food Process.*, 38(7), 77, 1977.
2. Commission on Biochemical Nomenclature, *Enzyme Nomenclature*, American Elsevier, New York, 1973, 308.
3. **Takasaki, Y. and Tanabe, O.**, Formation of fructose from glucose by bacteria. I. Properties of glucose isomerase, *Hakko Kyokaishi*, 20, 449, 1962.
4. **French, D.**, *Biochemistry of Carbohydrates — Biochemistry Series One*, Whelan, W. J., Ed., Butterworths, London, 1975, 281.
5. **Newkirk, W. B.**, Development and production of anhydrous dextrose, *Ind. Eng. Chem.*, 16, 760, 1936.
6. **Dale, J. K. and Langlois, D. P.**, U.S. Patent 2,201,609, 1940.
7. Commission on Biochemical Nomenclature, *Enzyme Nomenclature*, American Elsevier, New York, 1973, 212.

8. **Cantor, S. M. and Hobbs, K. C.**, U.S. Patent 2,354,664, 1944.
9. **Langlois, D. P. and Larson, D. H.**, U.S. Patent 2,746,889, 1956.
10. **Scallet, B. L. and Ehrenthal, I.**, U.S. Patent 3,285,776, 1966.
11. **Parrish, F. W.**, U.S. Patent 3,431,253, 1969.
12. **Nita, Y.**, Japanese Patent 3487, 1967.
13. **Mahler, H. R. and Cordes, E. H.**, *Biological Chemistry*, 2nd ed., Harper & Row, New York, 1974, 543.
14. **Akabori, S., Nehara, K., and Muramatsu, I.**, Biochemical formation of tetrose, pentose, and hexose, *J. Chem. Soc. Jpn.*, 73, 311, 1952.
15. **Cohen, S. S.**, Studies on D-ribulose and its enzymatic conversion to D-arabinose, *J. Biol. Chem.*, 201, 71, 1955.
16. **Green, M. and Cohen, S. S.**, Enzymatic conversion of L fucose to L-fuculose, *J. Biol. Chem.*, 219, 557, 1956.
17. **Palleroni, N. J. and Doudoroff, H.**, Mannose isomerase of *Pseudomonas* saccharophilia, *J. Biol. Chem.*, 218, 535, 1956.
18. **Marshall, R. O. and Kooi, E. R.**, Enzymatic conversion of D-glucose to D-fructose, *Science*, 125, 648, 1957.
19. **Marshall, R. O.**, U.S. Patent 2,950,228, 1960.
20. **Mitsuhashi, S. and Lampen, J. O.**, Conversion of D-xylose to D-xylulose in extracts of *Lactobacillus pentosus*, *J. Biol. Chem.*, 204, 1011, 1953.
21. **Hochster, R. M. and Watson, R. W.**, Enzymatic isomerization of D-xylose to D-xylulose, *Arch. Biochim. Biophys.*, 48, 120, 1953.
22. **Slein, M. W.**, Xylose isomerase from *Pasteurella pestis*, strain A-1122, *J. Am. Chem. Soc.*, 77, 1663, 1953.
23. **Littauer, V. Z., Volcani, B. E., and Bergmann, E. D.**, Observations on the metabolism of pentoses in *Eschericia coli*, *Biochim. Biophys. Acta*, 18, 523, 1955.
24. **Cohen, S. S. and Barner, H.**, Enzymatic adaptation in a thymine requiring strain of *Eschericia coli*, *J. Bacteriol.*, 69, 59, 1955.
25. **Takasaki, Y.**, U.S. Patent 3,616,221, 1971.
26. **Takasaki, Y., Kosugi, Y., and Kanabayashi, A.**, *Fermentation Advances*, Academic Press, New York, 1969, 561.
27. **Takasaki, Y.**, Japanese Patent 9823, 1970.
28. **Lloyd, N. E., Lewis, L. T., Logan, R. M., and Patel, D. N.**, U.S. Patent 3,817,832, 1974.
29. **Takasaki, Y. and Kamibayoshi, A.**, U.S. Patent 3,753,858, 1973.
30. **Lloyd, N. E., Lewis, L. T., Logan, R. M., and Patel, D. N.**, U.S. Patent 3,694,314, 1973.
31. **Thompson, K. N., Johnson, R. A., and Lloyd, N. E.**, U.S. Patent 3,788,945, 1974.
32. **Novo Terapeutisk Lab.**, Belgian Patent 797,984, 1973.
33. **Pazur, J. H.**, *Starch Chemistry and Technology*, Vol. I, Whistler, R. L. and Paschall, E. F., Eds., Academic Press, New York, 1965, 155.
34. **Abdullah, M., Fleming, I. D., Taylor, P. M., and Whelan, W. J.**, Action of glucoamylase, *Biochem. J.*, 89, 35, 1963.
35. **Miah, M. N. N. and Ueda, S.**, Multiplicity of glucoamylase of *Aspergillus oryzae*, *Starke*, 29, 235, 1977.
36. **Kooi, E. R., Harjes, C., and Gilkison, J. S.**, U.S. Patent 3,042,584, 1962.
37. **Abdullah, M., Calley, B. J., Lee, E. Y. C., Robyt, J., Wallenfels, K., and Whelan, W. J.**, The mechanism of carbohydrase action. II. Pullulanase, an enzyme specific for the hydrolysis of alpha-1→6 bonds in amylaceous oligo and polysaccharides, *Cereal Chem.*, 43, 111, 1966.
38. **Houge-Angeletti, R. A.**, Subunit structure and amino acid composition of xylose isomerase from *Streptomyces albus*, *J. Biol. Chem.*, 250, 7814, 1975.
39. **Lloyd, N. E., Khaleeluddin, K., and Lamm, W. R.**, Automated method for the determination of D-glucose isomerase activity, *Cereal Chem.*, 49, 544, 1972.
40. **Takasaki, Y.**, Studies on sugar isomerizing enzyme production and utilization of glucose isomerase from *Streptomyces* sp., *Agric. Biol. Chem.*, 30, 1247, 1966.
41. **Takasaki, Y.**, Kinetic and equilibrium studies on D-glucose-D-fructose isomerization catalyzed by glucose isomerase from *Streptomyces* sp., *Agric. Biol. Chem.*, 31, 309, 1967.
42. **Mildvan, A. S.**, *Bioinorganic Chemistry — Advances in Chemistry Series*, Gould, R. F., Ed., American Chemical Society, Washington, D.C., 1971, 404.
43. **Khaleeluddin, K., Sutthoff, R. F., and Nelson, W. J.**, U.S. Patent 3,834,940, 1974.
44. **Pigman, W. and Anet, E. F. L. J.**, *The Carbohydrates Chemistry and Biochemistry*, Vol. IA, 2nd ed., Pigman, W. and Horton, D., Eds., Academic Press, New York, 1972, 178.
45. **Wankat, P. C.**, Multicomponent cycling zone separations, *Ind. Eng. Chem. Fundam.*, 14, 96, 1975.

46. **Busbice, M. E. and Wankat, P. C.**, The pH cycling zone separation of sugars. Preparative separation technique for counter-current distribution and chromatography, *J. Chromatogr.*, 114, 369, 1975.
47. **Anon.**, *U.S. International Trade Commission Pub. 807*, U.S. Government Printing Office, Washington, D.C., 1977, 192.
48. **Anon.**, *F. O. Licht's Int. Sugar Rep.*, 109, No. 8, March 14, 1977, 7.
49. **Anon.**, *F. O. Licht's Int. Sugar Rep.*, 109, No. 8, March 14, 1977, 8.
50. **Bengtson, B. L. and Lamm, W. R.**, U.S. Patent 3,654,080, 1972.
51. **Lee, C. K., Hayes, L. E., and Long, M. E.**, U.S. Patent 3,645,848, 1972.
52. **Shieh, K. K., Lee, M. A., and Donnelly, B. J.**, U.S. Patent 3,834,988, 1974.
53. **Armbruster, F. C., Heady, R. E., and Cory, R.**, U.S. Patent Reissue 29, 152, 1977.
54. **Armbruster, F. C., Heady, R. E., and Cory, R. P.**, U.S. Patent 3,813,318, 1974.
55. **Armbruster, F. C., Heady, R. E., and Cory, R. P.**, U.S. Patent 3,957,587, 1976.
56. **Dworschack, R. G. and Lamm, W. R.**, U.S. Patent 3,666,628, 1972.
57. **Dworschack, R. G., Chen, J. C., Lamm, W. R., and Davis, L. G.**, U.S. Patent 3,736,232, 1973.
58. **Heady, R. E. and Jacaway, W. A., Jr.**, U.S. Patent 3,770,589, 1973.
59. **Diers, I. V.**, U.S. Patent 4,042,460, 1977.
60. **Lee, C. K. and Long, M. E.**, U.S. Patent 3,821,086, 1974.
61. **Long, M. E.**, U.S. Patent 3,935,069, 1976.
62. **Zienty, M. F.**, U.S. Patent 3,779,869, 1973.
63. **Amotz, S., Nielsen, T. K., and Thiessen, N. O.**, U.S. Patent 3,980,521, 1976.
64. **Tsumura, N. and Kasumi, T.**, U.S. Patent 4,001,082, 1977.
65. **Sipos, T.**, U.S. Patent 3,708,397, 1972.
66. **Lloyd, N. E. and Hurst, L. S.**, U.S. Patent 3,823,133, 1974.
67. **Fujita, Y., Matsumoto, A., Ishikawa, H., Hishida, T., Kato, H., and Takamisawa, H.**, U.S. Patent 4,008,124, 1977.
68. **Bouniot, A.**, U.S. Patent 3,990,943, 1976.
69. **Reynolds, J. H.**, U.S. Patent 3,705,084, 1972.
70. **Schmitt, E. E., Polistrina, R. A., and Forgione, P. S.**, U.S. Patent 3,809,605, 1974.
71. **Vieth, W. R., Mead, B., Wang, S. S., and Gilbert, S. G.**, U.S. Patent 3,843,446, 1974.
72. **Heady, R. E. and Jacaway, W. A., Jr.**, U.S. Patent 3,847,740, 1974.
73. **Messing, R. A.**, U.S. Patent 3,868,304, 1975.
74. **Eaton, D. L. and Messing, R. A.**, U.S. Patent 3,982,997, 1976.
75. **Bialousz, L. R., Herritt, E. R., Lartique, D. J., and Pitcher, W. H., Jr.**, U.S. Patent 3,965,035, 1976.
76. **Gregory, J. L. and Pitcher, W. H., Jr.**, U.S. Patent 4,002,576, 1977.
77. **Messing, R. A.**, U.S. Patent 3,804,719, 1974.
78. **Rochett, T. J., Doig, A. R., Komatsu, S., Evans, J., and O'Connell, J. J.**, U.S. Patent 3,802,909, 1973.
79. **Messing, R. A.**, U.S. Patent 3,850,751, 1974.
80. **Keyes, M. H.**, U.S. Patent 4,001,085, 1977.
81. **Butler, L. G.**, U.S. Patent 4,006,059, 1977.
82. **Messing, R. A.**, U.S. Patent 3,910,823, 1975.
83. **Nystrom, C. W.**, U.S. Patent 3,935,068, 1976.
84. **Stanley, W. L. and Olson, A. C.**, U.S. Patent 3,736,231, 1973.
85. **Goldstein, L. and Manecke, G.**, *Applied Biochemistry and Bioengineering*, Vol. 1, Wingard, L. B., Jr., Katchalski-Katzer, E., and Goldstein, L., Eds., Academic Press, New York, 1976, 43.
86. **Goldstein, L. and Manecke, G.**, *Applied Biochemistry and Bioengineering*, Vol. 1, Wingard, L. B., Jr., Katchalski-Katzer, E., and Goldstein, L., Eds., Academic Press, New York, 1976, 42.
87. **Dieter, J., Wolfgang, B., and Ulrich, B. H.**, U.S. Patent 3,775,253, 1973.
88. **Jaworek, D., Nelbock-Hochstetter, M., Beaucamp, K., Bergmeyer, H. V., and Karl-Heina, B.**, U.S. Patent 3,969,287, 1976.
89. **Smith, N. L.**, U.S. Patent 4,007,089, 1977.
90. **Axén, R. E. A. U. and Porath, J. O.**, U.S. Patent 3,645,852, 1972.
91. **Kagedal, S. L. and Johannes, S. H.**, U.S. Patent 3,788,948, 1974.
92. **Wood, L. L., Hartdegen, F. J., and Hahn, P. A.**, U.S. Patent 3,929,574, 1975.
93. **Zaborsky, O. R.**, U.S. Patent 3,830,699, 1974.
94. **Katchalski, E., Goldstein, L., and Blumberg, S.**, U.S. Patent 3,706,633, 1972.
95. **Wildi, B. S. and Wicks, L. E.**, U.S. Patent 4,013,514, 1977.
96. **Forgione, P. S.**, U.S. Patent 3,753,861, 1973.
97. **Epton, R., McLaren, J. V., and Thomas, T. H.**, U.S. Patent 3,761,357, 1973.
98. **Fergione, P. S.**, U.S. Patent 3,847,743, 1974.
99. **Olson, A. C. and Stanley, W. L.**, U.S. Patent 3,767,531, 1973.

100. **Wirth, P. C. and Tixier, R.**, U.S. Patent 3,836,433, 1974.
101. **Van Velzen, A. G.**, U.S. Patent 3,838,007, 1974.
102. **Stanley, W. L. and Wateers, G.**, U.S. Patent 3,909,358, 1975.
103. **Mehltretter, C. L. and Weakly, F. B.**, U.S. Patent 3,745,088, 1973.
104. **Messing, R. A.**, U.S. Patent 3,930,951, 1975.
105. **Royer, G. P.**, U.S. Patent 3,930,950, 1976.
106. **Kenyon, R. S., Garnett, J. L., and Liddy, M. J.**, U.S. Patent 3,981,775, 1976.
107. **Barker, S. A. and Gray, C. J.**, U.S. Patent 3,846,306, 1974.
108. **Goldstein, L. and Manecke, G.**, *Applied Biochemistry and Bioengineering*, Vol. 1, Wingard, L. B., Jr., Katchalski-Katzer, E., and Goldstein, L., Eds., Academic Press, New York, 1976, 77.
109. **Burns, R. A.**, U.S. Patent 3,953,292, 1976.
110. **Keyes, M. H.**, U.S. Patent 3,839,175, 1974.
111. **Emery, A. N., Barker, S. A., and Novais, J. M.**, U.S. Patent 3,841,969, 1974.
112. **Mathews, J. S.**, U.S. Patent 3,841,970, 1974.
113. **Keyes, M. H. and Semersky, F. E.**, U.S. Patent 3,860,486, 1975.
114. **Klibanov, A. M., Samokhiv, G. P., Martinek, K., and Berezin, I. V.**, A new mechanochemical method of enzyme immobilization, *Biotechnol. Bioeng.*, 19, 211, 1977.
115. **Thompson, K. N., Johnson, R. A., and Lloyd, N. E.**, U.S. Patent 3,909,354, 1975.
116. **Coughlin, R. W. and Charles, M.**, U.S. Patent 4,016,293, 1977.
117. **Coughlin, R. W. and Charles, M.**, U.S. Patent 4,048,018, 1977.
118. **Heady, R. E. and Jacaway, W. A., Jr.**, U.S. Patent 3,847,741, 1974.
119. **Hurst, T. L.**, U.S. Patent 4,026,764, 1977.
120. **Poulsen, P. B. R. and Zittan, L. E.**, U.S. Patent 4,025,389, 1977.
121. **Takasaki, Y.**, Japanese Patent 54,584, 1974.
122. **Takasaki, Y.**, Japanese Patent 42,885, 1974.
123. **Idaszak, L. R., Terranova, R. A., and Heady, R. E.**, U.S. Patent 3,956,065, 1976.
124. **Forgione, P. S., Polistrina, R. A., and Schmitt, T. T.**, U.S. Patent 3,791,927, 1974.
125. **Tamura, M., Ushiro, S., and Hasagawa, S.**, U.S. Patent 3,960,663, 1976.
126. **Cotter, W. P., Lloyd, N. E., and Hinman, C. W.**, U.S. Patent 3,623,953, 1971.
127. **Takasaki, Y. and Kamibayashi, A.**, U.S. Patent 3,715,276, 1973.
128. **Cory, R. P.**, U.S. Patent 3,910,821, 1975.
129. **Takasaki, Y.**, U.S. Patent 3,689,362, 1972.
130. **Takasaki, Y.**, Japanese Patent 21,386, 1977.
131. **Takasaki, Y.**, Japanese Patent 9,740, 1977.
132. **Hebeda, R. E. and Leach, H. W.**, U.S. Patent 3,922,201, 1975.
133. **Walon, R. G. P.**, U.S. Patent 4,009,074, 1977.
134. **Suekane, M., Hasegawa, S., Tamamura, M., and Ishikawa, V.**, U.S. Patent 3,935,070, 1976.
135. **Lauer, K., Budka, H., and Stoeck, G.**, U.S. Patent 3,686,117, 1972.
136. **Melaja, A. J.**, U.S. Patent 3,692,582, 1972.
137. **Sutthoff, R. F. and Nelson, W. J.**, U.S. Patent 4,022,637, 1977.
138. **Yoritomi, K., Yoshida, T., and Kikuchi, K.**, Japanese Patent 68,752, 1973.
139. **Kubota, H.**, Japanese Patent 09,737, 1976.
140. **Takasaki, Y.**, Japanese Patent 46,848, 1975.
141. **Kubota, H.**, Japanese Patent 9,738, 1976.
142. **Odawara, H., Noguchi, Y., and Ohno, M.**, U.S. Patent 4,014,711, 1977.
143. **Neuzil, R. W. and Priegnitz, J. W.**, U.S. Patent 4,024,331, 1977.
144. **Broughton, D. B., Bieser, W. J., Berg, R. C., Connel, F. D., Korous, D. L., and Neuzil, R. W,.** High Purity Fructose Via Continuous Adsorptive Separation, presented at Int. Biochemistry Symp., Lincolnshire, Ill., October 13, 1976.
145. **Kubo, T. and Takuki, R.**, U.S. Patent 3,666,647, 1972.
146. **Melaja, A. J., Forsberg, H.**, German Patent 2,209,243, 1972.
147. **Young, F. E. and Jones, F. T.**, U.S. Patent 2,588,449, 1952.
148. **Nakamura, S., Kurosawa, Y., Takezaki, Y., and Sawai, M.**, Japanese Patent 99,349, 1973.
149. **Lauer, K., Strahenburg, S., Stephan, P., and Stoeck, G.**, U.S. Patent 3,607,392, 1971.
150. **Nitsch, E.**, U.S. Patent 3,812,010, 1974.
151. **Yamauchi, T.**, U.S. Patent 3,929,503, 1975.
152. **Lundquist, J. T., Jr., Veltman, P. L., and Woodruff, E. T.**, U.S. Patent 3,956,009, 1976.
153. **Kutsch, T., Gosewinkel, W., and Stoeck, G.**, U.S. Patent 3,513,023, 1970.

Chapter 5

POTENTIAL AND USE OF IMMOBILIZED CARBOHYDRASES

Peter J. Reilly

TABLE OF CONTENTS

I. Introduction 114

II. α-Amylase 115
- A. Introduction 115
- B. Published Research 116

III. Glucoamylase 119
- A. Introduction 119
- B. Published Research 120
  - 1. Entrapment 120
  - 2. Adsorption 122
  - 3. Adsorption by Ion Exchange 123
  - 4. Covalent Attachment 125

IV. β-Amylase and Pullulanase 131
- A. Introduction 131
- B. Published Research 132
  - 1. Entrapment 132
  - 2. Adsorption 132
  - 3. Covalent Attachment 133

V. Invertase 135
- A. Introduction 135
- B. Published Research 136
  - 1. Entrapment 136
  - 2. Adsorption 138
  - 3. Covalent Attachment 140

VI. α-Galactosidase 141
- A. Introduction 141
- B. Published Research 141

VII. Conclusion 143

References 143

I. INTRODUCTION

In the previous chapter, we discussed the most successful of all immobilized enzymes. In the chapter to follow this, the reader will encounter a carbohydrase that may achieve commercial adoption. Here, in the following pages, is a group of immobilized enzymes that are similar to glucose isomerase and lactase in that they are active on carbohydrates but, with one exception, have not yet been successful industrially. This has not prevented members of this group of carbohydrases from being among the most studied of all immobilized enzymes. Glucoamylase has appeared in the technical literature approximately 100 times, and α-amylase and invertase about half as often. Three others, β-amylase, pullulanase, and α-galactosidase, have been much less studied in immobilized form, but the first two have commercial potential and the third has already seen industrial use.

It is no accident that four of these six enzymes are active on starch, for the hydrolysis of starch has been a major part of the food processing industry for many years. In this multistep process, α-amylase cleaves the interior bonds of the amylose and amylopectin molecules that make up starch, producing soluble products with the general designation of dextrins, while the other three attack specific linkages in the dextrin molecules: glucoamylase, α(1→4) and, more slowly, α(1→6) bonds to produce glucose; β-amylase, alternate α(1→4) links, to produce maltose and a limit dextrin; and pullulanase, α(1→6) bonds, to form straight-chain dextrins.

Acid can be employed to partially hydrolyze starch, but to produce relatively pure glucose glucoamylase is required in addition. To form syrups with fairly high maltose contents, enzymes high in β-amylase activity may be used. For very high proportions of maltose, a mixture of β-amylase and debranching enzyme such as pullulanase is necessary.

For some applications, such as the production of high-fructose syrups discussed in the previous chapter, partial acid hydrolysis of starch followed by glucoamylase treatment does not yield sufficiently pure glucose, and for those purposes, α-amylase must be used instead of acid.

None of these four starch-hydrolyzing enzymes has as yet achieved commercial use in immobilized form. Introduction of immobilized α-amylase has been precluded by the low cost of the native form, the ease of using it, and the difficulty of achieving high enzyme activity when both the enzyme and substrate are insoluble.

Immobilized glucoamylase, while much closer to commercial acceptance, must overcome the low cost of native glucoamylase as well as doubts whether it is sufficiently stable for the prolonged use under industrial conditions necessary to make any immobilized enzyme economically viable. In addition, most carefully controlled studies on immobilized glucoamylase have indicated that its yield of glucose is lower than that of the soluble form under identical conditions.

Reasons for not using immobilized β-amylase are somewhat similar. Crude amylases in soluble form may be employed at moderate cost to obtain syrups of fairly high maltose content. In addition, up to this point, no immobilized form of the enzyme has been found stable enough for prolonged use.

Pullulanase could be used in conjunction with β-amylase to achieve high maltose yields, and with glucoamylase to incrementally improve the yield of glucose. Unless those enzymes were employed in immobilized form, it is unlikely that pullulanase would be either.

The two other carbohydrases to be discussed here, invertase and α-galactosidase, both catalyze the cleavage of low molecular weight carbohydrates, as does lactase, which is covered in the next chapter.

Invertase cleaves β-fructofuranoside bonds and would most commonly be used industrially to hydrolyze sucrose to glucose and fructose. In relatively nonbuffered syrups, this reaction is more easily carried out with hydrochloric acid or alternatively with strong acid ion exchange resins. Only in highly buffered sugar solutions does invertase find a role, and this application is quite minor.

α-Galactosidase hydrolyzes the α-galactoside bonds of sugars such as raffinose (found in soybeans and in sugar beets grown and stored at lower temperatures) and stachyose (present in soybeans). This enzyme is used to reduce the raffinose content of a sugar beet molasses stream to aid sucrose crystallization. A potential further use is removal from soybean milk of the α-galactosides, which can cause flatulence upon ingestion. Of the six enzymes discussed in this chapter, α-galactosidase is the only one employed commercially in immobilized form, within the U.S. by Great Western Sugar in Billings, Mont. and Holly Sugar in Torrington, Wyo. under license from the Agency of Industrial Science and Technology of Japan.

In summary then, immobilized α-galactosidase has found some commercial acceptance, at least partially because the competing native form is expensive. Immobilized glucoamylase potentially is of industrial importance, as are β-amylase and pullulanase, if technical problems can be overcome. α-Amylase and invertase in immobilized form appear to have little chance for adoption, the first because of technical difficulties and the low cost of native enzyme, and the second because alternate processes severely limit the potential market.

Because, in general, this chapter covers potentialities rather than accomplished commercial successes, a full treatment of the technical literature will be given. It should be pointed out that in many, if not most cases, these enzymes have been immobilized for reasons quite unrelated to their potential industrial use. Nevertheless, the properties of these enzymes, when immobilized to various carriers, are of interest and may, in some cases, lead to fruitful research efforts.

II. α-AMYLASE

A. Introduction

α-Amylase [α(1→4) glucan 4-glucanohydrolase, EC 3.2.1.1] hydrolyzes interior α(1→4) linkages of amylose and amylopectin molecules. Because the enzyme is unable to cleave the α(1→6) bonds of amylopectin and because maltose is not attacked by α-amylase, an α-limit dextrin results on prolonged hydrolysis.[1] The enzyme, like other endohydrolases, breaks bonds with retention of configuration, a clue to the origin of its name.[1]

In present commercial use, α-amylase is employed to convert starch to dextrins, with sufficient hydrolysis occurring to make the products soluble and not susceptible to gelling upon cooling. At this point, the dextrins may be further broken down by glucoamylase to syrups of high glucose content.

Historically, the production of dextrins by α-amylase hydrolysis occurred in two steps, cooking of a starch slurry at temperatures of 100°C or higher to swell and gelatinize the starch granules, followed by prolonged enzymatic hydrolysis at temperatures of 80 to 95°C. More recently, upon discovery of more thermostable α-amylase varieties, the process has been streamlined, with starch slurry containing α-amylase being cooked and liquefied at slightly above 100°C by continuously passing the mixture through a jet cooker to which steam is added. After a short residence in a pressurized coil, the dextrin is held further for 2 to 4 hr for continued hydrolysis.

There are major drawbacks to the use of immobilized α-amylase. First, the native enzyme is relatively inexpensive, and immobilization would entail carrier and perhaps binding chemical costs not easily counterbalanced by savings in enzyme cost, especially since the native enzyme is expected to remain active for such a short time. Second and more important, α-amylase, at least in the early stages of hydrolysis, is attacking extremely large molecules that would not readily enter the close spaces in which most of the immobilized enzyme would be found. This accessibility problem (demonstrated in the following papers) is compounded by the problem that, even with the use of soluble starch which is already partly hydrolyzed, activities of immobilized α-amylase are often very low compared to the native form. To use immobilized α-amylase would require initial cooking and acid hydrolysis followed by filtering of the product to remove insoluble impurities. Unlike many processes with immobilized enzymes, this is more complicated than the procedure it would replace.

B. Published Research

α-Amylase has been entrapped (Table 1) and adsorbed (Table 2), but most work with the immobilized enzyme has dealt with its attachment to solid supports by covalent bonds (Table 3). In comparison to glucoamylase, immobilized α-amylase has been the subject of few studies yielding data useful for design purposes and, for that reason (and because of its low potential for commercialization), its treatment here will be brief.

In one of the earliest articles on any immobilized enzyme, Bernfeld and Wan[2] entrapped α-amylase in an *N,N'*-methylene-bisacrylamide/tetramethylethylenediamine copolymer, though little of the activity was retained. Better results were obtained with varying formulations of polyacrylamide gel irradiated with γ-rays.[3,4]

Another method of entrapment is behind ultrafiltration membranes. In this technique, enzyme and reactants cannot readily pass through the barrier, but product, being of lower molecular weight, can. Steady state operation may be difficult to attain because filtration rate may decrease as a gel layer forms on the membrane.[5] Use of an ultrafiltration reactor does allow stabilization of the enzyme by covalently coupling soluble polysaccharides to it.[6]

α-Amylase has been immobilized by adsorption to sand,[8] bentonite,[9] kaolin,[9] collagen,[11,12] and, after acylation with organic acids, by adsorption on Millipore® membranes.[10] With collagen, the enzyme is stabilized at low pHs[11] and high temperatures.[12]

Extensive Soviet research has dealt with immobilization by ion exchange,[13-19] and, in one case,[17] decay at low pHs was sharply decreased.

In a number of cases, the enzyme has been covalently attached to acrylic polymers through active groups incorporated in the support.[20-23,26] With Enzacryls® AA and AH, on the other hand, nitrous acid or thiophosgene was employed for activation.[24,25] Although little activity was retained on either Enzacryl®, at moderate temperatures some stabilization did occur with both as well as with Enzacryl® CHO.[26]

When α-amylase was bound with CNBr to Sephadexes G-25 and G-200 and Sepharoses 6B and 2B (carriers with successively greater exclusion limits), highest activity on 8000-dalton starch occurred with Sepharose 2B.[29] With phenyl maltoside, however, Sepharose 6B had greater activity. These results appeared to be caused not only by exclusion of substrate, but also by exclusion of enzyme, as Sephadex G-25 bound α-amylase only on bead surfaces and G-200 exhibited preferential binding there. pH-Activity profiles varied more after immobilization to the Sepharoses where more enzyme was bound in the interior of the carrier; in general, α-amylase attached to the Sepharoses was more stable to low pHs, high temperatures, or treatment with EDTA or urea. In most cases, native enzyme was less stable than enzyme bound to any of the carriers, and under all conditions that bound to Sepharose 6B was most stable.

In other work with Sepharoses, α-amylase attached to Sepharose 4B with CNBr

Table 1
α-AMYLASE IMMOBILIZED BY ENTRAPMENT

Organic gels

Carrier	Ref.
N,N′-Methylene-bisacrylamide/tetramethyl-ethylenediamine copolymer	2
Acrylamide/N,N′-methylene-bisacrylamide copolymer	3
Acrylamide/N,N′-methylene-bisacrylamide/sodium acrylate/ starch copolymer	4

Ultrafiltration reactors

Membrane	Ref.
Amicon PM10	5, 6
Amicon PM30	7

Table 2
α-AMYLASE IMMOBILIZED BY ADSORPTION

Carrier	Ref.
Sand	8
Bentonite or kaolin	9
Millipore® membranes	10
Collagen	11, 12
Ion Exchange Carriers	
KU-2	13
Amberlite® IRC-50	13
Amberlite® CG-50	13, 14, 15
Vofatite® L-150	16
KMT	17
Cationite gel	18
Dextran or starch modified with epichlorohydrin or polyethyleneamines	19

retained little activity but was stabilized after immobilization,[30] while α-amylase linked to ECD-Sepharose 2B by thiol-disulfide interchange exhibited highest retention of specific activity when the least amount of enzyme was bound.[31] Here again, the enzyme was more stable in immobilized form.

In general, cellulosic supports provided rather inhospitable environments for α-amylase, low amounts of bound protein or low retentions of activity resulting upon immobilization.[6,32-36] However, in some cases, coupling the enzyme to cellulosics stabilized it.[6,35,37,38]

Both Ledingham and Hornby[36] and Linko et al.[37] found that α-amylase linked to cellulosic carriers exhibited a higher degree of multiple attack than the native enzyme, and the former authors found the same with α-amylase coupled to poly(aminostyrene).[36] This suggests that the substrate was hindered by the support, leaving it more vulnerable to repeated hydrolysis by the same enzyme molecule and leading to the formation of more small carbohydrate fragments. These results were supplemented by those of Boundry et al.,[43] who observed that immobilized α-amylase yielded far less material below 35,000 daltons than did native α-amylase. However, these smaller molecules had a much lower degree of polymerization when produced by the immobilized enzyme than by the soluble one, similar to the previously cited work. In addition, though immobilized α-amylase attacked amylose at a higher rate than did the soluble enzyme, the relative rates were reversed with the larger amylopectin, suggesting again the influence of substrate exclusion and diffusion limitation.

α-Amylase has been covalently coupled to only a few inorganic supports, in contrast to the large amount of work conducted with organic carriers. In two cases low yields were obtained,[49,50] and in one case stability was decreased by immobilization.[47] Lai and Lan,[48] on the other hand, achieved moderately high coupling efficiency and increased stability with diazotized silica glass.

Table 3
α-AMYLASE IMMOBILIZED BY COVALENT ATTACHMENT

Carrier	Binding agent	Ref.
Nitrated methacrylic acid/methacrylic acid-*m*-fluoroanilide/divinylbenzene copolymer	—	20
Nitrated methacrylic acid/3 (or 4)-fluorostyrene/divinylbenzene copolymer	—	21
Acrylamide/*N,N'*-methylenebisacrylamide/dextrose/glycidyl acrylate/allyl glycidyl ether/eosin Y copolymer	—	22
Acrylamide/ethylene glycol dimethacrylate (or *N,N'*-methylene-bisacrylamide)/methacrylic acid anhydride	—	23
Enzacryl® AA	Nitrous acid or thiophosgene	24, 25
Enzacryl® AH	Nitrous acid	24, 25
Enzacryl® CHO	—	26
Poly *N*-acryloyl-4-and -5-aminosalicyclic acids	Titanium tetrachloride	27
4-Methacryloxybenzoic acid/divinylbenzene copolymer	*N*-ethoxycarbonyl-2-ethoxy-1,2-dehydroquinoline	28
Collagen	Glutaraldehyde crosslinking	12
Sephadex G-25 or G-250	Cyanogen bromide	9
Sepharose 2B or 6B	Cyanogen bromide	29
Sepharose 4B	Cyanogen bromide	30
ECD-Sepharose 2B	Epichlorohydrin, sodium thiosulfate, dithiothreitol, and 2,2'-dipyridyl disulfide	31
Cellulose	Titanium trichloride	32
	m-Diaminobenzene and nitrous acid	33
3-(*p*-Aminophenoxy)-2-hydroxypropyl ether of cellulose	Nitrous acid or thiophosgene	25, 34
CM-cellulose	Hydrazine and nitrous acid	35, 36
	Cyanogen bromide	37
AE-cellulose	Glutaraldehyde	38
p-Aminobenzyl cellulose	Nitrous acid	36
Soluble dextran, DEAE-dextran, and CM-cellulose	2-Amino-4,6-dichloro-s-triazine	6
Carboxylic polymer	Carbodiimides	39
Poly(allyl carbonate)	—	40
Poly(aminostyrene)	Nitrous acid	35, 36, 41, 42
Duolite® S-761	Glutaraldehyde crosslinking	43
Erythrocytes	*p*-Chloromercuribenzoate and isobutyl chloroformate on enzyme	44
	5,5'-Dithiobis (2-nitrobenzoic acid) or *m*-maleimidobenzoic acid plus isobutyl chloroformate on enzyme	45

Table 3 (continued)
α-AMYLASE IMMOBILIZED BY COVALENT ATTACHMENT

Carrier	Binding agent	Ref.
Inorganic Carriers		
Stainless steel	Titanium tetrachloride	46
Iron oxide	*m*-Diaminobenzene, nitrous acid, and titanium dioxide	47
Amino glass	Nitrous acid	48, 49
	Glutaraldehyde	49
Alkylamine silica	2,4-Tolylene diisocyanate or glutaraldehyde	50

III. GLUCOAMYLASE

A. Introduction

Glucoamylase [α(1→4) glucan glucohydrolase, EC 3.2.1.3], also known as amyloglucosidase and in past days as γ-amylase, is an exohydrolase that catalyzes the hydrolysis of α(1→4) bonds and, much more slowly, α(1→6) bonds in maltooligosaccharides from the nonreducing end of the molecule to produce β-glucose and a dextrin molecule containing one less glucose unit.[1] As a typical exohydrolase, glucoamylase is most active on oligimers containing five or more residues, with the hydrolysis of maltose (4-α-D-glucopyranosyl-D-glucose) being relatively slow.

Two factors prevent complete conversion of dextrin to glucose. Not only does maltose have a finite equilibrium concentration in the presence of glucose and glucoamylase, but even the purified preparations of the enzyme appear to have some transferase activity, catalyzing the slow formation of α(1→6) and α(1→3) bonds from β-glucose and an acceptor. This acceptor is most commonly another glucose molecule and, if so, the product is isomaltose (6-α-D-glucopyranosyl-D-glucose) or nigerose (3-α-D-glucopyranosyl-glucose).[51]

This second reaction causes the concentration of glucose to pass through a maximum at some elapsed time in a batch reaction or, if a column containing immobilized glucoamylase is employed, at some residence time determined by the reaction conditions. It is obviously a prime endeavor to make this maximum as high as feasible and to ensure that the hydrolysis is terminated at the maximum.

Because the reaction to produce isomaltose or nigerose requires two molecules of glucose, initial dextrin concentrations of industrial interest (those of 30% and above) lead to lower conversions to glucose than do lower dextrin concentrations. Varying other reaction parameters either has little effect on the maximum concentrations, as in the case of temperature, or leads to undesirable changes in enzyme activity and stability, as with pH.

In present industrial practice, glucoamylase is employed in native form for two purposes: in conjunction with fungal amylases to increase the DE (dextrose equivalent, the relative measure of the reducing power of a dextrin solution, based on pure dextrose [glucose] as 100) of an acid-hydrolyzed dextrin from 35—45 to 60—65, and to hydrolyze α-amylase liquefied dextrin solutions of DE 10—15 to as high a glucose content as possible. It is the second application where use of immobilized glucoamylase has excited the most interest, for when the native enzyme is employed, its concentration is low and the hydrolysis at 60°C and pH 4 to 5 requires 48 to 72 hr in large tanks. Use of immobilized glucoamylase, on the other hand, would drastically reduce the residence time because of the much higher enzyme concentrations possible and therefore would greatly reduce equipment size.

This major advantage is counterbalanced by two potential disadvantages. First, if the immobilized enzyme was to be used, its stability would have to be sufficiently high so that operation at 55 to 60°C, temperatures where microbial contamination is not a severe problem, could proceed for the long times necessary to make its use economical. To this point, no immobilized glucoamylase has demonstrated this stability, and so adoption of such a product awaits either a more stable variety or the discovery of methods to operate at lower temperatures for prolonged periods without serious microbial contamination.

Second, it is necessary that the immobilized enzyme gives the same high yields of glucose attained by the soluble enzyme, which at present has reached 95% or higher based on total carbohydrate. Use of immobilized glucoamylase under controlled conditions has usually been shown to give slightly lower yields, which would lead to a lower fructose content when the high-dextrose syrup produced was converted with glucose isomerase to a high-fructose syrup. Obviously, adoption of immobilized glucoamylase also awaits a commercially acceptable solution of this problem.

B. Published Research

Research has been conducted on glucoamylase immobilized in gels and ultrafiltration reactors and attached to solid carriers by adsorption, ion exchange, and covalent binding. Because glucoamylase has been so intensively studied, this discussion will concentrate on the factors that have so far prevented immobilized glucoamylase from being used commercially.

1. Entrapment

The enzyme has been entrapped in a variety of gels (Table 4) but often with rather disappointing retentions of activity. In acrylamide polymerized by γ-irradiation, activities entrapped ranged from 25% to 65% of those offered,[52,53] while poly(dimethyl acrylamide) retentions were between 38% and 59%.[54] In polyacrylamide cross-linked with *N,N'*-methylene-bisacrylamide and with starch added, the retained activity was only 6.5%,[3] but if sodium acrylate and calcium acrylate were added to the mixture, 71% could be attained.[4] Poly(hydroxyethyl acrylate) gave very low activities, and a mixture of acrylic acid and sodium acrylate could not be satisfactorily polymerized.[54] With vinyl alcohol polymerized by γ-irradiation, only 16 to 24% of the glucoamylase remained active, but if electron beam irradiation were used instead, retentions close to 50% could be attained, the values falling with increasing levels of radiation even though the amount of protein entrapped was increasing.[52,59] Much the same phenomenon occurred with vinylpyrrolidone polymerized by γ-irradiation.[60] Here the conversion of dextrin was similar to that with native glucoamylase, though the rate decreased and the Michaelis constant increased. Although no effect of immobilization on the optimum pHs for activity or stability was observed, the optimum temperature for a short assay decreased from 65°C to 60°C.[60]

When acrylamide and *N,N'*-methylene-bisacrylamide were copolymerized with ammonium persulfate using *N*-dimethylaminopropionitrile initiation, only 20% retention of activity occurred, and amylopectin was not degraded.[55] If, on the other hand, the polymerization took place in suspension with amylopectin, and the latter was later removed from the gel with free glucoamylase, retentions up to 98% were attained.[56] While the soluble enzyme could convert 96% of the 35% dextrin solution fed to it, immobilized glucoamylase was capable of only 93% conversion. There was no change in the pH-activity profile upon immobilization unless acrylic acid was added to the

Table 4
GLUCOAMYLASE IMMOBILIZED BY ENTRAPMENT

Carrier	Ref.
Organic gels	
Poly(acrylamide)	52—54
Poly(dimethyl acrylamide)	54
Acrylamide/ *N,N'*-methylene-bisacrylamide copolymer	55, 56, 57
Acrylamide/ *N,N'*-methylene-bisacrylamide/starch copolymer	3
Acrylamide/ *N,N'*-methylene-bisacrylamide/sodium acrylate/calcium acrylate/starch copolymer	4
Poly(2-hydroxyethyl acrylate)	54
Acrylic acid/sodium acrylate copolymer	54
Poly(vinyl alcohol)	52, 58, 59
Poly(*N*-vinylpyrrolidone)	60
Other entrapment methods	
Cellulose triacetate fibers	61
Lecithin/dicetyl phosphate liposomes	62

Membrane	Ref.
Ultrafiltration reactors	
Similar to Amicon UM2	63
Amicon UM2	64
Amicon PM10	5
Amicon PM30	7

polymerization mixture, when the profile was broadened because of the added buffering effect.[56] Unlike glucoamylase trapped in poly(vinyl-pyrrolidone), this form of the enzyme was more stable when immobilized, half-lives at 60°C with 1% starch approaching 45 hr and little activity loss occurring at 50°C in over 100 hr.[55,56]

With immobilization in gels, several problems must be overcome. If a low number of bonds in the gel are formed, the enzyme activity may approach that of the native form before entrapment, but gross leakage may result and the gel may be excessively soft. Greater cross-linking or more extensive bond formation, on the other hand, may yield gel of more nearly satisfactory mechanical properties, but often the enzyme is so tightly entrapped that significant activity is lost. In addition, access of the substrate becomes more and more limited as the gel becomes stiffer. The Beck and Rȧse[56] method to ensure greater access by formation of pores in the gel was an ingenious attempt to improve the activity of the immobilized enzyme, an especially important factor with glucoamylase, where the reactant is quite large and therefore especially susceptible to intraparticle diffusion limitation or even total exclusion.

Three other entrapement methods are worthy of note. Corno et al.[61] spun cellulose triacetate filters which incorporated glycoamylase but found that diffusion limitation was severe; retentions of activity were 40% or even less as enzyme concentration increased and that of substrate decreased. This effect of slow diffusion was also demonstrated in the large increases of the Michaelis constant over that of soluble glucoamylase, especially at higher temperatures. Despite these difficulties, a reactor packed with glucoamylase-impregnated threads could produce DEs up to 96 from 20 to 40% dextrin.

In a very different application, glucoamylase was entrapped in liposomes of lecithin and dicetyl phosphate.[62] Although only 5% or so of the enzyme offered was entrapped, little of the activity taken up was lost.

The third alternate method is the use of glucoamylase in reactors whose exits were blocked by ultrafiltration membranes.[5,7,63,64] Here the enzyme is in its native state, and the only mass transfer limitation is that on substrate and product passing through the membrane. High conversions can be attained, but filtration rates may be decreased with time because of formation of a gel layer adjacent to the membrane. This is less of a problem with glucoamylase than with α-amylase because the former more completely hydrolyzes the reactant.

2. Adsorption

Glucoamylase has been immobilized to a number of inorganic and a few organic adsorbents with varying degrees of success (Table 5). Usami and co-workers found that with acid clay, activity yields upon immobilization were 30 to 40%,[68,69] but that desorption under use was severe.[68] Immobilization decreased the specific activity rate of the enzyme[67,68] and sharply increased the Michaelis constant when soluble starch served as substrate.[67-69] The pH vs. activity profile was narrowed by immobilization, and the optimum pH was raised to above pH 5.[67,68] With 0.5 to 5.0% starch, the Michaelis-Menten equation was followed in column operations.[69]

With activated charcoal, the results were similar,[65,67-70] except that the activity after immobilization was above 50% and much less eluted with use. In addition, the pH optimum was not shifted as greatly as by acid clay.[68] With neither adsorbent did enzyme stability seem as high as exhibited by native glucoamylase.[69,70]

While little activity remained when glucoamylase was adsorbed on calcium phosphate gel, silica gel, or diatomaceous earth, high activities were achieved when the latter was mixed with acid clay. However, 2% starch was sufficient to elute the adsorbed glucoamylase.[65] Usami and Taketomi[65] had no success with alumina, but Solomon and Levin[71] found that the type employed was very important and that retentions of up to 90% could be attained.

A study of glucoamylase immobilization to hydroxylapatite[72] confirms much of the other work on adsorption to inorganic carriers. Approximately 10% of the amylase and 45% of the maltase activity were recovered on the hydroxylapatite, with the specific activity decreasing sixfold when assayed with starch but remaining roughly constant when assayed with maltose. The Michaelis constant for dextrin tripled while that for maltose doubled. It is evident, therefore, that immobilization of glucoamylase has caused more severe losses of activity on dextrin than on the smaller maltose, indicative of diffusion limitation.

There have been four papers describing the adsorption of glucoamylase to organic carriers,[73-76] two by hydrophobic interactions. To Sepharose 6B modified to yield 400 μmol hexyl groups per milliliter packed gel, the bond is very tight, with only a few detergents successfully removing the enzyme.[73] Adsorption is both faster and more complete in the presence of NaCl and, while some glucoamylase is desorbed upon reversion to a salt-free medium, the amount remaining bound is still higher than if no salt had been used during adsorption.[74] The Michaelis constant for adsorbed gluco-

Table 5
GLUCOAMYLASE IMMOBILIZED BY ADSORPTION

Carrier	Ref.
Acid clay	65—69
Activated charcoal	65, 67—70
Calcium phosphate gel, silica gel, and diatomaceous earth	65
Alumina	65, 71
Molecular sieve 4A	71
Hydroxylapatite	72
Hexyl-Sepharose 6B	73, 74
Phenylenediamine/glutaraldehyde copolymer	75
Collagen	76
Ion exchange carriers	
CM-Sephadex C-25	77
CM-cellulose	72, 77, 78
Amberlite® CG-50	65, 77, 79
DEAE-Sephadex A-50	80, 81
DEAE-cellulose	77, 78, 81—90
Guanidino-cellulose	90
p-Aminobenzenesulfonylethyl cellulose	91
Amberlite® CG-4B	77
Duolite® A-7	81
Amberlite® IR-45	82, 92, 93
Dowex® 1-X4, 1-X10	65, 82

amylase was only slightly higher than that of the native form, but the stability of the former decreased greatly in both buffer and 1% soluble starch.

Glucoamylase bound to a copolymer of phenylenediamine and glutaraldehyde gave high stability; little activity was lost in a column at 48°C and pH 4.6 over 3 months, while hydrolysis of 45% to 50% DE 42 syrup approached 100%.[75] The pH profile did not change appreciably upon immobilization. The mechanism of binding is unknown, and it is possible that covalent bonds are present here.

3. Adsorption by Ion Exchange

A number of ion exchange materials have served as carriers for glucoamylase (Table 5). We will first deal with the few cation exchangers that have been employed and then with the many more anionic carriers.

Miyamoto et al.[77] found that CM-Sephadex adsorbed 17 mg of enzyme per gram carrier with 73% retention of specific activity, but they were unsuccessful with CM-cellulose, as were Fukui and Nakagawa[72] and Kučera and Hanus.[78]

With Amberlite® CG-50 particles of 20 to 190 μm diameter, 0.065 g enzyme per gram carrier was adsorbed with 45% retention of specific activity.[77] Smaller particles (5 to 70 μm) took up glucoamylase faster, and larger ones (60 to 200 μm) absorbed slower than the 20 to 190 μm beads.[79] The activity of the three particle sizes on maltose and amylose followed the same pattern. In further confirmation of the effect of pore diffusion limitation, Miyamoto et al.[79] found that the activation energy decreased from

10.3 kcal/mol for native glucoamylase to 7.7 kcal/mol for the enzyme bound to the largest particles. Calculated activities for 0.003 *M* and 0.1 *M* maltose with all three particle sizes agreed well with the experimentally obtained values. As would be expected, less pore diffusion limitation occurred at higher maltose concentrations. When the largest particles were exposed to glucoamylase solutions for increasing lengths of time, the amount of enzyme bound increased from 30 to 100 mg/g, but the specific activity on maltose at 40°C decreased 60%.

Miyamoto et al.[79] solved by computer a second order material balance differential equation using varying Thiele moduli and obtained intraparticle concentrations of maltose at varying values of radius and bulk maltose concentrations. This yielded effectiveness factors as a function of the modulus and bulk concentration and demonstrated that increasing the first and decreasing the second led to increasing intraparticle diffusion limitation.

Glucoamylase has been attached to a number of anionic carriers, but the greatest number of attempts have employed DEAE-Sephadex[80,81] and DEAE-cellulose.[77,78,81-90] With the former, the optimum pH and temperature did not change, but the Michaelis constant with soluble starch increased fourfold upon immobilization.[80] In a stirred reactor, the adsorbed enzyme converted 30% dextrin to a glucose syrup of greater than DE 95. Another approach was to link an ethylene-maleic anhydride copolymer to glucoamylase and then to attach the conjugate to DEAE-Sephadex.[81] All was taken up and 60% remained active; with the native enzyme, all was adsorbed but just 35% retained activity.

When Bachler et al.[82] attached glucoamylase to DEAE-cellulose, 16 to 55% of the enzyme bound was active, the higher percentages occurring at lower loadings. A linear relationship was observed between enzyme offered and that adsorbed in active form, 17% of the former remaining active when immobilized. Glucoamylase was eluted from DEAE-cellulose when greater than 0.1 *M* acetate was present. Although the pH profile was sharpened, the optimum for the *Aspergillus* enzyme remained at pH 4.25. However, it was much less stable after immobilization. When adsorbed glucoamylase was employed in a stirred vessel to hydrolyze 5% DE 16 dextrin, DE 94 glucose syrup could be attained, the same as with native enzyme.[83] With 30% dextrin, though, the maximum DE with native glucoamylase was 92, and this fell even further when adsorbed enzyme was employed.

Another tack was to attach glucoamylase to DEAE-cellulose porous paper in an annular reactor.[84] As much as 0.5 g enzyme per gram paper could be adsorbed, of which 50% was active. Under different conditions, either film or pore diffusion limitation was evident.

Solomon and Levin's work with ethylene/maleic anhydride/glucoamylase (EMA-GA) conjugates attached to DEAE-Sephadex has just been discussed.[81] In the same project, they added both this material and succinylated glucoamylase to DEAE-cellulose. At 50°C EMA/GA/DEAE-cellulose was more stable than GA/DEAE-cellulose, and both the latter and EMA-GA were more stable than native glucoamylase. EMA/GA/DEAE-cellulose retained 25% of the native enzyme's activity when assayed on dextrin and 40% when tested with *p*-nitrophenyl-glucopyranoside. In addition, EMA/GA was less easily desorbed from DEAE-cellulose than glucoamylase, although below pH 4 accelerated desorption did occur, and above pH 5 the adsorbed enzyme was unstable. Succinylation of glucoamylase occurred with 40% yield loss, compared to total retention of activity when EMA/GA was formed; the succinylated enzyme could be bound to DEAE-cellulose with an overall activity loss of 78%.

With *Endomyces* glucoamylase, highest activities on DEAE-cellulose were attained at ionic strengths below 0.01.[87] Above this level there was a sharp drop in activity until a plateau at ionic strengths between 0.02 and 0.1. The optimum pH range for activity was 6.5 to 8, where approximately 60% of the original activity was bound, 70 to 74% of the specific activity being retained. The highest specific activity occurred when about 200 mg protein per gram carrier were bound, although it was possible to attach even higher levels of enzyme.

Chen and Tsao[89] have produced porous cellulose beads with excellent mechanical properties capable of binding high amounts of glucoamylase, especially when DEAE groups were added. Then activities to 9000 U/g were achieved, the immobilized enzyme being assayed with 10% maltose at 60°C.[90] When guanidino instead of DEAE groups were employed, activities were appreciably lower.

With *p*-aminobenzenesulfonylethyl cellulose, 37 to 64 mg protein per gram carrier could be bound with 20 to 30% retention of activity; however, the latter value could be increased to 44% with the proper ammonium sulfate treatment.[91] The optimum activity on starch was at pH 4.5 and 65 to 70°C. At 55°C and pH 4.5, 97.9% glucose on a dry basis could be obtained from 31 to 34% liquefied starch.

Glucoamylase could be adsorbed on strongly basic resins such as Amberlite® IR-45,[82,92,93] or Dowex® 1,[65,82] but on the former at least stabilities were low.[92,93] Optimum pH was slightly increased, while the Michaelis constant on soluble starch was 6.5 times that of native glucoamylase.[93]

4. *Covalent Attachment*

The chief method of immobilizing glucoamylase has been by covalent attachment, and a great number of carriers and binding agents have been used for that purpose (Table 6). This section will treat organic carriers first, followed by inorganic supports. In the discussion of the former, gel materials will be followed by cellulosics and then by carriers derived from petrochemicals.

Five different acrylic gels have been employed for the covalent immobilization of glucoamylase.[22,23,27,57,94,95] Poly(n-butyl-4-iodomethacrylate) yielded low activities,[94] but a formulation of acrylmide/dextrose/glycidyl acrylate/allyl glycidyl ether/eosin Y, when cross-linked with *N,N'*-methylene bisacrylamide (MBA), incorporated 62% of the glucoamylase offered when assayed with 30% DE 28 dextrin.[22] The amount of cross-linking greatly affected the product profile; as it increased, more oligosaccharides were left unreacted. Walton and Eastman[22] found that the best yields of glucose (approximately 94%) were obtained by hydrolyzing dextrin to DE 68 with immobilized α-amylase before passing it to the immobilized glucoamylase column (rather than using low DE dextrin where only 91 to 92% glucose could be obtained). This is opposite to experience with native glucoamylase. The immobilized enzyme decayed in two segments of radically different half-lives, with that of the shorter half-life increasing in proportion as the temperature increased. It appeared that the rapid decay was caused by cage collapse of the gel, and this was alleviated to some extent by greater crosslinking, or preferably by substitution of triacryloylhexahydro-s-triazine for MBA.

Copolymerization of glucoamylase in a mixture of acrylamide/MBA/acrylic acid/2,3-epoxypropyl ester led to six times the activity of an equal amount mechanically included in an acrylamide/MBA copolymer,[57] and a copolymer of acrylamide/methacrylic acid anhydride cross-linked with ethylene glycol dimethacrylate or MBA gave 100% retention of activity at pHs above 6. Below that point the beads shrank,[23] as they were anionic.[95]

In the only project with acrylic gels where added linking agents were employed, Kennedy and Epton[27] employed titanous and titanic complexes to link glucoamylase to poly(*N*-acryloylaminosalicylic) acids in up to 40% rentetion of activity and as high as 150 mg protein per gram carrier.

Table 6
GLUCOAMYLASE IMMOBILIZED BY COVALENT ATTACHMENT

Carrier	Binding agent	Ref.
Organic carriers		
Poly(*n*-butyl-4-iodomethacrylate) (Poliodal-4)	—	94
Acrylamide/*N,N'*-methylene-bisacrylamide (or triacryloylhexahydro-s-triazine)/dextrose/glycidyl acrylate/allyl glycidyl ether/eosin Y copolymer	—	22
Acrylamide/*N,N'*-methylene-bisacrylamide/acrylic acid/2,3-epoxypropyl ester copolymer	—	57
Acrylamide/ethylene glycol dimethacrylate (or *N,N'*-methylene-bisacrylamide)/methacrylic acid anhydride copolymer	—	23, 95
Poly *N*-acryloyl-4- and -5-aminosalicylic acids	Titanium tri- or tetrachloride	27
Sepharose 4B	Cyanogen bromide	78, 96
Agarose	Cyanogen bromide	97
Collagen	Methanol, hydrazine, and nitrous acid	98, 99
Gelatin	Glutaraldehyde	100
Cellulose	Metal chlorides	32, 101
	Cyanogen bromide	84, 90, 102
	Tolylene-2,4-diisocyanate	89, 90
	Hexamethylene diisocyanate, tolylene-2,4-diisocyanate and nitrous acid, glutaraldehyde, or 3-aminopropyltrithoxysilane and glutaraldehyde	90
	m-Diaminobenzene and nitrous acid	33, 103
3-(*p*-Aminophenoxy)-2-hydroxypropyl ether of cellulose	Thiophosgene or nitrous acid	104
Bromoacetyl or iodoacetyl cellulose	—	105
Dialdehyde cellulose	—	106
CM-cellulose	Hydrazine and nitrous acid	79, 106—110
	Hydrazine	106, 111
AE-cellulose	Glutaraldehyde or carbodiimide	112
DEAE-cellulose	2-Amino-4,6-dichloro-s-triazine	113, 114
Chitin	Glutaraldehyde	115
Duolite® S-30	Glutaraldehyde crosslinking	116
Nylon 66	Titanium tetrachloride	32, 101
Aminonylon	Glutaraldehyde	117
Poly(aminostyrene)	Nitrous acid	118
	Glutaraldehyde	119
Microorganisms	Titanium tetrachloride	32, 120
Inorganic carriers		
Stainless steel	Titanium tetrachloride	46
Alumina	Glutaraldehyde	71
Hornblende	Glutaraldehyde	121
	Titanium tetrachloride	121, 122

Table 6 (continued)
GLUCOAMYLASE IMMOBILIZED BY COVALENT ATTACHMENT

Carrier	Binding agent	Ref.
Enzacryl® TIO	*m*-Diaminobenzene and nitrous acid	122
Celite	*m*-Diaminobenzene and nitrous acid	103
Glass or silica	Metal chlorides	32, 101, 123
	Carbodiimide	123, 124
	Thiophosgene	125
	Nitrous acid	123, 124, 126—131
	Glutaraldehyde	123, 124, 126, 128, 131—138

There are two reports of Sepharose 4B being used for immobilization of glucoamylase,[78,96] one where it was coimmobilized with glucose oxidase.[96] The optimum pH in that system was 5.7, decreasing to pH 4.8 as the ratio of glucoamylase to glucose oxidase decreased and the former became more limiting.

Swanson et al.[97] have immobilized glucoamylase to small beads of 4% agarose and large beads of 10% agarose. With high concentrations of maltose and the large beads there was little intraparticle diffusion limitation, but with low concentrations the limitation on activity was severe. When the smaller particles were employed, diffusion limitation was not a problem even at low maltose concentrations, not only because the particle diameter was less but also because the smaller particles had less solid material with which to bind glucoamylase and, therefore, an appreciably lower enzyme activity. The authors pointed out that at low substrate concentrations where the reaction rate approached first order kinetics, it was possible to design an enzyme-carrier complex to have an arbitrary lower limit of effectiveness factor by setting the values for activity and particle size. In a system where substrate concentration is high, diffusion limitation should not be a problem. Swanson et al.[97] found this latter statement to be true with 50% w/v maltose. However, it would have been less true if the substrate had been a high molecular weight dextrin which has a lower diffusity than maltose and is attacked more readily by glucoamylase, or if the glucoamylase has been immobilized in higher concentrations.

Collagen has also served as a carrier,[98,99] with some increase in glucoamylase stability being noted.[99] After immobilization, Michaelis constants increased for maltose, suggesting diffusion limitation in the helioidal reactor that was being used. With gelatin containing bentonite, Celite, silica gel, or by itself, roughly 50% activity retention occurred upon linking with glutaraldehyde.[100] The immobilized enzyme was less stable in buffer than the native form, although 30% soluble starch stabilized it markedly.

A number of binding agents have been employed to link glucoamylase to cellulose. Among the metal chlorides, $TiCl_3$, $TiCl_4$, and $ZrCl_4$ appeared to be most effective, with approximately 100 mg/g of enzyme being bound with retentions of activities around 50%.[32,101] With *Aspergillus* glucoamylase at 50°C with 5% starch, there was no activity loss after 9 days. The carrier may be regenerated in high yield by addition of $TiCl_4$, followed by more enzyme.[32]

With cyanogen bromide, glucoamylase could be linked to cellulose with up to 51% retention of specific activity on maltose.[102] Activity of the immobilized enzyme on amylose was nil, however. Protein content could reach 10% on finely ground cotton linter pulp, although with wood pulp and microcrystalline cellulose it was lower. Another carrier was nonporous dialysis paper;[84] a third was porous cellulose beads.[90] In the latter case, Chen and Tsao found activity with CNBr linkage nearly as high as when the enzyme was adsorbed to DEAE-porous cellulose beads.[90] Other methods to bind glucoamylase to the beads, such as tolylene-2,4-diisocyanate by itself[89,90] and after diazotization,[90] glutaraldehyde both direct and after silanization,[90] and hexamethylene diisocyanate,[90] gave lower yields.

Binding with Bismark Brown, produced by diazotization of *m*-diaminobenzene coated on cellulose, led to low activities and low stability.[103] Activities were also low when glucoamylase was attached to 3-(*p*-aminophenoxy)-2-hydroxypropyl ether of cellulose with thiophosgene or nitrous acid, but here stabilities in buffer were slightly higher than that of the native enzyme.[104]

A seemingly successful carrier was cellulose with bromoacetyl or iodoacetyl active groups.[105] Maeda and Suzuki reported that specific activities were retained almost completely, based on maltose assays, when glucoamylase was bound to either carrier after it had been dissolved in dimethyl sulfoxide and reprecipitated; the protein content of the complex was approximately 4%. Specific activity to amylose fell to approximately 15% of the native enzyme, however. Smaller particle sizes led to higher activities and to Michaelis constants closer to those of the native enzyme on maltose and amylose, again emphasizing the effect of diffusion limitation. There was no increase of stability upon immobilization.

In another linkage to an active cellulose carrier, 46% retention of specific activity was attained and protein was bound in nearly 1% concentration to the carrier, dialdehyde cellulose.[106]

Carboxymethylcellulose has been employed as a carrier for glucoamylase with mixed results. Maeda and Suzuki[107] achieved binding through the azide with 90% retention of specific activity and with 22 mg protein coupled per gram carrier, but the conjugate was less stable than native glucoamylase. Michaelis constants for maltose and amylose increased moderately over values for soluble glucoamylase. In a second paper,[108] bound specific activity was 75% of the soluble value for maltose and 70% for liquefied starch. With 35% liquefied starch, DE 98 syrup could be obtained originally, and DE 97 after 28 days' operation at 40°C. At 50°C, 33% of the initial activity remained after 28 days.

Kučera and Hanus[78] achieved activities and protein binding equivalent to those on Sepharose 4B where glucoamylase was coupled with CNBr, or on DEAE-cellulose with ionic bonding, when they attached glucoamylase to highly hydrated CMC-azide gel. Michaelis constants for all three were lower than that for the native enzyme because of partition effects,[78,110] and stabilities were similar. With the CMC-glucoamylase conjugate, rates peaked at maltooctaose and decreased at higher chain lengths, although when the enzyme was free, reaction rates were constant for d.p. 6 to 7 and above.[109,110]

Christison[111] oxidized glucoamylase in approximately 80% yield with periodate and attached the enzyme through the aldehyde groups on the enzyme's carbohydrates to CMC-hydrazide. He obtained 14% yield, based on the activity offered, and the specific activity of the immobilized enzyme was 41.5% of the soluble form. The amount of protein bound was 23 mg per gram carrier. On the other hand, Roth et al.[106] obtained low activities when they attempted to bind glucoamylase to CMC-azide or hydrazide or when they oxidized the enzyme with periodate.

While AE-cellulose gave a conjugate less stable than native glucoamylase,[112] DEAE-cellulose produced one that was more stable.[114] With the DEAE-cellulose immobilized enzyme packed in a column, there was no loss in activity in 100 hr with 53% DE 56 feed at 55°C. With maltose, the integrated Michaelis-Menten equation was followed up to moderate conversions.[113] Dextrins of higher DE values more easily reached the enzyme than less hydrolyzed material. Approximately 25% of the specific activity of the enzyme was recovered when glucoamylase was linked to DEAE-cellulose, and about 100 mg protein per gram was bound.[114] The Michaelis constant for maltose increased by a factor of 4 upon immobilization, but the activation energy remained constant at 11.4 kcal/mol.

Glucoamylase has been successfully linked to a number of carriers derived from petrochemical feedstocks. It was adsorbed to Duolite® S-30 phenol-formaldehyde resin and cross-linked with glutaraldehyde;[116] in addition, it was attached to nylon 66 with $TiCl_4$.[32,101] Inman and Hornby[117] treated nylon tubes to increase the surface area and liberate free amino groups, and then used glutaraldehyde to link glucoamylase either in series or together with glucose oxidase. In this manner, they could determine the concentration of maltose.

With aminopolystyrene, binding has been by diazo linkage[118] and with glutaraldehyde.[119] In the former,[118] up to 15 mg protein per gram carrier could be attached, but at that level, specific activity was sharply diminished. With gluteraldehyde, the pH profile for activity was centered at 4.5 but became very sharp.[119] Meanwhile, the energy of activation decreased to 7.9 kcal/mol, roughly half the value with soluble glucoamylase, possibly because of slow diffusion of the substrate dextrin.

A last mention of covalent coupling to an organic substance deals with not a standard carrier but with microorganisms. Glucoamylase has been attached to brewers yeast, *B. subtilis*, and *E. coli* with $TiCl_4$.[32,120]

While there are nearly as many reports of glucoamylase being bound covalently to inorganic supports as to organic carriers, many fewer have been employed. Indeed, the overwhelming choice of those working with inorganic materials has been to use various forms of glass or silica. We will end our discussion of immobilized glucoamylase with these applications, but will first deal with other inorganic carriers.

Stainless steel was employed as a support mainly because its high density made it suitable in fluidized beds.[46] With alumina, glutaraldehyde linkage caused an increase in glucoamylase stability, but the pH range where this occurred was quite narrow.[71]

Johnson and Costelloe[121] immobilized the enzyme to hornblende, a ferromagnesium silicate, with either glutaraldehyde or $TiCl_4$. In the former case, activities were roughly the same whether the carrier was silanized or not, but activity was increased fourfold if the enzyme were further cross-linked with glutaraldehyde after being attached to silanized hornblende. They found higher activities with $TiCl_4$ linkage than with glutraldehyde and silanized carriers, and in a second paper,[122] even higher activities were found when glucoamylase was bound to Enzacryl® TIO with *m*-diaminobenzene and nitrous acid. On either hornblende or Enzacryl® TIO, stabilities of glucoamylase at 50°C in tap water or buffer were lower than those of native enzyme.[121,122]

By far the greatest bulk of research with glass and silica supports has concerned immobilization with nitrous acid or glutaraldehyde, with a fewer number of attempts with metal chlorides, thiophosgene, and carbodiimides.

Moderately high activities could be attained when glucoamylase was coupled to ground borosilicate glass with either $TiCl_3$ or $TiCl_4$, and lower activities with $SnCl_4$.[32,101] In the first two cases, 70 to 80 mg protein per gram carrier were attached with about 50% retention of specific activity; in the last, 30 mg/g were bound with 30% retention. Both specific activity and storage stability were high after immobilization on various silicate carriers, with titanium salts.[123]

The first coupling of glucoamylase to a silica by azo linkage was Weetall and Havewala's work with arylamine porous glass.[126] The immobilized enzyme had a pH optimum displaced to 5 and a Michaelis constant that increased with dextrin concentration, but was roughly equal to that of the native enzyme. The activation energy was slightly decreased by immobilization, from 16.26 to 13.77 kcal/mol. In a continuously stirred tank reactor, with the immobilized enzyme in a cage extensively hydrolyzing the 30% DE 15 to 25 dextrin feed, half-lives ranged from 4.2 days at 60°C and 22.6 days at 45°C to 33 days at 37°C. Increasing pore diameter decreased the enzyme activity that could be bound because the surface area decreased, but if the pore diameter became too small, the enzyme was excluded from the interior of the particle, a finding confirmed by Nakhapetyan et al.[130] using borosilicate glass with pores of 250 to 1750 Å. There appeared to be an optimum at about a pore diameter of 550 Å.[126]

A similar system to that of Weetall and Havewala studied by Marsh et al.[128] was found to be affected by pore diffusion limitation if particle size increased from 37—74 μm to 177—420 μm. At the latter size, the effectiveness factor was only 0.56 with maltose. Here also the pH optimum was 5.

There are few differences between the properties of glucoamylase immobilized to silica or glass by azo linkage or by glutaraldehyde, except in the latter case where the pH optimum is closer to that of the native enzyme.[126,128] In general, activities bound on similar supports by the two methods were similar,[126] as were stabilities, which in both cases were greater than native glucoamylase.[133] As before, the effect of slow intraparticle diffusion was noticeable under certain conditions.

The most complete study of the engineering properties of immobilized glucoamylase is that of Lee et al.,[136] who coupled the enzyme from *Aspergillus niger* to alkylamine porous silica with glutaraldehyde and measured the activity and stability of the complex under different conditions at both laboratory and pilot plant scale. At approximately 40°C, pH 4.5, and 30% dextrin, the immobilized enzyme consistently yielded approximately 1 DE unit or 1% less glucose than native glucoamylase when the residence time was chosen for maximum yield. Feed DEs above approximately 22 led to lower maximum glucose and DE values, but with the immobilized enzyme, feeds that had been hydrolyzed to a lesser extent required greater residence times. As would be expected, dextrin produced by α-amylase hydrolysis immediately before treatment with glucoamylase gave higher maximum values of DE and glucose than did dextrin produced by acid or a mixture of acid and enzyme hydrolysis, since the concentration of maltulose and other materials not easily hydrolyzed by glucoamylase was lower.

Lee et al.[136] found that decay of their immobilized glucoamylase was first order, in contrast to many preparations where two isozymes of varying stabilities were present. Half-lives at pH 4.4 in 30% dextrin ranged from 7.5 hr at 70°C to 581 hr at 55°C, yielding an activation energy for decay of 64 kcal/mol. At 38°C, immobilized glucoamylase lost no activity after 80 days of continuous pilot plant operation. Microbial contamination was generally of the order of 50 organisms/mℓ; when it became severe after process upsets, it could be controlled by washing with saturated aqueous chloroform.

The authors explained the lower yield with the immobilized enzyme as being caused by a glucose gradient within the carrier pores, caused by slow outward diffusion and allowing accelerated production of isomaltose and other reversion products. They also confirmed by calculation of effectiveness factors that the reduced yield at constant residence times (or conversely, the requirement for higher residence times) at low feed DEs was caused by slow intraparticle diffusion of reactants.

This work was carried on in a second paper from the same laboratory[138] where, by calculation of the intraparticle concentrations of each of the members of the maltooligosaccharide series from glucose to maltodecaose throughout a spherical particle, the strong effect of slow pore diffusion of low DE materials on their hydrolysis rate was confirmed. Two other experimental phenomena were attributed by simulation to slow pore diffusion: the increased production of isomaltose and other reversion products previously mentioned, and a lower production of maltose at intermediate residence times by immobilized glucoamylase. This latter result is caused by a higher concentration of maltose and lower concentration of maltotriose in the carrier than in the bulk solution; this allows maltose to be depleted because of its low production and high conversion to glucose by the immobilized enzyme compared to the native form. Experimentation with very small immobilized glucoamylase particles increased final yields of glucose and intermediate concentrations of maltose to levels equal to that with soluble glucoamylase.

Lee[138a] experimentally reduced the concentration of glucose available for reversion reactions by employing a mixture of immobilized glucoamylase and glucose isomerase to convert dextrin to high-fructose syrup. Because the pH employed was intermediate between the activity and stability optima of the two enzymes, the activity and stability of neither were particularly attractive, but lower amounts of reversion products were formed.

This discussion of immobilized glucoamylase has concentrated largely on methods of obtaining high enzyme stabilities and on the effects of pore diffusion limitation; because of the difficulties of comparing activities when different assays were employed, this aspect has been deemphasized. While it is not yet clear that the stability of immobilized glucoamylase is sufficient for industrial operation, certainly it is very close.

The lower yield with the immobilized enzyme is clearly caused by slow intraparticle diffusion. This can be ameliorated in a number of ways: smaller carrier particles, lower immobilized enzyme activities, and lower feed dextrin concentrations. Each causes difficulties: higher pressure drops in column operation, larger column sizes, and greater evaporation requirements, respectively. Perhaps the best solution is the use of carrier particles porous only to a limited depth. This allows moderately high surface areas but does not allow the very severe gradients in glucose and oligosaccharide concentrations, possible with deep pores, to be set up.

IV. β-AMYLASE AND PULLULANASE

A. Introduction

β-Amylase [α(1→4) glucan maltohydrolase, EC 3.2.1.2] catalyzes the hydrolysis of α(1→4) glycosidic bonds in maltodextrins, attacking from the nonreducing end of the molecule.[1] As with other exohydrolases, configuration of the resulting product, in this case β-maltose, is opposite to the bond being broken, hence the derivation of the enzyme's popular name.[1]

The enzyme cannot hydrolyze the α(1→6) branch points in amylopectin and its derivatives nor attack maltotriose, and for that reason, breakdown of regular starch to maltose is not complete, a β-limit dextrin resulting. With normally constituted maize dextrin of low DE, this limit is attained at roughly DE 55, and if more complete hydrolysis is desired, a debranching enzyme such as pullulanase must be used. As would be expected, β-amylase exhibits negligible transferase ability, not forming α(1→6) bonds to any extent, nor transferring maltose to acceptors.[1]

The main commercial use of β-amylase would be to produce high-maltose syrups from dextrin, primarily for use as brewing adjuncts. However, the native enzyme from plants, the only source, is expensive, and therefore much of its role has been usurped by cheaper fungal amylases. Nevertheless, syrups of higher maltose content are of interest and, if an immobilized β-amylase of high stability could be produced, it might well allow the enzyme to gain markets presently denied it.

In immobilized form, β-amylase has been comparatively lightly researched, with the sweet potato and barley malt enzymes nearly exclusively studied.

Pullulanase (pullulan 6-glucanohydrolase, EC 3.2.1.41) will be discussed with β-amylase because, to this point, all reported research on its immobilization has been conducted as part of joint projects with the latter. This has occurred because pullulanase can hydrolyze $\alpha(1\rightarrow6)$ bonds in molecules as small as 6^2-α-maltosylmaltose,[139] smaller than those found in β-limit dextrins, and therefore pure maltose can potentially be produced from dextrins by mixtures of β-amylase and pullulanase. On the other hand, isoamylase, the other direct debranching enzyme that could possibly be employed, cannot completely hydrolyze β-limit dextrins.[139]

The more likely use of immobilized pullulanase may be in conjunction with immobilized glucoamylase, for it is possible with mixtures of these two enzymes to attain dextrose syrups of marginally higher DE from dextrin than with glucoamylase alone. To this point, however, no research using mixtures of immobilized glucoamylase and pullulanase has reached the published literature.

B. Published Research

Three different techniques have been employed to immobilize β-amylase: entrapment in polyacrylamide gels and by ultrafiltration membranes (Table 7), adsorption on inorganics and by hydrophobic interactions to Sepharose derivatives (Table 8), and covalent attachment to a wide range of organics and to alkylamine porous silica (Table 9).

1. Entrapment

The earliest reported immobilization of the enzyme was by Bernfeld and Wan,[2] who entrapped sweet potato β-amylase in a gel of acrylamide, *N,N'*-methylene-bisacrylamide, and $AlNH_4(SO_4)_2$ whose polymerization was initiated by potassium persulfate. Although nearly all the activity was taken up, only about 6.5% was retained upon immobilization. No improvement resulted when barley β-amylase was entrapped in a gel of acrylamide and *N,N'*-methylene-bisacrylamide radiopolymerized with γ-radiation.[3]

Another type of immobilization was attempted by Marshall and Whelan,[7] who used a mixture of β-amylase and pullulanase in an ultrafiltration reactor with an Amicon PM30 membrane to obtain maltose from potato starch. Pullulanase alone gave maltotriose from pullulan.

When sweet potato β-amylase was mixed with waxy maize starch containing 98.4% amylopectin flowing through an 18,000 dalton cutoff tubular membrane, performance was superior to the equivalent nonpermeable reactor, even though permeation of products decreased with time because of buildup of a gel layer.[140] With a crude barley malt enzyme mixture containing both α-amylase and β-amylase, the performance of the permeable reactor was better only at long operation times.[141]

2. Adsorption

In the only report of β-amylase adsorbed to inorganic carriers, Velikanov et al.[9] bound the enzyme to bentonite and kaolin. The optimum temperature was 30°C for both soluble and adsorbed forms, though activity remained up to 70°C for the latter. Energies of activation were higher for the immobilized enzyme, as were Michaelis constants through a range from 20°C to 40°C. K_m reached a minimum value for soluble and both immobilized forms at 30°C.

A very thorough study of β-amylase bound to cross-linked hexyl-Sepharose 6B by hydrophobic interactions was carried out by Caldwell et al.[73,142-144] Approximately 45

Table 7
β-AMYLASE IMMOBILIZED BY ENTRAPMENT

Gel or membrane	Ref.
Poly(acrylamide)	2.3
Amicon PM30 ultrafiltration membrane	7
Calgon 18,000 dalton cutoff ultrafiltration membrane	140, 141

Table 8
β-AMYLASE IMMOBILIZED BY ADSORPTION

Carrier	Ref.
Bentonite and kaolin	9
Allyl-ECD-Sephrose 6B	142
Hexyl-ECD-Sepharose 6B	73, 142—144

Table 9
β-AMYLASE IMMOBILIZED BY COVALENT ATTACHMENT

Carrier	Linking agent	Ref.
Sephadex G-200 (treated with *p*-nitrophenyl glycidyl ether)	Thiophosgene	145
3-(*p*-Aminophenoxy)-2-hydroxypropyl ether of cellulose	Nitrous acid or thiophosgene	25, 104
Enzacryl® AA	Nitrous acid or thiophosgene	24, 25
Enzacryl® AH	Nitrous acid	24
Bio-Gel® CM100	1-Cyclohexyl-3-(2-morpholinoethyl)-carbodiimide metho-p-toluene sulfonate	146—148
ECD-Sepharose 4B or 6B	Cyanogen bromide, 4,4′-methylene dianiline (or 1,6-diaminohexane), acetaldehyde, and cyclohexyl isocyanide	149
AH- and CH-Sepharose 4B	1-Ethyl-3-(3-dimethylaminopropyl)-carbodiimide hydrochloride	150
Agarose	Cyanogen bromide	150
Cellulose	Tolylene-2,4-diisocyanate	151
Aminoethyl cellulose or poly(amino-styrene)	Ethyl adipimidate	152
Alkylamine porous silica	Glutaraldehyde	153

mg sweet potato β-amylase could be immobilized for every gram of wet carrier, but activities were not proportional to enzyme loading because of increasing diffusion limitation and protein-protein interactions.[73,143] About 30% remained active at low concentrations, but only 10% or less remained active when high concentrations of enzymes were offered.[143] The enzyme was most highly adsorbed in the presence of high salt concentrations,[73] but could be washed off by some detergents.[142] After reconditioning with butanol or a number of other organics, β-amylase could be readsorbed with full activity.[142] The enzyme was most highly adsorbed and stable at intermediate levels of hexyl substitution[144] but nevertheless was less stable than the native form when stored in buffer at 25°C to 45°C.[143] Immobilized β-amylase was stabilized by soluble starch and, in addition, apparent decay was lower when high concentrations were immobilized because of the effect of intraparticle diffusion limitation. Diffusion limitation also increased the Michaelis constant eightfold when the enzyme was immobilized.[143]

3. *Covalent Attachment*

The largest amount of work with β-amylase immobilization has concerned covalent linking to organic carriers. The first attempt was a failure, Sephadex G-200 treated with *p*-nitrophenyl glycidyl ether and activated with thiophosgene leading to 23 mg per gram carrier bound but with no activity.[145] However, soon thereafter, Barker and coworkers were successful in binding sweet potato *β*-amylase to Enzacryl® AA[24,25] and to a cellulose derivative.[25,104] This latter derivative, the 3-(*p*-aminophenoxy)-2-hydroxypropyl ether, accepted 17.5 mg protein per gram carrier with 25% retention of specific activity when activated with thiophosgene, and 42.4 mg protein per gram carrier with 17% retention of activity after nitrous acid activation.[104] With Enzacryl® AA, protein loading was 32 mg per gram carrier after nitrous acid treatment and 26 mg per gram after thiophosgene, but specific activity retentions were only 1.5% and 0.8%, respectively.[24] Enzacryl® AH activated with nitrous acid gave an inactive enzyme.[24] At 40°C and 50°C and pH 4.8 in acetate buffer, the cellulose derivatives were more stable than the native enzyme,[104] while the Enzacryl® AA conjugates were less stable;[24] at 20°C, the latter were more stable than either cellulose-bound enzymes.[25]

This work was followed by that of Mårtensson, who with Mosbach immobilized pullulanase to Bio-Gel® CM100,[154] and then bound *β*-amylase to the same carrier[146] and studied mixtures of the two immobilized enzymes.[147,148] Pullulanase from *Aerobacter aerogenes* was bound with a carbodiimide linkage to the extent of 19 mg/g of dry polymer, 34% of that offered, with 43% retention of specific activity. No increase of stability occurred upon immobilization.[154] With barley *β*-amylase, 6.8 mg enzyme per gram carrieer was bound, 40% of that offered.[146] Activity on the resin leveled out with increasing enzyme offered at values much below the maximum binding capacity, probably because the inherently high activity of the enzyme led to diffusion limitation. While coimmobilized glutathione or serum albumin did not improve binding efficiency, operational stability of the bound enzyme was much increased, half-lives with 1% starch at pH 6 and 45°C increasing from 18 hr with no addition to 90 hr with serum albumin and 110 hr with glutathione.

When *β*-amylase was mixed with pullulanase and pullulan and bound to Bio-Gel® CM100 by carbodiimide, optimum conversions of soluble starch to maltose were attained at roughly six times the activity of *β*-amylase as pullulanase.[147] Based on amounts added, coupling yields were 40% for *β*-amylase and 38% for pullulanase, with retentions of specific activity being 22% and 32%, respectively. With 1% starch, optimum pH was near 6 and temperature was near 45°C. Under these conditions, the immobilized mixture was appreciably more stable than the corresponding soluble mixture and had an operational half-life with 0.5% starch of over 30 days.

Mårtensson[148] also tested the mixed immobilized enzyme system in stirred tanks and found that increasing concentrations of starch stabilized both enzymes. When *β*-amylase was limiting, its energy of activation was 5.5 kcal/mol, same as that of the native enzyme. The Michaelis constant for starch increased sixfold over that of soluble *β*-amylase, probably indicating the effect of diffusion limitation.

A successful method of covalently immobilizing barley *β*-amylase to Sepharose was demonstrated by Vretblad and Axén,[149] who treated the carrier with cyanogen bromide, either 1,6-diaminohexane or 4,4′-methylane dianiline, and acetaldehyde followed by cyclohexyl isocyanide. Up to 35% of the activity offered was taken up, the percentage falling with increasing *β*-amylase added. Relative activity retentions were 20 to 45%.

The immobilized enzyme had an optimum pH range from 4 to 5, compared to 4.5 to 5.5 for the native enzyme, and was less sensitive to high ionic strength. *β*-Amylase when immobilized was far more stable at 4°C in buffer and also at elevated temperatures when processing 0.1% starch at pH 4.8 in a column. An interesting point was that while the native enzyme appeared to follow an Arrhenius relationship for activity at varying temperatures, the immobilized form did not (activity increasing linearly with temperature).[149]

When Schell et al.[150] attached β-amylase to CH- and AH-Sepharose with a carbodiimide bond, approximately 35% of the specific activity was retained upon immobilization, with 0.19 and 0.13 mg protein per milliliter carrier being bound, respectively. Only 6.7% was retained when the enzyme was immobilized to cyanogen bromide activated agarose. With the latter conjugate, the pH optimum was the same as the native form, but activity on the basic side increased greatly. The Michaelis constant was the same for soluble and agarose-bound β-amylase.

Maeda et al.[151] coupled soybean β-amylase to porous cellulose beads with tolylene-2,4-diisocyante. When the beads were sufficiently small (less than 420 μm), over 60% of the enzyme bound was active, and the protein concentration on the carrier exceeded 10%. Even at these low particle sizes, Michaelis constants for the bound enzymes were much higher than those of the native form, indicating pore diffusion limitation. Half-lives in column operation at pH 4.8 and 25% DE10 dextrin were approximately 20 days at 50°C and 40 days at 40°C. Optimum pHs for activity and stability were not altered by immobilization, nor was energy of activation.

Cook and Ledingham[152] employed yet two other organic polymers, AE-cellulose and poly(aminostyrene), as carriers for barley β-amylase. In both cases, the linking agent was ethyl adipimidate. The Michaelis constant for soluble starch with AE-cellulose bound β-amylase was less than that of the native enzyme, while that with poly(aminostyrene) was greater, as the positive charge on the former at pH 4.8 enhanced the ability of the weak positive charge of the enzyme's active center to attract the weakly negatively charged substrate. The increase with polyaminostyrene was caused by the hydrophobic carrier repelling the hydrophilic substrate.

The two carriers also greatly affected β-amylase's action pattern, with AE-cellulose causing a shift from multiple chain attack to single chain attack because the enhanced positive charge of the active center was more effective in retaining the substrate amylose. On the other hand, poly(aminostyrene) caused even greater multiple chain attack, the substrate chain being repelled by the carrier.[152]

The only published research on the covalent immobilization of the enzyme to an inorganic carrier is that of Hon and Reilly,[153] who employed glutaraldehyde to link barley and wheat β-amylase to alkylamine porous silica. When enzyme was offered to carrier in a ratio of 1:10, 60% of the protein was taken up, yielding an immobilized activity of 460 U per gram dry carrier, assayed at pH 4.8 and 35°C with 30% dextrin. When lower ratios were offered, higher percentages of protein were taken up and higher proportions of the offered activity appeared on the carrier.

Energies of activation were between 6 and 7 kcal/mol for immobilized barley β-amylase and approximately 4 kcal/mol for the enzyme from wheat. Both types of immobilized enzyme reached maximum activity with 30% dextrin between pHs 4 and 5, activities at lower pHs being difficult to measure because of rapid decay. They both exhibited decay featuring two first order segments, with half-lives at pH 4.8 for immobilized barley β-amylase ranging from approximately 300 min at 60°C to 15 min at 70°C for the first segment, and 1800 min at 60°C to 35 min at 70°C for the second segment. Highest stability for both barley and wheat β-amylase was at pHs between 5 and 6.

V. INVERTASE

A. Introduction

Invertase (β-fructofuranoside fructohydrolase, EC 3.2.1.26) is active only on terminal unsubstituted β-D-fructofuranosyl residues.[155] Specificity on the aglycone residue is low, however, as would be expected with a glycoside hydrolase.[155] The enzyme has substantial transferase activity, fructose being added to acceptor sugar and alcohol molecules.

Invertase, or β-fructofuranosidase, is one of the most studied of all enzymes, being the subject of both the classic kinetics research of Michaelis and Menten[156] and of the first immobilization efforts.[157-159]

However, of the enzymes covered in this chapter, invertase along with α-amylase have the lowest probability of achieving commercial use in immobilized form. While there is a major market for invert sugar, it is easier to hydrolyze pure sucrose solution (which has little buffering capacity) with hydrochloric acid or cation exchange resins or to supply the market with high fructose corn syrups made by isomerization of nearly pure glucose solutions. A minor application in the inversion of strongly buffering sucrose molasses may exist.

The numerous publications on immobilization which are discussed on the following pages are evidence more of its suitability as a model enzyme for experimental purposes than as a potential large-scale commercial product. As with α-amylase, because of the low probability of commercial use, treatment of immobilized invertase will be relatively brief.

B. Published Research

As with the enzymes already discussed in this chapter, invertase has been immobilized to a wide range of carriers by a large number of techniques (Tables 10 to 12). As was found previously, these different immobilization methods have led to widely varying activities and stabilities.

1. Entrapment

Use of poly(acrylamide) gel led to low retained activity[52,53,165] and, at least in some cases, to an increase in Michaelis constant at larger particle sizes[161] and at the more viscous solutions characteristic of high sucrose concentrations.[162] Stability of the enzyme trapped in poly(acrylamide) gel was roughly the same as the native form.[164,165]

Other acrylic gels have also been employed. Neither poly(dimethyl acrylamide) nor poly(2-hydroxyethyl acrylate) gave high retentions of activity, and an acrylic acid/sodium acrylate copolymer could not be gelled at all.[54] However, poly(acrylamide) stiffened with various cross-linking agents and related gels could not only stabilize invertase[3,168,171] but also gave fairly high activity.[3,167,169,171] As before, diffusional effects were sometimes evident.[167,170,172] Two interesting methods to form particles with better engineering properties were used by Kawashima and Umeda,[167] who polymerized frozen drops of acrylamide and additives suspended in hexane or other solvents to obtain gel beads, and by Krosing et al.,[169] who allowed the gel to form within the interior of porous silica beads.

Polymerizing vinyl alcohol by electron beam irradiation[52,59] was more successful in retaining immobilized invertase activity than was use of γ-irradiation,[52,58] though neither method was particularly promising.

Invertase can also be entrapped in fibers of cellulose triacetate[173-175] or poly(γ-methyl glutamate)[173] or less successfully in those of cellulose diacetate, ethyl cellulose, or poly(vinyl chloride).[173] The enzyme in cellulose triacetate was extremely stable at room temperature or below, half-lives in the order of years being possible.[173-175] As more enzyme was entrapped, specific activities decreased,[173,175] as they did when fiber porosity was low.[175] Stability, on the other hand, increased as porosity decreased, as the enzyme had less chance to escape.[175]

A variation of entrapment methods was to immobilize whole cells of invertase-containing *Saccharomyces pastorianus* in spherical agar particles.[176,177] The beads were

Table 10
INVERTASE IMMOBILIZED BY ENTRAPMENT

Organic gels

Carrier	Ref.
Poly(acrylamide)	52, 53, 160—165
Poly(dimethyl acrylamide)	54
Acrylamide/*N,N'*-methylene-bis-acrylamide/*N,N,N',N'*-tetramethylenediamine	166
Acrylamide/*N,N'*-methylene-bis-acrylamide/starch copolymer	3
Acrylamide/*N,N'*-methylene-bis-acrylamide/sodium acrylate/calcium acrylate copolymer	167, 168
Acrylamide/itaconic acid anhydride copolymer	169
Acrylic acid/sodium acrylate copolymer	54
Poly(2-hydroxyethyl acrylate)	54
Hydroxyethyl methacrylate/ethylene glycol dimethacrylate copolymer	170
Poly(ethylene glycol dimethacrylate)	171, 172
Poly(vinyl alcohol)	52, 58, 59
Poly(*N*-vinylpyrrolidone)	60
Cellulose triacetate fibers	173—175

Other entrapment methods

Carrier	Ref
Cellulose diacetate, ethyl cellulose, poly(vinyl chloride), or poly(γ-methyl glutamate)	173
Agar/yeast cells	176, 177

Ultrafiltration reactors

Membrane	Ref.
Amicon UM10	178
Amicon PM30, XM100A, XM300	179
Amicon XM100A	180
Lumen of Romicon P10 fiber	181

employed in an air-agitated fluidized bed that acted as a CSTR[176,177] and in a fixed-bed column.[177] In both reactors, reaction rate was strongly dependent on particle size, and in the latter, low liquid flow rates led to film diffusion limitation.

Ultrafiltration membranes have been employed to entrap invertase in several configurations. In a reactor whose exit was closed by an Amicon UM10 membrane, the enzyme was completely stable at 50°C in the presence of sucrose when attached to bentonite but lost 95% of its activity when in native form.[178] It was also operated in a CSTR ultrafiltration reactor with a series of membranes with larger exclusion limits,[179] and in a thin-channel reactor with an Amicon XM100A membrane.[180] In a fourth paper, invertase was held in the lumens of the hollow fibers in a Romicon P10 car-

Table 11
INVERTASE IMMOBILIZED BY ADSORPTION

Carrier	Ref.
Charcoal or aluminum hydroxide	157—159
Bentonite	178, 182
Tannic acid	183, 184
Collagen	185—187
Sepharose CL 4B/Anti-human serum albumin adsorbing enzyme coupled to human serum albumin	188
Ion exchange carriers	
CM-celullose	166
DEAE-cellulose	90, 176, 189—192
DEAE-Sephadex	193
Guanidino-cellulose	90, 194
Diethylaminoacetylcellulose	195
o-Phenosulfonic, *p*-phenosulfonic, and salicylic resins	196
Rohm and Haas resins	197
Bio-Rad AG3-X4A	198
Ion exchange resins and celluloses/fungal spores	199

tridge.[181] There was significant diffusion limitation in this system, as the substrate had to pass through the membrane to the solution in the lumen to reach the enzyme.

2. Adsorption

As mentioned at the beginning of this section, invertase was the enzyme employed in the earliest attempts at immobilization. Over 60 years ago, Griffin and Nelson[157,158] used aluminum hydroxide and charcoal as adsorbents. In a later paper, Nelson and Hitchock[159] eliminated film diffusion by decreasing the amount of adsorbed invertase.

In much more recent work, invertase was bound to bentonite by adsorption.[178,182] Activity was rather low,[182] but stability was greater than that of the soluble enzyme, though not as great as when the enzyme was fixed to bentonite with cyanuric chloride.[178,182]

An insoluble invertase-tannic acid complex, which could be dissociated by treatment with urea and sucrose, was more resistant to proteins[183,184] and to inhibition by glucose, fructose, and PCMB, than was soluble invertase.[183] There was no significant stabilization after immobilization.[183,184]

Collagen has been studied extensively as a carrier for invertase, both when the enzyme was adsorbed after soaking[185-187] and when it was bound by soaking or electrodeposition followed by tanning with a cross-linking agent,[203-205] presumably covalently attaching at least some of the invertase to the carrier. Characteristically, there was a sharp decrease in activity as loosely bound enzyme was washed off, followed by operation at constant activity.[185,186,203] In two cases, invertase was coimmobilized with other enzymes, first with glucose oxidase[187] and then with glucose oxidase and mutarotase.[205]

A totally different concept of binding invertase by adsorption was that of Mattiasson,[188] who covalently bound the enzyme to human serum albumin (HSA) with glutaraldehyde and anti-HSA to Sepharose CL 4B with CNBR. The heat of reaction when substrate encountered enzyme-HSA complexed to carrier-anti-HSA could be sensed

Table 12
INVERTASE IMMOBILIZED BY COVALENT ATTACHMENT

Carrier	Binding agent	Ref.
Organic carriers		
Nitrated methacrylic acid/ methacrylic acid *m*-fluoroanilide/ divinylbenzene copolymer	—	200
Poly(hydroxymethacrylate)	Cyanogen bromide	201
Poly(4-formyl-2-methoxyphenyl methacrylate) or 4-formyl-2-methoxyphenyl methacrylate/allyl alcohol copolymer	—	202
Collagen	Chromium sulfate tanning	203
	Glutaraldehyde cross-linking	204, 205
Cellulose	Titanium trichloride or titanium tetrachloride	101
	Tolylene-2,4-diisocyanate or hexamethylene diisocyanate	90
AE-cellulose	Glutaraldehyde	192
Duolite® S-30	Glutaraldehyde cross-linking	116, 206
Aminonylon	Glutaraldehyde	117
Poly(aminostyrene)	Nitrous acid	207
Amberlite® IR-45	Nitrous acid	203
Inorganic carriers		
Bentonite	Cyanuric chloride, thionyl chloride, or sulfonyl chloride	178, 182, 208
Brick	Thionyl chloride or sulfonyl chloride	178, 208
Glass	Thionyl chloride or sulfonyl chloride	178
	Nitrous acid	209, 210
	Titanium tetrachloride	211—213
	Glutaraldehyde	212, 213
Sand	Titanium tetrachloride	211
Biotite or muscovite	Titanium tetrachloride or glutaraldehyde	212
Hornblende	Titanium tetrachloride	212—214
	Glutaraldehyde	212, 213
	Nitrous acid	213
Pyroxene	Titanium tetrachloride or glutaraldehyde	213
Enzacryl® TIO	Glutaraldehyde	213
	Titanium tetrachloride	214
Enzacryl® ALO	Glutaraldehyde	213
Magnetite	Glutaraldehyde cross-linking	215
Oxides of tin, titanium, and aluminum	—	216

with a thermistor, and in that way, the substrate sucrose could be assayed. Old enzyme could be replaced by washing with glycine at pH 2.2.

A number of ion exchange materials have been employed to immobilize invertase. With CM-cellulose, which bound invertase below pH 5.5, stability was increased.[166] With DEAE-cellulose, on the other hand, stability decreased upon immobilization,[189-191] and the optimum pH for activity decreased significantly.[189,190,192] Desorption at fairly low ionic strengths was a serious problem,[176,190,192] as the optimum pH of operation was well below the pH where strong adsorption would occur. Attached to a guanidino ionic group, the enzyme was much more stable than when soluble and had a wide pH optimum.[194] When bound to diethylaminoacetyl-cellulose, it had the same stability as the native enzyme at elevated temperatures but was less stable when stored.[195] It was also less easily desorbed from this carrier than from DEAE-cellulose, but unfortunately retained activity upon immobilization was less than 50%.

When adsorbed to a series of petroleum-based ion exchange resins, invertase retained highest activity (about 50% overall) on the macroreticular varieties.[197] Between pH 3 and 5.9, there was little desorption from anionic or cationic carriers with 0.1 *M* acetate, but operation outside these limits or at high temperatures caused severe loss. As would be expected, pH optima decreased from that of the soluble enzyme with anionic resins and increased with cationic carriers.

With the weak anionic resin AG3-4XA, binding efficiency was low, and product inhibition was decreased.[198] Intraparticle diffusion limitation increased in importance as particle diameter increased.

A survey of ion exchange resins and celluloses found ECTEOLA-cellulose was most efficient at adsorbing invertase-containing fungal spores.[199]

3. Covalent Attachment

As noted with other enzymes, the largest proportion of work with immobilized invertase has been with covalent binding. With the polyaldehydes vanacryl and covanacryl, Brown and Joyeau[202] achieved 14% to 35% retention of specific activity upon immobilization, with the amount fixed approximately 10 mg/g.

Attachment of invertase to cellulose with titanium salts was unsatisfactory,[101] while with diisocyanates, lower activities were obtained than with ionic binding to DEAE-cellulose or guanidino-cellulose.[90] Poor results were also obtained when an attempt was made to couple invertase to AE-cellulose with glutaraldehyde.[192]

Several petroleum-based materials have been used as carriers to which invertase has been linked covalently: Duolite® S-30 phenol-formaldehyde resin,[116,206] aminonylon,[117] poly(aminostyrene),[207] and Amberlite® IR-45.[203] In the first case, the enzyme was adsorbed to the carrier and then cross-linked with glutaraldehyde.[116,206] In the second, invertase was immobilized with glucose oxidase on the inside of hollow tubes so that sucrose could be determined.[117] With poly(aminostyrene) beads, the maximum rate was sharply decreased, and with both beads and hollow tubes, the Michaelis constant was increased over that of the soluble enzyme.[207] The pH-activity curve for the immobilized enzyme was sharpened, perhaps because of the increased hydrophobicity of the microenvironment.[207] Saini et al.[203] observed the opposite behavior with Amberlite® IR-45, a weak-base polystyrene resin. The enzyme was bound to the carrier in low concentration but high efficiency; however, its stability decreased upon immobilization.

With inorganic carriers, the bulk of the published research has dealt with carriers that are minerals, with some work on synthesized materials such as glass. On neither brick nor glass was binding with thionyl or sulfonyl chloride particularly effective,[178] nor were significantly better results obtained when the enzyme was coupled to bentonite with cyanuric chloride.[182] Greater activity was attained when cyanuric chloride was replaced with sulfonyl or thionyl chloride.[208]

Invertase coupled to amino porous glass by azo linkage has lower stability than the soluble form of the enzyme,[209] but appreciable activity can be obtained both by this binding method[209] and by linkage with titanium salts.[211-213] In some cases, an equally good linking agent was glutaraldehyde after silanization of the carrier to provide an amino group.[212,213] Though hornblende-invertase conjugates had low activities because the surface area of the carrier was low, stabilities were roughly the same as those of the soluble enzyme.[212-214] Other inorganic materials that served as carriers for invertase were sand,[211] biotite,[212] muscovite,[212] pyroxene,[213] Enzacryl® TIO and ALO,[213,214] magnetite,[215] and hydrated oxides of tin, titanium, and aluminum from the corresponding chlorides.[216] In the latter case, invertase was bound to tin oxide with almost complete retention of activity (titanium and aluminum oxides gave low activities), but stability was lower than that of the soluble enzyme.[216]

VI. α-GALACTOSIDASE

A. Introduction

Of the six enzymes discussed in this chapter, α-galactosidase (α-D-galactoside galactohydrolase, EC 3.2.1.22) is the only one whose immobilized form is currently used commercially. This is as good an indication as any that intensity of published research is not necessarily correlated with potential commercial success, for immobilized α-galactosidase has been the subject of remarkably few reports in the scientific literature.

α-Galactosidase, also known as melibiase after one of its substrates, is active only on α-D-galactopyranosyl residues.[217] Like invertase a glycoside hydrolase, it has low specificity towards the aglycon residue and substantial transferase activity. Industrially, its present importance lies in its ability to hydrolyze raffinose [*O*-α-D-galactopyranosyl-(1→6)-*O*-β-D-fructofuranosyl-(2→1)-α-D-glucopyranoside] in sugar beet molasses to galactose and sucrose, and it is this application for which immobilized α-galactosidase is employed.

Raffinose is found in appreciable quantities in sugar beets that have been grown and stored at lower temperatures. In the processing of sugar beet juice, raffinose concentrations of 4.5% or greater seriously retard the crystallization of sucrose, and these levels may be reached when molasses is treated with finely divided lime in a Steffen house and the apprecipitated di- and trisaccharides recycled to concentration and crystallization steps. The raffinose level may be reduced by placing an α-galactosidase treatment step between the crystallizers and the Steffen house.

Another potential use for the enzyme is in the hydrolysis of raffinose and stachyose [*O*-α-D-galactopyranosyl-(1→6)-*O*-α-D-galactopyranosyl-(1→6)*O*-α-D-glucopyranosyl-(1→2)-β-D-fructofuranoside] in soybean milk to galactose and sucrose, for both cause flatulence in some humans when ingested. Since it is possible to wash the sugars out of defatted soy flour or flake, soy protein concentrates are not a candidate for α-galactosidase treatment.

B. Published Research

α-Galactosidase has been immobilized in three different fashions: by use of mycelial pellets of *Morteriella vinacea* containing the enzyme, by entrapment in polyacrylamide gel or hollow fibers, and by covalent attachment to nylon.

The first method is the one now employed industrially, after extensive research, starting in the 1960s, at the Fermentation Research Institute, Inage, Chiba Prefecture, Japan. Suzuki et al.[218] showed that *M. vinacea* must be grown on media containing galactose, lactose, melibiose, or raffinose for α-galactosidase to be formed. The enzyme in mycelial suspension has an optimum pH of 4 and an optimum temperature of 50°C in a 15-min assay on raffinose. Though the mycelia are not particularly suitable for continuous operation, α-galactosidase is not easily extracted.

Pellets with better engineering properties than the mycelia are produced instead by increasing lactose concentration in the medium, decreasing glucose, replacing urea with ammonium sulfate, and adding calcium carbonate to maintain the pH at 5.6.[219] High agitation or low medium viscosity were effective in reducing pellet size and increasing α-galactosidase activity as intraparticle diffusion limitation of substrate was reduced. It appeared, however, that large pellets were more resistant than smaller ones or than mycelia to extraction of enzyme under alkaline conditions.

A third paper from the Fermentation Research Institute further delineated the influence of slow substrate and product diffusion through the pellets.[220] Though the K_m for *p*-nitrophenyl-α-D-galactoside (PNPG) remained constant at 0.43 m*M* for varying pellet diameters and was the same as the enzyme in cell-free suspension, effectiveness factors were markedly affected by pellet size and enzyme activity. When diffusion limitation was severe, no maximum activity at intermediate substrate concentration was found, even though one would be expected because of the strong inhibition exerted by PNPG.

Product galactose inhibition was ameliorated by intraparticle diffusion limitation, with apparent effectiveness factors increasing with increasing galactose concentration even though reaction rates decreased.[220] Because of the strong product inhibition, rates were first order in reactant concentration throughout the batch reaction, with rate coefficients decreasing with increasing initial PNPG concentration.

Thananunkul et al.[221] entrapped α-galactosidase from disrupted mycelia of *M. vinacea* in a polyacrylamide/*N,N'*-methylene-bisacrylamide gel, formed after initiation with ammonium persulfate and *p*-dimethylaminopropionitrile, with 60 to 65% retention of the relative activity. The optimum pH on melibiose decreased from 4.5 to 4 for the enzyme in the homogenized mycelia upon entrapment, with some broadening of the pH profile, while the optimum temperature in a 60-min assay increased from 50°C to 55°C. Despite this apparent increase in heat stability, half-lives at 50°C and pH 5 remained at approximately 1 day. However, after the first day, the entrapped enzyme was more stable. A fluidized bed containing α-galactosidase-gel complex hydrolyzed 60% of the raffinose and stachyose in soybean milk but at very low flow rates.

Two different papers report the entrapment of α-galactosidase in hollow fibers.[181,222] Smiley et al.[222] placed dialyzed enzyme solution obtained by extraction of *Aspergillus awamori* grown on moist wheat bran in an Amicon DC30 dialyzer. When soybean milk or whey at 45°C was circulated around the outside of the fibers, the reaction went to completion in 4 hr. All monosaccharides were produced, as the crude enzyme solution contained invertase.

A similar paper by Korus and Olson[181] concerned the use of *Bacillus stearothermophilus* α-galactosidase in Amicon X50 or Romicon P10 hollow fiber cartridges. While the former was unsuitable because of rapid enzyme leakage through the fiber walls, the latter, after preconditioning with bovine serum albumin, retained 82% of the initial enzyme activity with raffinose as substrate. After 7 days at 37°C, the activity had decreased 10%, though under those conditions, native enzyme was totally stable. The authors calculated a Thiele modulus of 23, indicating moderate diffusion limitation. This explains the noncomplete retention of activity but means that the actual stability was lower than that measured experimentally.

Only two articles[223,224] have appeared on the covalent immobilization of α-galactosidase, both for where it was attached to nylon. In the first, Reynolds[223] bound the *Bacillus stearothermophilus* enzyme in 1% raffinose to precipitated nylon floc with dimethyladipimidate hydrochloride, yielding 8 mg protein bound per gram of catalyst. Activity, assayed on 1.03 m*M* PNPG at pH 7 and 25°C, was 16.2U/g catalyst, and 96% yield was obtained. The pH optimum of the immobilized α-galactosidase was approximately 7 for PNPG, raffinose, stachyose, and melibiose. Neither soluble nor immobilized enzyme lost any activity in 15 days in 0.1 *M* phosphate buffer at 40°C and pH 7, and the latter kept full activity after 7 days' contact with diluted beet sugar molasses containing 1.5% raffinose. Immobilized α-galactosidase could be freed of contaminating *Leuconostoc mesenteroides* with formaldehyde.

In other work where covalent bonding was attempted, Faulstich et al.[224] immobilized green coffee bean α-galactosidase to nylon net that had been treated to obtain the *N*-hydroxysuccinimido ester. Of the 40 mg protein per square centimeter offered, 10% was bound, yielding a specific activity of 4.1 U/mg of protein, when assayed with PNPG at 25°C in 0.1*M* phosphate buffer at pH 6.5. This was 38% retention of activity.

VII. CONCLUSION

There is no reason at this point to repeat statements made earlier about the commercial potential of the enzymes discussed in this chapter. Instead, I would like to close by emphasizing two points that perhaps have become apparent in the last few pages.

The first, which has previously been stressed, is that diffusional limitations play an extremely important role in the potential use of immobilized enzymes, particularly those where the substrate is of high molecular weight. Certainly that is true with the starch-degrading enzymes treated here, where activity, incidence of multiple attack, and selectivity are all affected by slow diffusion of substrates and products. While these effects can be eliminated in many systems by decreasing the amount of enzyme bound or the particle size of the carrier, this is not always feasible, especially when small particle diameter would cause excessive column pressure drop. Therefore, there are situations where high enzyme activities immobilized on carriers can lead to difficult engineering decisions and not altogether satisfactory compromises before an immobilized enzyme supplants the corresponding soluble form.

The second point I would like to cover here is the absolutely random nature of the changes in enzyme activity and stability caused by immobilization. The ability to forecast enzyme properties after immobilization is an art not yet mastered by those in the field, and that problem is the major justification for the thorough coverage of the literature of immobilized carbohydrases presented in this chapter. To this point, whether coupling of a particular enzyme to a particular carrier will lead to high or low stabilities or retained activities is not safely predictable, and, until it is, we will be testing many carriers with many methods of binding before choosing the most economically feasible enzyme-carrier systems.

REFERENCES

1. **Thoma, J. A., Spradlin, J. E., and Dygart, S.,** Plant and animal amylases, in *The Enzymes*, Vol. V, 3rd ed., Boyer, P. D., Ed., Academic Press, New York, 1971, chap. 6.
2. **Bernfeld, P. and Wan, J.,** Antigens and enzymes made insoluble by entrapping them into lattices of synthetic polymers, *Science*, 142, 678, 1963.
3. **Kawashima, K. and Umeda, K.,** Immobilization of enzymes by radiopolymerization of acrylamide, *Biotechnol. Bioeng.*, 16, 609, 1974.
4. **Kawashima, K. and Umeda, K.,** Immobilization of enzymes by radiocopolymerization of monomers, *Biotechnol. Bioeng.*, 17, 599, 1975.
5. **Butterworth, T. A., Wang, D. I. C., and Sinskey, A. J.,** Application of ultrafiltration for enzyme retention during continuous enzymatic reaction, *Biotechnol. Bioeng.*, 12, 615, 1970.
6. **Wykes, J. R., Dunnill, P., and Lilly, M. D.,** Immobilization of α-amylase by attachment to soluble support materials, *Biochim. Biophys. Acta*, 250, 522, 1971.

7. **Marshall, J. J. and Whelan, W. J.**, A new approach to the use of enzymes in starch technology, *Chem. Ind. (London)*, 701, 1971.
8. **Steenhoek, I. and Kooiman, P.**, Adsorption of bacterial α-amylase on quartz and denaturation of adsorbed enzyme by shaking, *Enzymologia*, 35, 335, 1968.
9. **Velikanov, L. L., Velikanov, N. L., and Zvyagintsev, D. G.**, Effect of temperature on the activity of free and adsorbed enzymes, *Pochvovedenie*, 1971(3), 62.
10. **Urabe, I. and Okada, H.**, Immobilization of acylated bacterial α-amylase, in *Fermentation Technology Today*, Terui, G., Ed., Society of Fermentation Technology, Osaka, 1972, 367.
11. **Suzuki, S., Karube, I., and Watanabe, Y.**, Electrolytic preparation of enzyme-collagen films, in *Fermentation Technology Today*, Terui, G., Ed., Society of Fermentation Technology, Osaka, 1972, 375.
12. **Strumeyer, D. H., Constantinides, A., and Freudenberger, J.**, Preparation and characterization of α-amylase immobilized on collagen membranes, *J. Food Sci.*, 39, 498, 1974.
13. **Köstner, A. and Pappel, K.**, Adsorption of amylolytic enzymes on ion exchangers. I. Comparison of ion exchangers, *Tr. Tallin. Politekh. Inst. Ser. A*, 300, 33, 1971; *Chem. Abstr.*, 77, 66513, 1972.
14. **Pappel, K. and Köstner, A.**, Adsorption of amylolytic enzymes on ion exchangers. III. Adsorption of *Aspergillus oryzae* amylase on the cation exchanger Amberlite GC-50/II, *Tr. Tallin. Politekh. Inst. Ser. A*, 300, 51, 1973; *Chem. Abstr.*, 79, 102072, 1973.
15. **Köstner, A., Pappel, K., and Kask, K.**, Adsorption of amylolytic enzymes of mold fungi on ion exchangers, *Fermenty Mikroorganizmov*, p. 245, 1973; *Chem. Abstr.*, 80, 106831, 1974.
16. **Angelova, M., Angelov, T., and Grigorov, I.**, Bound α-amylase from *Aspergillus oryzae, Prilozhna Mikrobiol.*, 6, 54, 1975; *Chem. Abstr.*, 84, 161184, 1976.
17. **Ivanova, G. P., Mirgorodskaya, O. A., Moskvichev, B. V., and Samsonov, G. V.**, Stability of α-amylase of *Bacillus subtilis*, immobilized by a gel of the ion exchanger KMT, *Appl. Biochem. Microbiol. (U.S.S.R.)*, 12, 23, 1976.
18. **Kiseleva, E. M., Mirgorodskaya, O. A., Momot, N. N., Avizhenis, V. Yu., Moskvichev, B. V., and Samsonov, G. V.**, Denaturation of α-amylase from *Bacillus subtilis* in acid medium, *Prikl. Biokhim. Mikrobiol.*, 13, 577, 1977.
19. **Il'in, V. A.**, Chemistry of the sorption of α-amylase by gel sorbents, *Zh. Fiz. Khim.*, 51, 2060, 1977.
20. **Manecke, G. and Günzel, G.**, Application of a nitrated copolymer from methacrylic acid and methacrylic acid-m-fluoroanilide to the preparation of enzyme resins, to the resolution of racemic compounds, and to tanning experiments, *Makromol. Chem.*, 51, 199, 1962.
21. **Manecke, G. and Förster, H.-J.**, Reactive polystyrene-based polymers as carriers for proteins and enyzmes, *Makromol. Chem.*, 91, 136, 1966.
22. **Walton, H. M. and Eastman, J. E.**, Insolubilized amylases, *Biotechnol. Bioeng.*, 15, 951, 1953.
23. **Krämer, D. M., Lehmann, K., Plainer, H., Reisner, W., and Sprössler, B. G.**, Enzymes covalently bound to acrylic gel beads. I. Interaction of hydropholic anionic gel beads with biomacromolecules, *J. Polym. Sci. Polym. Symp.*, 47, 77, 1974.
24. **Barker, S. A., Somers, P. J., Epton, R., and McLaren, J. V.**, Cross-linked polyacrylamide derivatives (Enzacryls) as water-soluble carriers of amylolytic enzymes, *Carbohydr. Res.*, 14, 287, 1970.
25. **Barker, S. A., Somers, P. J., and Epton, R.**, Recovery and re-use of water-insoluble amylase derivatives, *Carbohydr. Res.*, 14, 323, 1970.
26. **Epton, R., McLaren, J. V., and Thomas, T. H.**, Water-insolubilisation of glycoside hydrolases with poly(acryloyl aminoacetaldehyde dimethyl acetal) (Enzacryl Polyacetal), *Carbohydr. Res.*, 22, 301, 1972.
27. **Kennedy, J. F. and Epton, J.**, Poly(N-acryloyl-4- and -5-aminosalicyclic acids), III. Uses as their titanium complexes for the insolubilisation of enzymes, *Carbohydr. Res.*, 27, 11, 1973.
28. **Bartling, G. J., Chattopadhyay, S. K., Brown, H. D., Barker, C. W., and Vincent, J. K.**, A convenient new method for enzyme immobilization, *Biotechnol. Bioeng.*, 16, 1425, 1974.
29. **Horigome, T., Kasai, H., and Okuyama, T.**, Stability of taka-amylase A immobilized on various sizes of matrix, *J. Biochem.*, 75, 299, 1974.
30. **Murao, S., Inui, M., and Arai, M.**, Preparation of immobilized bacterial α-amylase of liquefying type and its properties, *Hakko Kogaku Kaishi*, 55, 75, 1977.
31. **Carlsson, J., Axén, R., and Unge, T.**, Reversible, covalent immobilization of enzymes by thiol-disulphide interchange, *Eur. J. Biochem.*, 59, 567, 1975.
32. **Emery, A. N., Hough, J. S., Novais, J. N., and Lyons, T. P.**, Some applications of solid phase enzymes in biological engineering, *Chem. Eng. (London)*, 258, 71, 1972.
33. **Gray, C. J., Livingstone, C. M., Jones, C. M., and Barker, S. A.**, A new and convenient method for enzyme insolubilisation using diazotized m-diaminobenzene, *Biochim. Biophys. Acta*, 341, 457, 1974.

34. **Barker, S. A., Somers, P. J., and Epton, R.**, Preparation and properties of α-amylase chemically coupled to microcrystalline cellulose, *Carbohydr. Res.*, 8, 491, 1968.

35. **Fukushi, T. and Isemura, T.**, Regeneration of the native three-dimensional structure of *Bacillus subtilis* α-amylase and its formation in biological systems, *J. Biochem.*, 64, 283, 1968.

36. **Ledingham, W. M. and Hornby, W. E.**, The action pattern of water-insoluble α-amylases, *FEBS Lett.*, 5, 118, 1969.

37. **Linko, Y.-Y., Saarinen, P., and Linko, M.**, Starch conversion by soluble and immobilized α-amylase, *Biotechnol. Bioeng.*, 17, 153, 1975.

38. **Galich, I. P., Tsyperovich, A. S., Kolesnik, L. A., and Tsesarskaya, V. D.**, Preparation of immobilized α-amylase and its properties, *Ukr. Biokhim. Zh.*, 48, 480, 1976; *Chem. Abstr.*, 85, 173219, 1976.

39. **Torchilin, V. P.**, Preparation and some properties of immobilized α-amylase, *Bioorg. Khim.*, 1, 991, 1975; *Chem. Abstr.*, 84, 40256, 1976.

40. **Kennedy, J. F., Barker, S. A., and Rosevear, A.**, Use of a poly(allyl carbonate) for the preparation of active, water insoluble derivatives of enzymes, *J. Chem. Soc. Perkin Trans. I*, p. 2568, 1972.

41. **Grubhofer, N. and Schleith, L.**, Modified ion-exchange resins and specific adsorbents, *Naturwissenshaften*, 40, 508, 1953.

42. **Grubhofer, N. and Schleith, L.**, Coupling of proteins on diazotized polyaminostyrene, *Hoppe-Seyler's Z. Physiol. Chem.*, 297, 108, 1954.

43. **Boundry, J. A., Smiley, K. L., Swanson, C. L., and Hofreiter, B. T.**, Exoenzymic activity of alpha-amylase immobilized on a phenol-formaldehyde resin, *Carbohydr. Res.*, 48, 239, 1976.

44. **Chen, L.-F. and Richardson, T.**, Enzyme derivatives containing reactive groups. Immobilization of alpha-amylase on human erythrocytes, *Pharmacol. Res. Commun.*, 6, 273, 1974.

45. **Chen, L.-F. and Richardson, T.**, Enzyme derivatives containing reactive groups. II. Immobilization of alpha-amylase on human erythrocytes using 5,5′-dithiobis(2-nitrobenzoic acid) and m-maleimidobenzoic acid adducts, *Pharmacol. Res. Commun.*, 6, 581, 1974.

46. **Hasselberger, F. X., Allen, B., Paruchuri, E. K., Charles, M., and Coughlin, R. W.**, Immobilized enzymes. Lactase bonded to stainless steel and other dense carriers for use in fluidized bed reactors, *Biochem. Biophys. Res. Commun.*, 57, 1054, 1974.

47. **Kennedy, J. F., Barker, S. A., and White, C. A.**, Immobilization of α-amylase on polyaromatic and titanium compounds incorporating a magnetic material, *Stärke*, 29, 240, 1977.

48. **Lai, T.-S. and Lan, M.-J.**, Some properties of covalently bound α-amylase and the effect of α-amylolysis on starch saccharification, *J. Chin. Biochem. Soc.*, 5, 7, 1976; *Chem. Abstr.*, 85, 173272, 1976.

49. **Nakhapetyan, L. A., Sosedova, O. N., Menyailova, I. I., Vainer, L. M., Zhandov, S. P., and Koromal'di, E. V.**, α-Amylase immobilized by covalent binding to modified porous glass, *Fermentn. Spirt. Prom-st.*, 6, 39, 1976; *Chem. Abstr.*, 85, 190665, 1976.

50. **Tertykh, V. A., Yanishpol'skii, V. V., Chuiko, A. A., Galich, I. P., Tsyperovich, A. S., and Koval'chuk, T. A.**, Immobilization of α-amylase on the surface of highly dispersed silica, *Dopov. Akad. Nauk Ukr. RSR Ser. B, Geol. Khim. Biol. Nauki*, p. 651, 1977.

51. **Hehre, E. J., Okada, G., and Genghof, D. S.**, Configurational specificity: unappreciated key to understanding enzymatic reversions and *de novo* glycosidic bond synthesis. I. Reversal of hydrolysis by α-, β- and glucoamylases with donors of correct anomeric form, *Arch. Biochem. Biophys.*, 135, 75, 1969.

52. **Maeda, H. and Yamauchi, A.**, New technology in the use of radiation. Application to enzyme insolubilization, *Genshiryoku Kogyo*, 19, 49, 1973; *Chem. Abstr.*, 80, 79640, 1974.

53. **Maeda, H., Yamauchi, A., and Suzuki, H.**, Preparation of immobilized enzymes by γ-ray irradiation, *Biochim. Biophys. Acta*, 315, 18, 1973.

54. **Maeda, H., Suzuki, H., Yamauchi, A., and Sakimae, A.**, Preparation of immobilized enzymes from acrylic monomer under γ-ray irradiation, *Biotechnol. Bioeng.*, 17, 119, 1975.

55. **Gruesbeck, C. and Rase, H. F.**, Insolubilized glucoamylase enzyme system for continuous production of dextrose, *IEC Prod. Res. Dev.*, 11, 74, 1972.

56. **Beck, S. R. and Rase, H. F.**, Encapsulated enzyme: a glucoamylase copolymer system, *IEC Prod. Res. Dev.*, 12, 260, 1973.

57. **Jaworek, D.**, New immobilization techniques and supports, in *Enzyme Engineering*, Vol. 2, Pye, E. K. and Wingard, L. B., Jr., Eds., Plenum Press, New York, 1974, 105.

58. **Maeda, H., Suzuki, H., and Yamauchi, A.**, Preparation of immobilized enzymes using poly(vinyl alcohol), *Biotechnol. Bioeng.*, 15, 607, 1973.

59. **Maeda, H., Suzuki, H., and Yamauchi, A.** Preparation of immobilized enzymes by electron-beam irradiation, *Biotechnol. Bioeng.*, 15, 827, 1973.

60. **Maeda, H., Suzuki, H., Yamauchi, A., and Sakimae, A.**, Preparation of immobilized enzymes by N-vinylpyrrolidone and the general properties of the glucoamylase gel, *Biotechnol. Bioeng.*, 16, 1517, 1974.
61. **Corno, C., Galli, G., Morisi, F., Bettonte, M., and Stopponi, A.**, Glucoamylase trapped into cellulosic fibres. Properties and use, *Stärke*, 24, 420, 1972.
62. **Gregoriadis, G., Leathwood, P. D., and Ryman, B. E.**, Enzyme entrapment in liposomes, *FEBS Lett.*, 14, 95, 1971.
63. **Stavenger, P. L.**, Putting semipermeable membranes to work, *Chem. Eng. Prog.*, 67(3), 30, 1971.
64. **Madgavkar, A. M., Shah, Y. T., and Cobb, J. T.**, Hydrolysis of starch in a membrane reactor, *Biotechnol. Bioeng.*, 19, 1719, 1977.
65. **Usami, S. and Taketomi, N.**, Activities of enzyme adsorbed on adsorbents. I, *Hakko Kyokaishi*, 23, 267, 1965; *Chem. Abstr.*, 63, 11939e, 1965.
66. **Usami, S., Yamada, T., and Kimura, A.**, Activities of enzymes adsorbed on adsorbents. II. Stability of adsorbed enzyme, *Hakko Kyokaishi*, 25, 513, 1967; *Chem. Abstr.*, 68, 84499, 1968.
67. **Kimura, A., Shirasaki, H., and Usami, S.**, Kinetics of enzymes adsorbed on adsorbent, *Kogyo Kagaku Zasshi*, 72, 489, 1969.
68. **Usami, S. and Shirasaki, H.**, Kinetics of enzymes adsorbed on adsorbent, *J. Ferment. Technol.*, 48, 506, 1970.
69. **Usami, S. and Inoue, S.**, Kinetics and continuous reaction of glucoamylase adsorbed on adsorbent, *Asahi Garasu Kogyo Gijutsu Shoreikai Kenkyu Hokoku*, 25, 39, 1974; *Chem. Abstr.*, 83, 203225, 1975.
70. **Usami, S., Matsubara, M., and Noda, J.**, Activities of enzymes adsorbed on adsorbents. III. Continuous reaction by the glucoamylase-activated charcoal complex, *Hakko Kyokaishi*, 29, 195, 1971; *Chem. Abstr.*, 76, 123528, 1972.
71. **Solomon, B. and Levin, Y.**, Adsorption of amyloglucosidase on inorganic carriers, *Biotechnol. Bioeng.*, 17, 1323, 1975.
72. **Fukui, T. and Nakagawa, C.**, Properties and utilization of glucoamylase bound with carboxymethylcellulose and hydroxyapatite, *Dempun Kagaku*, 19(2), 51, 1972; *Chem. Abstr.*, 83, 117524, 1975.
73. **Caldwell, K. D., Axén, R., and Porath, J.**, Reversible immobilization of enzymes to hydrophobic agarose gels, *Biotechnol. Bioeng.*, 18, 433, 1976.
74. **Caldwell, K. D., Axén, R., Bergwall, M., and Porath, J.**, Immobilization of enzymes based on hydrophobic interaction. II. Preparation and properties of an amyloglucosidase adsorbate, *Biotechnol. Bioeng.*, 18, 1589, 1976.
75. **Krasnobajew, V. and Böniger, R.**, Application possibilities of PAG-immobilized enzymes in the starch industry, *Chimia*, 29, 123, 1975.
76. **Eskamani, A., Chase, T., Jr., Freudenberger, J., and Gilbert, S. G.**, Determination of protein immobilized on solid support by tryptophan content, *Anal. Biochem.*, 57, 421, 1974.
77. **Miyamoto, K., Fujii, T., and Miura, Y.**, On the insolubilized enzyme activities studied using adsorbents and ion exchangers, *J. Ferment. Technol.*, 49, 565, 1971.
78. **Kučera, J. and Hanus, J.**, Preparation of carboxymethyl-cellulose gels and their use for immobilization of amyloglucosidase, *Collect. Czech. Chem. Commun.*, 40, 2536, 1975.
79. **Miyamoto, K., Fujii, T., Tamaoki, N., Okazaki, M., and Miura, Y.**, Intraparticle diffusion in the reaction catalyzed by immobilized glucoamylase, *J. Ferment. Technol*, 51, 566, 1973.
80. **Anon.**, Immobilized enzymes. I. The preparation of immobilized glucoamylase by adsorption on DEAE-Sephadex, *Wei Sheng Wu Hseuh Pao*, 13, 25, 1973.
81. **Solomon, B. and Levin, Y.**, Studies on adsorption of amyloglucosidase on ion-exchange resin, *Biotechnol. Bioeng.*, 16, 1161, 1974.
82. **Bachler, M. J., Strandberg, G. W., and Smiley, K. L.**, Starch conversion by immobilized glucoamylase, *Biotechnol. Bioeng.*, 12, 85, 1970.
83. **Smiley, K. L.**, Continuous conversion of starch to glucose with immobilized glucoamylase, *Biotechnol. Bioeng.*, 13, 309, 1971.
84. **Emery, A., Sorenson, J., Kolarik, M., Swanson, S., and Lim, H.**, An annular bound-enzyme reactor, *Biotechnol. Bioeng.*, 16, 1359, 1974.
85. **Gembicka, D., Janicki, J., and Nowacka, I.**, Method of obtaining and characteristics of glucoamylase combined with DEAE-cellulose, *Rocz. Technol. Chem. Zywn.*, 24, 193, 1974; *Chem. Abstr.*, 83, 24328, 1975.
86. **Gembicka, D., Nowacka, I., and Janicki, J.**, Application of glucoamylase bound with DEAE-cellulose for hydrolysis of starch by continuous method, *Acta Aliment. Pol.*, 1, 33, 1975; *Chem. Abstr.*, 83, 181405, 1975.
87. **Oreshkin, E. N., Nakhapetyan, L. A., and Vainer, L. M.**, Preparation of immobilized glucoamylase by ion exchange sorption on DEAE-cellulose, *Appl. Biochem. Microbiol.*, 10, 753, 1974.

88. **Monosov, E. Z., Oreshkin, E. N., and Nakhapetyan, L. A.**, Study on the localization of glucoamylase during ion exchange immobilization on DEAE-cellulose, *Biol. Nauki (Moscow)*, 19, 91, 1976; *Chem. Abstr.*, 85, 30112, 1975.
89. **Chen, L. F. and Tsao, G. T.**, Physical characteristics of porous cellulose beads as supporting material for immobilized enzymes, *Biotechnol. Bioeng.*, 18, 1507, 1976.
90. **Chen, L. F. and Tsao, G. T.**, Chemical procedures for enzyme immobilization of porous cellulose beads, *Biotechnol. Bioeng.*, 19, 1463, 1977.
91. **Li, K.-H., Chang, S.-K., Sun, W.-R., Ku, S.-F., Yang, L.-W., Li, S.-F., and Yang, K.-Y.**, Immobilized enzymes. II. Preparation of a p-aminobenzenesulfonylethyl cellulose-glucoamylase complex, *Wei Sheng Wu Hsueh Pao*, 13, 31, 1973.
92. **Park, Y. K. and Lima, D. C.**, Continuous conversion of starch to glucose by an amyloglucosidase-resin complex, *J. Food Sci.*, 38, 358, 1973.
93. **Park, Y. K.**, Enzymic properties of a fungal amyloglucosidase-resin complex, *J. Ferment. Technol.*, 52, 140, 1974.
94. **Brown, E., Racois, A., Joyeau, R., Bonte, A., and Rioual, J.**, Preparation and properties of insoluble chymotrypsin derivatives, *C. R. Acad. Sci. Ser. C*, 273, 668, 1971.
95. **Krämer, D. M., Lehmann, K., Plainer, H., Reisner, W., and Sprössler, B. G.**, Enzymes covalently bound to acrylic gel beads. II. Practical application of hydrolases covalently bound to hydrophilic anionic gel beads, *J. Polym. Sci. Polym. Symp.*, 47, 89, 1974.
96. **Gestrelius, S., Mattiasson, B., and Mosbach, K.**, Studies on pH-activity profiles of an immobilized two-enzyme system, *Biochim. Biophys. Acta*, 276, 339, 1972.
97. **Swanson, S. J., Emery, A., and Lim, H. C.**, Pore diffusion in packed-bed reactors containing immobilized glucoamylase, *AIChE J.*, 24, 30, 1978.
98. **Brillouet, J.-M., Coulet, P. R., and Gautheron, D. C.**, Thin layer-flow reactor with amyloglucosidase bound to collagen membranes, *Biotechnol. Bioeng.*, 18, 1821, 1976.
99. **Brillouet, J.-M., Coulet, P. R., and Gautheron, D. C.**, Chemically activated collagen for amyloglucosidase attachment. Use in a helicoidal reactor, *Biotechnol. Bioeng.*, 19, 125, 1977.
100. **Solomon, B. and Levin, Y.**, Studies on the binding of amyloglucosidase to inert proteins, *Biotechnol. Bioeng.*, 16, 1393, 1974.
101. **Barker, S. A., Emery, A. N., and Novais, J. M.**, Enzyme reactors for industry, *Process Biochem.*, 6(10), 11, 1971.
102. **Maeda, H. and Suzuki, H.**, Studies on water-insoluble enzyme. IV. Activities of glucoamylase chemically fixed to cyanogen bromide-activated cellulose, *Agric. Biol. Chem.*, 36, 1839, 1972.
103. **Gray, C. J. and Livingston, C. A.**, Properties of enzymes immobilized by the diazotized m-diaminobenzene method, *Biotechnol. Bioeng.*, 19, 349, 1977.
104. **Barker, S. A., Somers, P. J., and Epton, R.**, Preparation and stability of exoamylolytic enzymes chemically coupled to microcrystalline cellulose, *Carbohydr. Res.*, 9, 257, 1969.
105. **Maeda, H. and Suzuki, H.**, Studies on the water-insoluble enzyme. III. Preparation and general properties of glucoamylase bound to halogenacetyl cellulose, *Agric. Biol. Chem.*, 36, 1581, 1972.
106. **Roth, P., Feist, U., Flemming, Ch., Gabert, A., and Täufel, A.**, Synthesis and properties of immobilized enzymes. VII. Coupling of glucoamylase to dialdehyde cellulose, carboxymethylcellulose hydrazide, and carboxymethylcellulose azide, *Acta Biol. Med. Ger.*, 36, 179, 1977.
107. **Maeda, H. and Suzuki, H.**, Studies on the water-insoluble enzyme. I. General properties of CM-cellulose glucoamylase, *Nippon Nogei Kagaku Kaishi*, 44, 547, 1970.
108. **Maeda, H., Miyado, S., and Suzuki, H.** Studies on the water-insoluble enzyme. II. Continuous saccharification of liquid starch, *Hakko Kyokaishi*, 28, 391, 1970; *Rept. Ferm. Res. Inst.*, 42, 17, 1972.
109. **Kučera, J.**, Continuous hydrolysis of soluble starch by immobilized amyloglucosidase (E.C.3.2.1.3), *Collect. Czech. Chem. Commun.*, 41, 2978, 1976.
110. **Kučera, J. and Hanus, J.**, Kinetics of starch splitting by immobilized amyloglucosidase, *Folia Microbiol. (Prague)*, 21, 209, 1976.
111. **Christison, J.**, Preparation of immobilized glucoamylase, *Chem. Ind.*, 215, 1972.
112. **Kvesitadze, G. I., Tokhadze, Z. V., Dvali, M. Sh., Bregvadze, Ts. R., and Svanidze, R. S.**, Immobilization of glucoamylase on aminoethylcellulose, *Isv. Akad. Nauk Gruz. SSR, Ser. Biol.*, 2, 82, 1976; *Chem. Abstr.*, 85, 74269, 1976.
113. **Wilson, R. J. H. and Lilly, M. D.**, Preparation and use of insolubilized amyloglucosidase for the production of sweet glucose liquors, *Biotechnol. Bioeng.*, 11, 349, 1969.
114. **O'Neill, S. P., Dunnill, P., and Lilly, M. D.**, A comparative study of immobilized amyloglucosidase in a packed bed reactor and a continuous feed stirred tank reactor, *Biotechnol. Bioeng.*, 13, 337, 1971.
115. **Stanley, W. L., Watters, G. G., Kelly, S. H., and Olson, A. C.**, Glucoamylase immobilized on chitin with glutaraldehyde, *Biotechnol. Bioeng.*, 20, 135, 1978.

116. **Olson, A. C. and Stanley, W. L.**, Lactase and other enzymes bound to a pheno-formaldehyde resin with glutaraldehyde, *J. Agric. Food Chem.*, 21, 440, 1973.
117. **Inman, D. J. and Hornby, W. E.**, Preparation of some immobilized linked enzyme systems and their use in the automated determination of disaccharides, *Biochem. J.*, 137, 25, 1973.
118. **Ledingham, W. M. and Ferreira, M. do S. S.**, The preparation and properties of amyloglucosidase chemically attached to polystyrene beads, *Carbohydr. Res.*, 30, 196, 1973.
119. **Baum, G.**, Enzyme immobilization on macroreticular polystyrene, *Biotechnol. Bioeng.*, 17, 253, 1975.
120. **Hough, J. S. and Lyons, T. P.**, Couplings of enzymes onto microorganisms, *Nature (London)*, 235, 389, 1972.
121. **Johnson, D. B. and Costelloe, M.**, Glucoamylase immobilization on hornblende, *Biotechnol. Bioeng.*, 18, 421, 1976.
122. **Flynn, A. and Johnson, D. B.**, Immobilized glucoamylase: studies on hornblende and Enzacryl TIO, *Int. J. Biochem.*, 8, 501, 1977.
123. **Kvesitadze, G. I., Gvalia, T. Sh., Svanidze, R. S., Tokhadze, Z. V., and Nutsubidze, N. N.**, Immobilization of *Aspergillus* glucoamylase on inorganic carriers, *Bioorg. Khim.*, 3, 836, 1977.
124. **Kvesitadze, G. I., Tokhadze, Z. V., Dvali, M. Sh., Bregvadze, Ts. R., and Fraikina, T. Ya.**, Immobilization of glucoamylase on silicate carriers, *Isv. Akad. Nauk Gruz. SSR, Ser. Biol.*, 1, 154, 1975; *Chem. Abstr.*, 83, 174764, 1975.
125. **Flemming, Ch., Gabert, A., and Wand, H.**, Synthesis and properties of insolubilized enzymes. V. Covalent coupling of trypsin, glucoamylase, peroxidase, aminoacrylase, and alkaline phosphatase to γ-isothiocyanatopropyl-diethoxysilyl glass, *Acta Biol. Med. Ger.*, 32, 135, 1974.
126. **Weetall, H. H. and Havewala, N. B.**, Continuous production of dextrose from corn starch. A study of reactor parameters necessary for commercial application, in *Enzyme Engineering*, Wingard, L. B., Jr., Ed., Wiley-Interscience, New York, 1972, 241.
127. **Weetall, H. H., Havewala, N. B., Garfinkel, H. M., Buehl, W. M., and Baum, G.**, Covalent bond between the enzyme amyloglucosidase and a porous glass carrier. The effect of shearing, *Biotechnol. Bioeng.*, 16, 169, 1974.
128. **Marsh, D. R., Lee, Y. Y., and Tsao, G. T.**, Immobilized glucoamylase on porous glass, *Biotechnol. Bioeng.*, 15, 483, 1973.
129. **Lai, T.-S. and Hsu, F.-F.**, Insolubilization of glucoamylase by covalent attachment to porous silica glass beads, *J. Chin. Biochem. Soc.*, 4, 36, 1975; *Chem. Abstr.*, 84, 175866, 1976.
130. **Nakhapetyan, L. A., Menyailova, I. I., Zhandov, S. P., Koromal'di, E. V., and Antonov, V. K.**, Immobilization of glucoamylase on porous glass, *Fermentn. Spirt. Prom-st.*, 1975(1), 37; *Chem. Abstr.*, 82, 120914, 1975.
131. **Nakhapetyan, L. A., Menyailova, I. I., Zhandov, S. P., and Koromal'di, E. V.**, Properties of glucoamylase immobilized on inorganic supports, *Fermentn. Spirit. Prom-st.*, 1976(5), 35; *Chem. Abstr.*, 85, 139200, 1976.
132. **Marsh, D. R. and Tsao, G. T.**, A heat transfer study on packed bed reactors of immobilized glucoamylase, *Biotechnol. Bioeng.*, 18, 339, 1976.
133. **Nakhapetyan, L. A. and Menyailova, I. I.**, Hydrolysis of concentrated starch solutions using immobilized glucoamylase, *Fermentn. Spirt. Prom-st.*, 2, 41, 1976; *Chem. Abstr.*, 85, 18957, 1976.
134. **Kvesitadze, G. I., Tokhadze, Z. V., Dvali, M. Sh., and Bregvadze, Ts. R.**, Immobilization of glucoamylase on silica gel by the glutaraldehyde method, *Isv. Akad. Nauk Gruz. SSR, Ser. Khim.*, 1, 396, 1975; *Chem. Abstr.*, 84, 117766, 1976.
135. **Swanson, S. J., Emery, A., and Lim, H. C.**, Cycle time and carrier life in immobilized-glucoamylase reactors, *J. Solid-Phase Biochem.*, 1, 119, 1976.
136. **Lee, D. D., Lee, Y. Y., Reilly, P. J., Collins, E. V., Jr., and Tsao, G. T.**, Pilot plant production of glucose with glucoamylase immobilized to porous silica, *Biotechnol. Bioeng.*, 18, 253, 1976.
137. **Chang, H. N. and Reilly, P. J.**, Experimental operation of immobilized multienzyme systems in optimal and suboptimal configurations, *Biotechnol. Bioeng.*, 20, 243, 1978.
138. **Lee, D. D., Lee, G. K., Reilly, P. J., and Lee, Y. Y.**, Effect of pore diffusion limitation on the products of enzymatic dextrin hydrolysis, *Biotechnol. Bioeng.*, 22, 1, 1980.
138a. **Lee, G. K.**, Ph.D. Dissertation, Iowa State University, Ames, 1978.
139. **Lee, E. Y. C. and Whelan, W. J.**, Glycogen and starch debranching enzymes, in *The Enzymes*, Vol. 5, 3rd ed., Boyer, P. D., Ed., Academic Press, New York, 1971, chap. 7.
140. **Closset, G. P., Cobb, J. T., and Shah, Y. T.**, Study of performance of a tubular membrane reactor for an enzyme catalyzed reaction, *Biotechnol. Bioeng.*, 16, 345, 1974.
141. **Tachauer, E., Cobb, J. T., and Shah, Y. T.**, Hydrolysis of starch by a mixture of enzymes in a membrane reactor, *Biotechnol. Bioeng.*, 16, 545, 1974.

142. Caldwell, K. D., Axén, R., and Porath, J., Utilization of hydrophobic interaction for the formation of an enzyme reactor bed, *Biotechnol. Bioeng.*, 17, 613, 1975.
143. Caldwell, K. D., Axén, R., Bergwall, M., and Porath, J., Immobilization of enzymes based on hydrophobic interaction. I. Preparation and properties of β-amylase adsorbate, *Biotechnol. Bioeng.*, 18, 1573, 1976.
144. Caldwell, K. D., Axén, R., Bergwell, M., Olsson, I., and Porath, J., Immobilization of enzymes based on hydrophobic interaction. III. Adsorbent substituent density and its impact on the immobilization of β-amylase, *Biotechnol. Bioeng.*, 18, 1605, 1976.
145. Axén, R. and Porath, J., Chemical coupling of enzymes to cross-linked dextran (Sephadex), *Nature (London)*, 210, 367, 1966.
146. Mårtensson, K., Preparation of an immobilized two-enzyme system, β-amylase-pullulanase, to an acrylic copolymer for the conversion of starch to maltose. I. Preparation and stability of immobilized β-amylase, *Biotechnol. Bioeng.*, 16, 567, 1974.
147. Mårtensson, K., Preparation of an immobilized two-enzyme system, β-amylase-pullulanase, to an acrylic copolymer for the conversion of starch to maltose. II. Cocoupling of the enzymes and use in a packed bed column, *Biotechnol. Bioeng.*, 16, 579, 1974.
148. Mårtensson, K., Preparation of an immobilized two-enzyme system, β-amylase-pullulanase, to an acrylic copolymer for the conversion of starch to maltose. III. Process kinetic studies on continuous reactors, *Biotechnol. Bioeng.*, 16, 1567, 1974.
149. Vretblad, P. and Axén, R., Preparation and properties of an immobilized barley β-amylase, *Biotechnol. Bioeng.*, 15, 783, 1973.
150. Schell, H. D., Cornoiu, I., and Mateescu, M. A., Covalent binding of β-amylase to CNBr-activated agarose, to CH- and AH-Sepharose 4B, and the catalytic properties of the resulting conjugate, *Rev. Roum. Biochim.*, 11, 199, 1974.
151. Maeda, H., Tsao, G. T., and Chen, L. F., Preparation of immobilized soybean β-amylase on porous cellulose beads and continuous maltose production, *Biotechnol. Bioeng.*, 20, 383, 1978.
152. Cook, D. and Ledingham, W. M., The effect of immobilization on the action pattern of β-amylase, *Biochem. Soc. Trans.*, 3, 996, 1975.
153. Hon, C. C. and Reilly, P. J., Properties of β-amylase immobilized to alkylamine porous silica, *Biotechnol. Bioeng.*, 21, 505, 1979.
154. Mårtensson, K. and Mosbach, K., Covalent coupling of pullulanase to an acrylic copolymer using a water-soluble carbodiimide, *Biotechnol. Bioeng.*, 14, 715, 1972.
155. Lampen, J. O., Yeast and *Neurospora* invertases, in *The Enzymes*, Vol. 5, 3rd ed., Boyer, P. D., Ed., Academic Press, New York, 1971, chap. 10.
156. Michaelis, L. and Menten, M. L., The kinetics of invertase action, *Biochem. Z.*, 49, 333, 1913.
157. Griffin, E. G. and Nelson, J. M., The influence of certain substances on the activity of invertase, *J. Am. Chem. Soc.*, 38, 722, 1916.
158. Nelson, J. M. and Griffin, E. G., Adsorption of invertase, *J. Am. Chem. Soc.*, 38, 1109, 1916.
159. Nelson, J. M. and Hitchcock, D. J., The activity of adsorbed invertase, *J. Am. Chem. Soc.*, 43, 1956, 1921.
160. Kreen, M., Köstner, A., and Kask, K., Preparation of invertase immobilized on polyacrylamide gel, *Tr. Tallin. Politekh. Inst. Ser. A.*, 300, 21, 1971; *Chem. Abstr.*, 79, 63040, 1973.
161. Kreen, M., Köstner, A., and Kask, K., Kinetics of invertase immobilized in polyacrylamide gel, *Tr. Tallin. Politekh. Inst.*, 331, 117, 1973; *Chem. Abstr.*, 82, 53355, 1975.
162. Kreen, M., Köstner, A., and Kask, A., Use of bound invertase for sucrose hydrolysis, *Tr. Tallin. Politekh. Inst.*, 331, 131, 1973; *Chem. Abstr.*, 82, 41857, 1975.
163. Krosing, V., Jarvet, J., Siimer, E., and Köstner, A., Kinetics of hydrolysis of sucrose aqueous solutions catalyzed by an invertase immobilized in polyacrylamide gel, *Tr. Tallin. Politekh. Inst.*, 367, 17, 1974; *Chem. Abstr.*, 83, 128164, 1975.
164. Krosing, V., Jarvet, J., Siimer, E., and Köstner, A., Thermal inactivation of soluble invertase and invertase immobilized in a polyacrylamide gel, *Tr. Tallin. Politekh. Inst.*, 383, 17, 1975; *Chem. Abstr.*, 85, 89269, 1976.
165. Usami, S. and Kuratsu, Y., Preparation and properties of water-insoluble saccharase immobilized on polyacrylamide gel, *J. Ferment. Technol.*, 51, 789, 1973.
166. Nakagawa, H., Arao, T., Matsuzawa, T., Ito, S., Ogura, N., and Takahana, H., Comparision of the properties of tomato β-fructofuranosidase embedded within a polyacrylamide gel and adsorbed on CM-cellulose, *Agric. Biol. Chem.*, 39, 1, 1975.
167. Kawashima, K. and Umeda, K., Studies on immobilized enzymes by radiopolymerization. III. A method for preparing bead-shaped immobilized enzymes, *Agric. Biol. Chem.*, 40, 1143, 1976.
168. Kawashima, K. and Umeda, K., Studies on the immobilized enzymes by radiopolymerization. IV. Preparation of immobilized invertase and its characteristics, *Agric. Biol. Chem.*, 40, 1151, 1976.

169. **Krosing, A., Treimann, R., and Köstner, A.**, Immobilization of invertase in reinforced polyacrylamide gel, *Tr. Tallin. Politekh. Inst.*, 402, 15, 1976; *Chem. Abstr.*, 86, 167135, 1977.
170. **O'Driscoll, K. F., Izu, M., and Korus, R.**, Gel entrapment of enzymes, *Biotechnol. Bioeng.*, 14, 847, 1972.
171. **Fukui, S., Tanaka, A. Iida, T., and Hasegawa, E.** Application of photo-crosslinkable resin to immobilization of an enzyme, *FEBS Lett.*, 66, 179, 1976.
172. **Tanaka, A., Yasuhara, S., Fukui, S., Iida, T., and Hasegawa, E.**, Immobilization of invertase using crosslinkable resin oligimers and properties of the immobilized enzyme, *J. Ferment. Technol.*, 55, 71, 1977.
173. **Dinelli, D.**, Fibre-entrapped enzymes, *Process Biochem.*, 7(8), 9, 1972.
174. **Dinelli, D. and Morisi, F.**, Fiber-entrapped enzymes, in *Enzyme Engineering*, Vol. 2, Pye, E. K. and Wingard, L. B., Jr., Eds., Plenum Press, New York, 1974, 293.
175. **Marconi, W., Gulinelli, S., and Morisi, F.**, Properties and use of invertase trapped in fibers, *Biotechnol. Bioeng.*, 16, 501, 1974.
176. **Toda, K. and Shoda, M.**, Sucrose inversion by immobilized yeast cells in a complete mixing reactor, *Biotechnol. Bioeng.*, 17, 481, 1975.
177. **Toda, K.**, Intraparticle mass transfer study with a packed column of immobilized microbes, *Biotechnol. Bioeng.*, 17, 1729, 1975.
178. **Monsan, P. and Durand, G.**, New preparation of enzyme fixed on inorganic supports, *C. R. Acad. Sci. Ser. C*, 273, 33, 1971.
179. **Boudrant, J. and Cheftel, C.**, Continuous sucrose hydrolysis using soluble immobilized invertase, *Biochimie*, 55, 413, 1973.
180. **Bowski, L. and Ryu, D. Y.**, Determination of invertase activity during ultrafiltration, *Biotechnol. Bioeng.*, 16, 697, 1974.
181. **Korus, R. A. and Olson, A. C.**, The use of α-galactosidase and invertase in hollow fiber reactors, *Biotechnol. Bioeng.*, 19, 1, 1977.
182. **Monsan, P. and Durand, G.**, Preparation of insolubilized invertase by fixation or bentonite, *FEBS Lett.*, 16, 39, 1971.
183. **Negoro, H.**, Continuous inversion of sucrose by using insoluble sucrase, *J. Ferment. Technol.*, 48, 689, 1970.
184. **Negoro, H.**, Continuous inversion of sucrose by using insoluble saccharase. II. Enzymatic properties of soluble fractions derived from insoluble saccharase treated with urea, *J. Ferment. Technol.*, 50, 136, 1972.
185. **Vieth, W. R., Gilbert, S. G., and Wang, S. S.**, Performance of collagen-invertase complex membrane in a biocatalytic module, *Trans. N.Y. Acad. Sci. Ser. 2*, 34, 454, 1972.
186. **Wang, S. S. and Vieth, W. R.**, Collagen-enzyme complex membranes and their performance in biocatalytic modules, *Biotechnol. Bioeng.*, 15, 93, 1973.
187. **Fernandes, P. M., Constantinides, A., Vieth, W. R., and Venkatasubramanian, K.**, Enzyme engineering. V. Modeling and optimizing multienzyme reactor systems, *Chem. Technol.*, 5, 438, 1975.
188. **Mattiasson, B.**, A general enzyme thermistor based on specific reversible immobilization using the antigen-antibody interaction. Assay of hydrogen peroxide, penicillin, sucrose, glucose, phenol and tyrosine, *FEBS Lett.*, 77, 107, 1977.
189. **Suzuki, H., Ozawa, Y., and Maeda, H.**, Hydrolysis of sucrose by insoluble yeast invertase, *Agric. Biol. Chem.*, 30, 807, 1966.
190. **Suzuki, H., Ozawa, Y., Maeda, H., and Tanabe, O.**, Studies on the water-soluble enzyme. Hydrolysis of sucrose by insoluble yeast invertase, *Rep. Ferment. Res. Inst.*, 31, 11, 1967.
191. **Usami, S., Noda, J., and Goto, K.**, Preparation and properties of water-insoluble saccharase, *J. Ferment. Technol.*, 49, 598, 1971.
192. **Lilly, M. D., O'Neill, S. P., and Dunnill, P.**, Bioengineering of immobilized enzymes, *Biochimie*, 55, 985, 1973.
193. **Kirstein, D. and Besserdich, H.**, Specific yields of tank-type enzyme reactors, *Chem. Tech. (Leipzig)*, 29, 259, 1977.
194. **Dickensheets, P. A., Chen, L. F., and Tsao, G. T.**, Characteristics of yeast invertase immobilized on porous cellulose beads, *Biotechnol. Bioeng.*, 19, 365, 1977.
195. **Maeda, H., Suzuki, H., and Sakimae, A.**, Preparation of immobilized invertase, *Biotechnol. Bioeng.*, 15, 403, 1973.
196. **Rosetti, M.**, The dynamics of invertase adsorption on ion-exchange resins, *Farmacia, 11, 469, 1963; Chem. Abstr.*, 61, 893b, 1964.
197. **Boudrant, J. and Cheftel, C.**, Continuous hydrolysis of sucrose by invertase adsorbed in a tubular reactor, *Biotechnol. Bioeng.*, 17, 827, 1975.

198. **Kobayashi, T. and Moo-Young, M.**, Kinetics and mass transfer behavior of immobilized invertase on ion-exchange resin beads, *Biotechnol. Bioeng.*, 15, 47, 1973.
199. **Johnson, D. E. and Ciegler, A.**, Substrate conversion by fungal spores entrapped in solid matrixes, *Arch. Biochem. Biophys.*, 130, 384, 1969.
200. **Manecke, G. and Singer, S.**, Reactions on copolymers of methacrylic acid fluoranilide, *Makromol. Chem.*, 39, 13, 1960.
201. **Besserdich, H., Kahrig, E., Krenz, R., and Kirstein, D.**, Kinetic studies on substrate inhibition of free and bound invertase, *J. Mol. Catal.*, 2, 361, 1977.
202. **Brown, E. and Joyeau, R.**, Immobilized enzymes. VI. Preparation of invertase and glucose oxidase immobilized on vanacryls, *Makromol. Chem.*, 175, 1961, 1974.
203. **Saini, R., Vieth, W. R., and Wang, S. S.**, Preparation and characterization of several forms of immobilized invertase, *Trans. N.Y. Acad. Sci. Ser. 2*, 34, 664, 1972.
204. **Venkatasubramanian, K. and Vieth, W. R.**, Effect of pressure on the hydrolysis of sucrose immobilized on collagen, *Biotechnol. Bioeng.*, 15, 583, 1973.
205. **Satoh, I., Karube, I., and Suzuki, S.**, Enzyme electrode for sucrose, *Biotechnol. Bioeng.*, 18, 269, 1976.
206. **Finley, J. W. and Olson, A. C.**, Automated method for measuring added sucrose in sweetened cereal products with immobilized invertase, *Cereal Chem.*, 52, 500, 1975.
207. **Filippusson, H. and Hornby, W. E.**, Preparation and properties of yeast β-fructofuranosidase chemically attached to polystyrene, *Biochem. J.*, 120, 215, 1970.
208. **Monsan, P. and Durand, G.**, Continuing use of invertase fixed on solid supports. Possibilities of use in agricultural and food industries, *Ann. Technol. Agric.*, 21, 555, 1972.
209. **Mason, R. D. and Weetall, H. H.**, Invertase covalently coupled to porous glass: Preparation and characterization, *Biotechnol. Bioeng.*, 14, 637, 1972.
210. **Marrazzo, W. N., Merson, R. L., and McCoy, B. J.**, Enzyme immobilized in a packed bed reactor: Kinetic parameters and mass transfer effects, *Biotechnol. Bioeng.*, 17, 1515, 1975.
211. **Thornton, D., Francis, A., Johnson, D. B., and Ryan, P. D.**, The immobilization of lactoperoxidase and β-fructofuranosidase on glass and on sand, by the metal-link method, *Biochem. Soc. Trans.*, 2, 137, 1974.
212. **Thornton, D., Byrne, M. J., Flynn, A., and Johnson, D. B.**, The immobilization of enzymes on inorganic supports, *Biochem. Soc. Trans.*, 2, 1360, 1974.
213. **Thornton, D., Flynn, A., Johnson, D. B., and Ryan, P. D.**, The preparation and properties of hornblende as a support for immobilized invertase, *Biotechnol. Bioeng.*, 17, 1679, 1975.
214. **Flynn, A. and Johnson, D. B.**, The immobilization of invertase on hornblende and on Enzacryl-TIO: Optimization and stability studies, *Int. J. Biochem.*, 8, 243, 1977.
215. **Van Leemputten, E. and Horisberger, M.**, Immobilization of enzymes on magnetic particles, *Biotechnol. Bioeng.*, 16, 385, 1974.
216. **Laurinavicyus, V. A. and Kulys, Yu. Yu.**, Immobilization of enzymes on hydrated oxides of transition metals and aluminum, *Appl. Biochem. Microbiol.*, 13, 346, 1977.
217. **Barman, T. E.**, *Enzyme Handbook*, Vol. 2, Springer-Verlag, New York, 1969.
218. **Suzuki, H., Ozawa, Y., Oota, H., and Yoshida, H.**, Studies on the decomposition of raffinose by α-galactosidase of mold. I. α-Galactosidase formation and hydrolysis of raffinose by the enzyme preparation, *Agric. Biol. Chem.*, 33, 506, 1969.
219. **Kobayashi, H. and Suzuki, H.**, Studies on the decomposition of raffinose by α-galactosidase of mold. II. Formation of mold pellet and its enzyme activity, *J. Ferment. Technol.*, 50, 625, 1972.
220. **Kobayashi, H. and Suzuki, H.**, Kinetic studies of α-galactosidase-containing mold pellets on PNPG hydrolysis, *Biotechnol. Bioeng.*, 18, 37, 1976.
221. **Thananunkul, D., Tanaka, M., Chichester, C. O., and Lee, T.-C.**, Degradation of raffinose and stachyose in soybean milk by α-galactosidase from *Mortierella vinacea*. Entrapment of α-galactosidase within polyacrylamide gel, *J. Food Sci.*, 41, 173, 1976.
222. **Smiley, K. L., Hensley, D. E., and Gasdorf, H. J.**, Alpha-galactosidase production and use in a hollow-fiber reactor, *Appl. Environ. Microbiol.*, 31, 615, 1976.
223. **Reynolds, J. H.**, An immobilized α-galactosidase continuous flow reactor, *Biotechnol. Bioeng.*, 16, 135, 1974.
224. **Faulstich, H., Schäfer, A., and Weckauf-Bloching, M.**, α- and β-galactosidases bound to nylon nets, *FEBS Lett.*, 481, 226, 1974.

Chapter 6

APPLICATIONS OF LACTASE AND IMMOBILIZED LACTASE

Robert W. Coughlin and Marvin Charles

TABLE OF CONTENTS

I. Abstract 154

II. Introduction 154

III. Sources of Lactases 155

IV. Applications to Milk 155

V. Applications to Whey 158
 A. Foods and Beverages 158
 B. Fermentation of LH Whey 159

VI. Technology of Lactase Application 163
 A. Immobilized Lactase 164

VII. Regulatory Aspects 169

Acknowledgment 171

References 171

I. ABSTRACT

This review treats the sources of the enzyme lactase and its use to hydrolyze the lactose present in milk and whey. The properties, advantages, applications, and potential use of such lactose-hydrolyzed (LH) milk or whey are also covered, including the manufacture of fermented food products from LH milk and use of LH whey as a medium for the commercial production of ethanol and other commodities. A significant portion of the paper is devoted to a review of the technology of applying immobilized lactase to the processing of milk and whey. It concludes with a brief discussion of some FDA regulatory aspects.

II. INTRODUCTION

The enzyme lactase (ECN 3.2.1.23), so named because it catalyzes the hydrolysis of milk sugar (lactose), is more formally designated β-D-galactosidase (or β-D-galactoside galactohydrolase) owing to its general ability to accelerate hydrolysis of the β-D-galactoside bond between D-galactose and an alcohol. Because the major source of lactose is milk the significant practical applications of the enzyme lactase are in the treatment of milk and milk products in order to convert the disaccharide lactose therein into its simple constituent monose sugars: glucose and galactose. The structural formulas for lactose, glucose, and galactose are shown in the rendering of the hydrolysis reaction

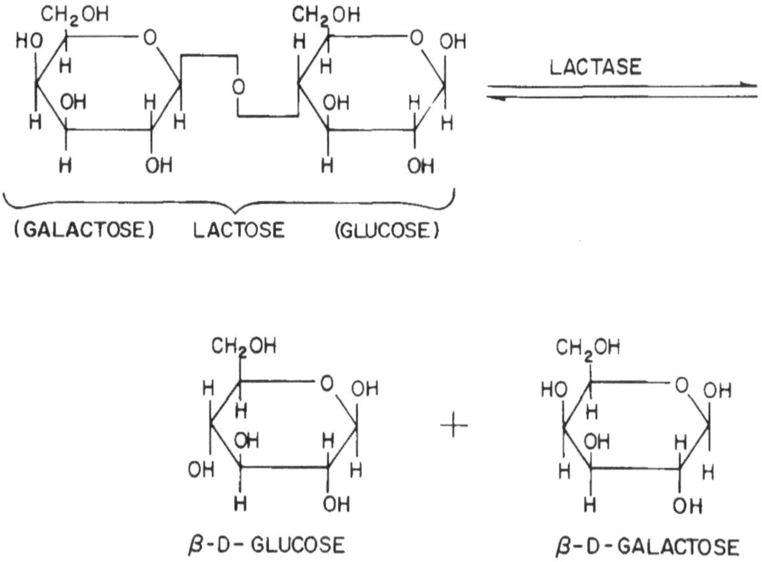

in Figure 1. One of the major economic incentives for conducting this hydrolysis is the increased sweetness conferred by the glucose and galactose as compared to lactose; this is evident from a comparison of the relative sweetness of these sugars in reference to sucrose as shown in Table 1.

Although milk and whey (the clear supernatant remaining after coagulation of curd to make cheese) are the largest-volume dairy products rich in lactose, they nevertheless contain only about 4 to 5% lactose by weight. Because of this the economic value of the sweetness added by lactose hydrolysis in some cases may be insufficient to satisfy the cost of enzymatic hydrolysis. As discussed below there are other technologically favorable effects of such hydrolysis such as:

Table 1
RELATIVE SWEETNESSES OF 10% AQUEOUS SUGAR SOLUTIONS

Sugar	Sweetness (%)
Sucrose	100
Lactose	40
Glucose	75
Galactose	70

1. Glucose is a more readily fermentable sugar than lactose.
2. Glucose and galactose are more soluble than lactose.
3. A large segment of the human population cannot digest lactose and these people are troubled by the consumption of lactose-containing products.
4. If lactose hydrolysis can convert whey from a waste stream to a utilizable product, concomitant savings in waste treatment costs can be realized.

III. SOURCES OF LACTASES

Although lactase activity is found in plants (especially grains and leguminous seeds) and animals, the only sources of expected commercial viability are microorganisms. In particular, the two major commercially available lactases are that from the fungus *Aspergillus niger* (Wallerstein Laboratories) and that from the yeast *Saccharomyces lactis* (Enzyme Development Corporation). The *S. lactis* lactase has a pH optimum of 6.8 to 7.0 which suits it ideally for application to milk (pH 6.6) and sweet whey (pH 6.1) whereas the *A. niger* enzyme has a pH optimum of 4.0 to 4.5 which makes it well suited for treating acid whey. The *S. lactis* enzyme deactivates rapidly below pH 6 and therefore is unsuitable for acid whey whereas the *A. niger* enzyme, although stable over a wide pH range, shows relatively low enzymatic activity above about pH 4.5 and therefore is unsuitable for treating milk or sweet whey. In general the pH optima of microbial lactases fall in the following ranges: from fungi (pH 2.5 to 4.5); from yeasts (pH 6.0 to 7.0) and from bacteria (pH 6.5 to 7.5). It seems that little effort has been expended to develop and exploit bacterial lactases.

For treating large volumes of liquids such as milk and whey, reactors containing immobilized-enzyme catalysts are best suited from the standpoints of economics and process technology. To date, *A. niger* lactase has been successfully bound to a variety of solid supports[6] and such preparations have shown operational half-lives of as much as 100 days, depending on conditions. On the other hand, efforts to immobilize the *S. lactis* enzyme, although successful, have produced catalyst products possessing operational half-lives too short for practical consideration.[8, 9]

IV. APPLICATIONS TO MILK

Lactose-hydrolyzed (LH) milk may be consumed directly or may be used to produce products as shown schematically in Figure 2. Essentially all the work with LH milk has been conducted using soluble *S. lactis* enzyme of compatible pH optimum because no practically promising immobilized forms of this enzyme have yet been prepared.

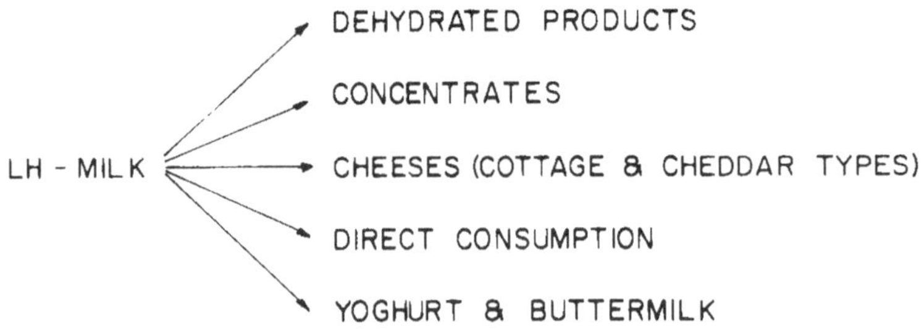

Nevertheless, the potentially useful and beneficial aspects of LH milk are sufficiently large that they represent a strong inducement to the development of suitable long-lived immobilized enzymes from yeasts or bacteria and amenable to the almost neutral pH of milk; for this reason such beneficial aspects of using LH milk are summarized here.

Hydrolysis of the lactose in milk causes several changes of potential value in the manufacture and marketing of dairy products. These changes include: increased sweetness, inceased solubility of the carbohydrate sugars, broader fermentation possibilities, more ready fermentation of these sugars, and reduced lactose concentration with associated diminished possibility of lactose crystallization. Such effects can be harnessed not only for increasing product quality but also to permit the development of a new market — viz. those consumers who are deficient in lactase (i.e., lactose intolerant) and who normally eschew dairy products. Milk intolerance, which seems to be widespread in non-Caucasians, is attributed to low levels of intestinal lactase and is marked by impairment of the normal digestive process and gastrointestinal discomfort such as flatulence, cramps, and diarrhea.[10,11] In testing the organoleptic properties of LH milk, Holsinger and Roberts[15] found that trained judges assessed milk with 30, 60, and 90%, respectively, of its lactose hydrolyzed as equivalent to a control milk containing, respectively, 0.3, 0.6, and 0.9% added sucrose. Similarly Paige et al.[12] reported that Negro adolescents found milk with 90% of its lactose hydrolyzed an acceptable beverage although 56% of these subjects judged it to be sweeter than an unhydrolyzed control.

LH milk has demonstrated superior performance of promising economic advantage when used to make yoghurt, cottage cheese, and cheddar cheese. Not only is yoghurt from LH milk preferred over conventional yoghurt in taste tests (owing to increased sweetness) but the time required to reach the desired pH was reduced by about 40 min;[13] this is shown in Figure 3. In the same article[13] it is reported that in cottage cheese manufacture from LH milk the time to reach desired pH is reduced by about 135 min and curd yields are improved by 5 to 12% due to lessened curd shattering during cooking; Figure 4 shows the faster rate of acid development when LH milk is used for cottage cheese manufacture. In related work[14] it was shown that by using LH milk cheddar cheese could be manufactured which at 3 to 4 months had the flavor, body, and texture of 6 to 8 month-old cheese; also for cheddar cheese faster acid development led to reduced times for renneting, curd cooking, and cheddaring. Although similar increases in rate of acid development were also observed[13] when buttermilk was prepared from LH milk, the increased sweetness of the buttermilk was found to be organoleptically objectionable so it presently appears improbable that LH milk could serve as a suitable raw material for buttermilk production.

Frozen concentrates of milk usually tend to thicken and coagulate when stored owing to crystallization of lactose. Holsinger and Roberts,[15] however, found that the use of LH milk not only increased the storage stability of such concentrates but that the reconstituted concentrates showed no significant difference in flavor score in comparison with an unhydrolyzed control milk with sucrose added. Spray-drying LH milk

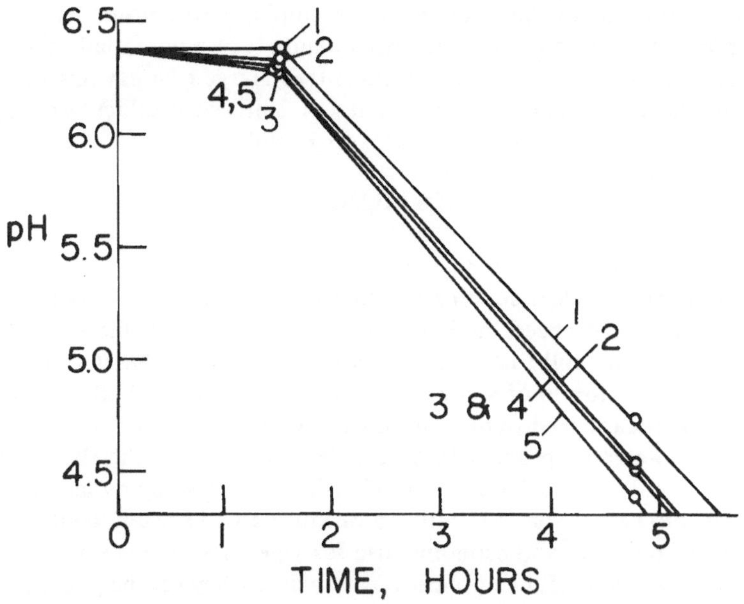

FIGURE 3. Acid development in the manufacture of yoghurt (45 C.) as a function of time for (1) 0, (2) 25, (3) 50, (4) 75, and (5) 90% hydrolyzed lactose milks.

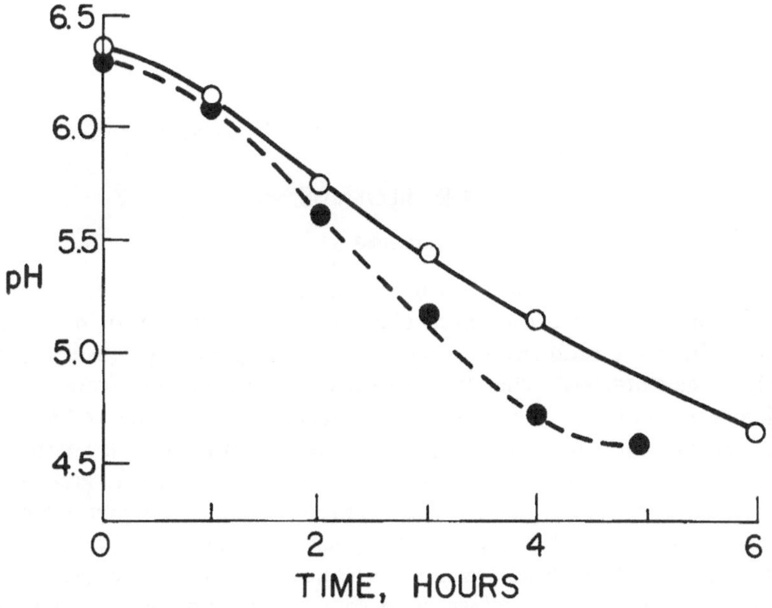

FIGURE 4. Acid development in the manufacture of cottage cheese (30C.) as a function of time for control milk (solid line) and 90% hydrolyzed lactose skim milk (dashed line).

has led to some problems with the product powder tending to stick to metal surfaces unless the apparatus surfaces are held below 60°C and the powder is cooled rapidly after collection.[16]

Aside from the suitability of LH milk itself for the lactase-deficient consumer, it is evident that products made therefrom, such as cheese and yogurt, will also be low in lactose and therefore might be suitable for developing new consumers of these dairy products among the lactose intolerant. This advantage of a potentially expanded market is an incentive quite aside from the potentially large-cost savings that might be realized from shortened manufacturing times of cultured products or certain other technological advantages inherent in the use of LH milk.

V. APPLICATIONS TO WHEY

A. Foods and Beverages

Whey is the relatively clear supernatant that remains after the coagulated casein is separated from milk for cheese making; the whey contains all the soluble minerals, proteins, and sugars of milk; its lactose content is 4.2 to 4.4%. Because little more than half of the 32 billion lb of whey produced annually in the U.S. is utilized, there appear to be significant environmental and economic incentives for converting this high-BOD waste water by-product into useable products and thereby returning this whey fraction of milk to the food chain. Modification of whey by lactose hydrolysis (or production of cheese from LH milk to produce an LH whey) could lead to new pathways toward practical and economic use for whey that is presently a waste product. Figure 5 shows some of the applications to which whey can be put. Essentially all

FIGURE 5.

of the whey which does already find use is so-called sweet whey (pH 6 to 7) which results from the manufacture of hard cheeses; acid whey (approximate pH 3 to 4) which results from the production of softer cheeses is presently considered essentially exclusively a waste material. The potential for utilizing acid whey seems greatest in an application where not only the sweetness produced by lactose hydrolysis but also its acidic properties would be beneficial, e.g., in the manufacture of sherbets or fruit-based drinks. In general, the increased solid solubility (in addition to greater sweetness) caused by lactase treatment can permit the preparation of a noncrystallizing, high-solids whey concentrate which might find application in producing a number of foods including ice cream. Whey has been used to make a variety of beverages ranging from liquid breakfasts to beers and wines, with some of these having protein contents as high as 3.5%. One of these called Rivella® (a sparkling, clear infusion of alpine herbs) has achieved commercial status with 20 to 30 million *ℓ* sold annually, mainly in Europe. Beverages made from whey have been reviewed exhaustively in a recent paper.[17]

Using LH wheys and lactose solutions Guy[18] prepared LH syrups by deproteinizing, demineralizing, and concentrating to 60% solids. These syrups showed good humectant properties, retained satisfactory solubilities, did not support mold growth, and as a humectant in caramel offered the same stabilizing effect in retarding sucrose crystallization as did invert sugar. These syrups were as sweet as concentrations of sucrose above 40% solids. Whey has been used in caramels[19] for years and in many other confections.[20] Guy et al. have also evaluated LH whey as an ingredient in ice cream[16] and in water ice.[21] In the case of ice cream[16] it was found that the addition of LH whey as 25% of the serum solids of the ice cream permitted sucrose concentrations to be reduced by 10%. Lowenstein[22] found that such ice cream formulations produced flavor scores equivalent to those of a control made with nonfat dry milk. In the case of the water ice,[21] cottage cheese whey was employed and provided a useful acidulation in addition to sweetness and other contributions. LH whey has also been used in making sherbets[23] containing 5% whey solids and it appears that LH acid whey could well supply the total acid requirements of sherbets.

B. Fermentation of LH Whey

Raw whey has found only limited use as an industrial fermentation substrate because most organisms of commercial interest cannot metabolize lactose to any practically significant extent. The growth of *Kluyveromyces fragilis* for animal feed is the only known large-scale use of dairy whey for fermentation processes. LH whey, on the other hand, will be an attractive substrate for the many commercially important species capable of metabolizing both glucose and galactose when strains and fermentation conditions are found which can overcome the usual problems associated with the growth of organisms on mixtures containing glucose and a less metabolizable carbohydrate such as galactose. The major problems include:

1. Galactose metabolism is usually slower than glucose metabolism.
2. A prolonged diauxic lag is often associated with the switch from primarily glucose metabolism to primarily lactose metabolism.
3. The products of galactose metabolism may not be the same as those of glucose metabolism.

O'Leary et al.[24] investigated the production of ethanol from LH whey by *K. fragilis* (NRRL Y-1109) and by *Saccharomyces cerevisiae* (ATCC 834). *K. fragilis* produced approximately 2% ethanol in 120 hr from an LH whey medium containing approximately 2% glucose, 2% galactose, and 1% lactose. As is illustrated in Figures 6 and 7, glucose was metabolized rapidly during the first 24 hr of fermentation, galactose was metabolized slowly throughout, and the alcohol production rate decreased markedly once all of the glucose had been consumed. When the same organism was grown on nonhydrolyzed whey there was no such decrease and the total fermentation time was decreased by 50 hr. Galactose-adapted *S. cerevisiae* produced approximately 1.7% alcohol in 120 hr when grown on the same LH whey medium but exhibited a diauxic lag of almost 50 hr between the disappearance of glucose and the initial metabolism of galactose (Figures 8 and 9). Furthermore, O'Leary et al.[24] found that if this strain was not adapted to galactose prior to the main fermentation on LH whey, galactose was not metabolized and only 0.85% alcohol was produced (Figure 10).

While these results are not particularly encouraging, it should be noted that O'Leary et al.[24] investigated only a small number of parameters for only two organisms and that there are many possibilities for improvement. Also, the results of their control experiments differed considerably from results reported by others[25-27] studying the fermentation of nonhydrolyzed whey by various yeasts.

Roland and Alm[28] have produced several wines by fermenting various combinations of grape juice concentrates and ultrafiltered LH whey with selected strains of *S. cerevisiae*. Fermentations were conducted at 18°C for 7 to 13 days (O'Leary et al.[24] used 30°C) and in all cases yielded between 10 and 12% alcohol. The case in which no grape juice was used is of particular interest. LH whey was fed to the fermentor on a daily

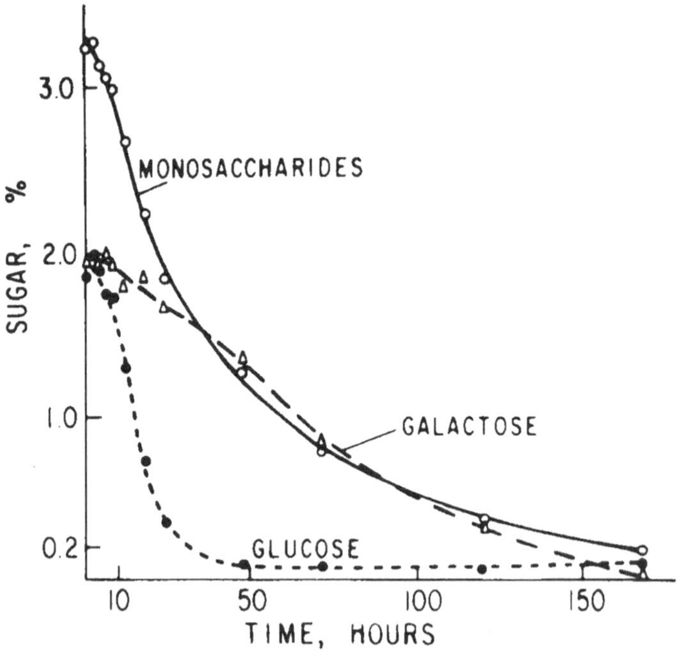

FIGURE 6. Sugar utilization patterns for *K. fragilis* grown on LH acid whey.[26]

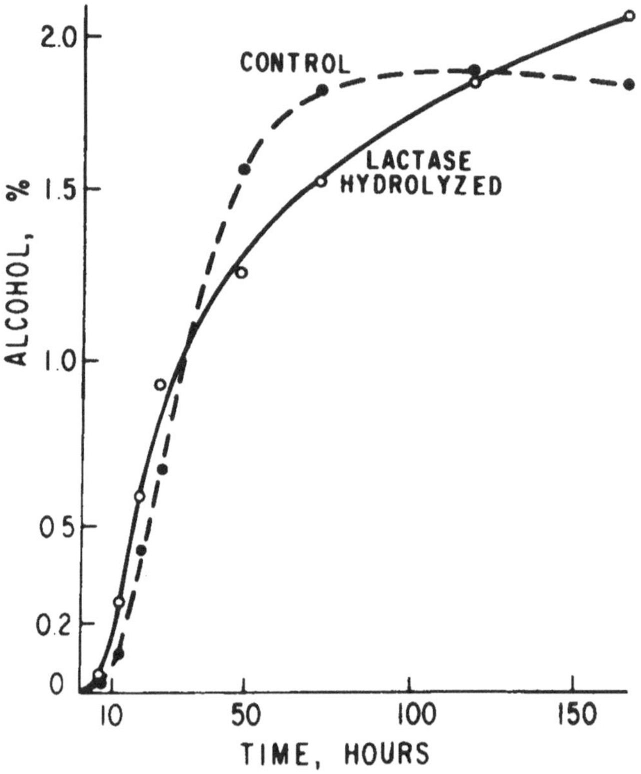

FIGURE 7. Alcohol production by *K. fragilis* grown on acid whey (control) and LH acid whey.[26]

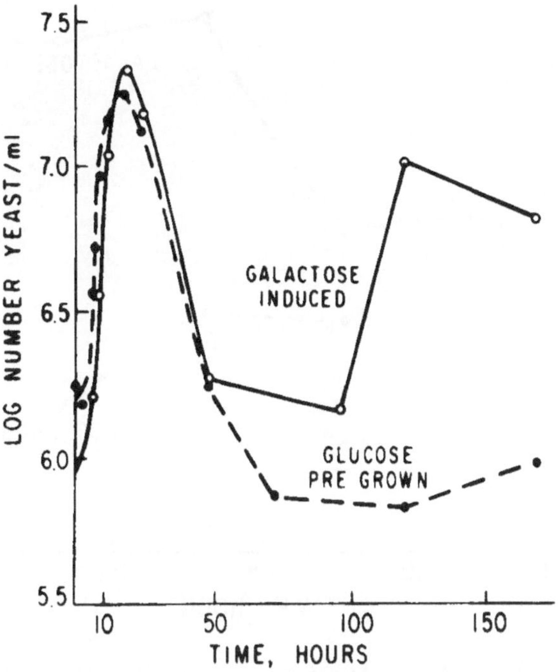

FIGURE 8. Growth of *S. cerevisiae* on LH whey.[26]

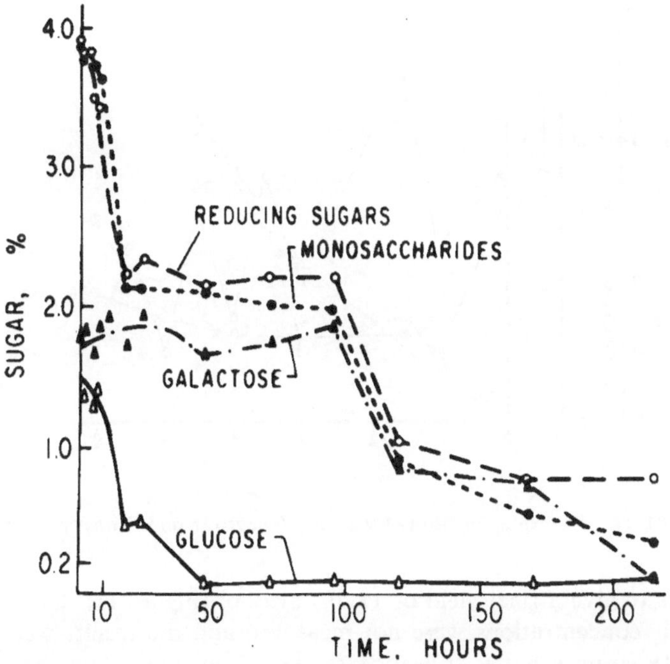

FIGURE 9. Sugar utilization pattern of galactose-pregrown *S. cerevisiae* grown on LH acid whey.[26]

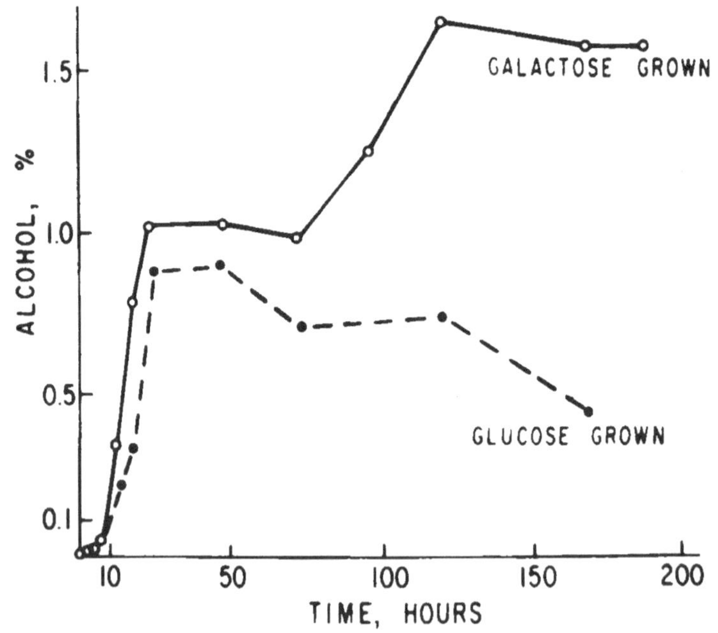

FIGURE 10. Alcohol production by *S. cerevisiae* grown on LH acid whey.[26]

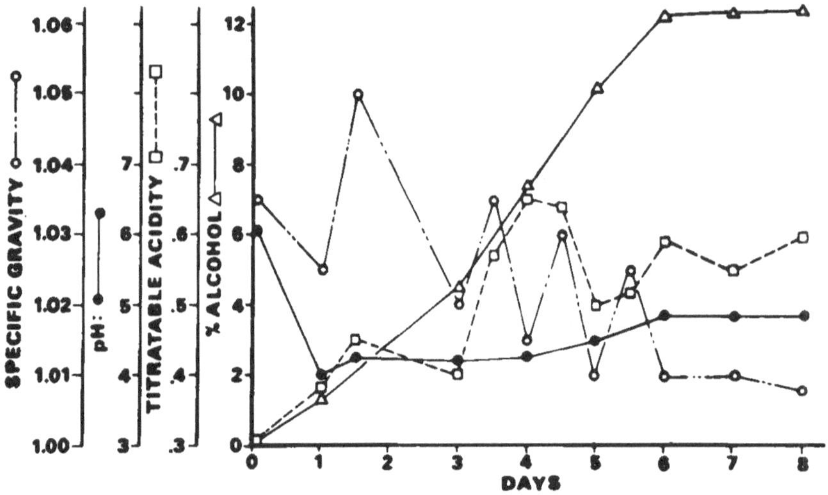

FIGURE 11. Fermentation history for *S. cerevisiae* grown on LH whey permeate.[30]

basis for 6 days to give a final yield of 12.5% alcohol (Figure 11). Unfortunately the individual sugar concentrations were not measured and the results were reported in terms of specific gravity but it still appears clear from their data that most of the galactose was consumed without the pronounced diauxic lag reported by O'Leary et al.[24] despite the fact that Roland and Alm carried their strains on glucose agar and grew their seed cultures in glucose mineral broth.

Charles and Radjai[29,30] cultured *Xanthomonas campestris* (NRRL B1459A) on ultrafiltered LH whey to produce xanthan gum in concentrations ranging from 1.5 to

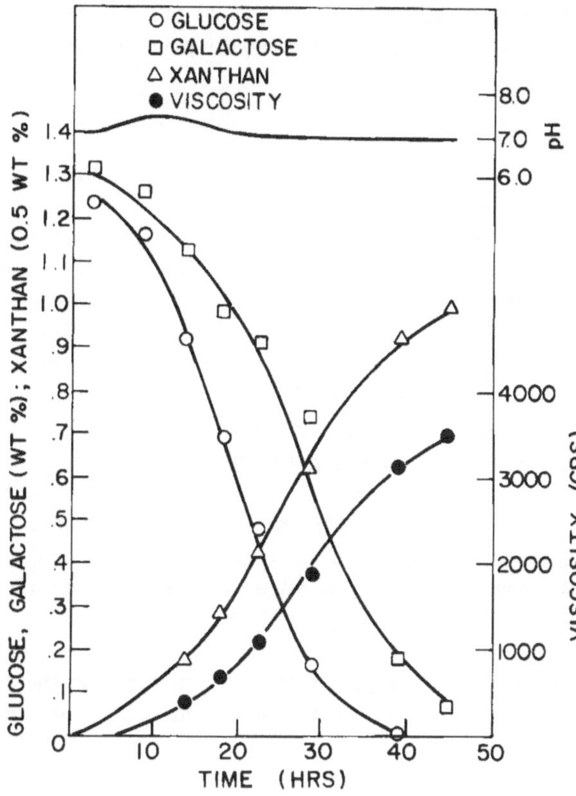

FIGURE 12. Xanthan gum production and sugar utilization pattern; *X. campestris* grown on LH acid whey (2-fold dilution).[32]

3.5% depending on the extent of dilution of the hydrolyzate. No obvious diauxesis was observed, fermentation times were comparable to those reported for media containing glucose as the only carbohydrate, and yields based on total sugar were as good as those obtained with conventional commercial glucose media. A typical fermentation history is given in Figure 12.

Lucas[31] used LH whey to produce an oriental-type soy sauce flavor by inoculating pasteurized LH whey with a selected strain of *S. cerevisiae* and fermenting the culture aerobically at 26°C for 12 to 24 hr; alternatively, the fermentation was conducted using a mixture of *S. cerevisiae* and a lactase-producing organism such as *S. lactis.*

Current technological and political trends make it clear that in many instances it will be possible to supply LH whey fermentation media under logistically favorable conditions at lower costs than those of comparable glucose media, and hence fermentations of LH whey deserve much greater consideration than they have received.

VI. TECHNOLOGY OF LACTASE APPLICATION

Soluble lactase,[32-34] freely dispersed nonviable lactase-containing microorganisms,[35-37] gel-entrapped lactase-containing microorganisms (probably viable),[38] fiber-entrapped lactase-containing organisms,[60] fungal spores,[38] and immobilized lactase[40,61] have all been studied for commercial hydrolysis of lactose in both milk and whey. Of these, immobilized-lactase technology has been studied and developed to the greatest extent and, in all likelihood, will be the method of choice for large-scale industrial applications unless there is a drastic reduction in the price of the enzyme. Only immobilized lactase will be discussed here.

A. Immobilized Lactase

A summary of selected reported studies of immobilized-lactase hydrolysis processes is given in Tables 2 and 3. For technical, design, operational, and economic considerations, tubular reactors containing particulate catalysts appear to be the best choice for large-scale hydrolysis of lactose in milk or whey. Such reactors are the basis of the most highly developed of the proposed lactose hydrolysis technologies and are being used successfully for other large-scale immobilized enzyme processes already commercialized (e.g., isomerization of glucose, resolution of racemic mixtures of amino acids).

The three tubular reactor processes which appear most promising are those developed by Olson and Stanley,[40,41] the Corning® Group,[44-46] and Coughlin and Charles.[47-53] All three use relatively inexpensive, highly active, very stable catalysts incorporating *Aspergillus niger* lactase and having pH optima between 3 and 4.5. All are very well suited for acid whey but only one[45] has been suggested for sweet whey or milk, but even in this case the evidence favoring the application of the *A. niger* enzyme to near neutral feed streams (of milk or sweet whey) is not compelling. Indeed, it appears that no commercially promising catalyst suitable for the hydrolysis of milk lactose has been developed yet.

The first two processes[40,41,44-46] employ fixed-bed reactors and therefore are suitable only for filtered or ultrafiltered whey. Furthermore, microbial contamination has been found to cause plugging problems even when deproteinized whey is processed.[4,40,41,43,45,47] Also, pressure drops across fixed-bed reactors containing small catalyst particles can be quite high even in the absence of blockages caused by particulates and microorganisms. The fluidized-bed reactor developed by Coughlin and Charles[47-53] is not subject to plugging or pressure drop problems regardless of the nature of the feed. However, the fluidized-bed reactor must be almost twice as long as a comparable fixed-bed and furthermore the nonplugging liquid distributors employed by Coughlin and Charles must be further developed to avoid serious departure from plug flow.[63]

Ford and Pitcher[46] have calculated that for the Corning process hydrolysis cost will be 1 to 4 ¢/lb of lactose hydrolyzed (1974 prices). The effects of plant capacity, catalyst cost, and degree of conversion on the cost of hydrolysis are illustrated in Figures 13, 14, and 15. These estimates assume the use of a catalyst (*A. niger* lactase bound covalently to either controlled pore glass or titania) having an initial activity of 300 IU /g (35°C) and a half-life of 62 days (35°C) in a fixed-bed reactor the temperature of which is raised continuously from 35°C to 50°C over a period of 559 days to compensate for loss in enzymic activity. It is also assumed that the whey will be first demineralized and ultrafiltered. The cost of demineralization, 2 to 5 ¢/lb, was not included in the estimate of hydrolysis cost. Ford and Pitcher stated that the additional cost is justified by their finding that deionized, deproteinized whey has an average half-life of 60 days in comparison to an average half-life of only 10 days for deproteinized whey. (The authors also have evidence that there are some negative effects of whey protein and whey salts on half-life but that the effects are smaller than those reported by Ford and Pitcher.) They also suggested that in some cases (primarily food applications) demineralization will be required to make the hydrolyzate saleable. If concentration is necessary, there will be an additional cost of 2 to 5 ¢/lb of lactose.

Paruchuri[47] has calculated that for the fluidized-bed process, the hydrolysis cost will be 8 to 9 ¢/lb of lactose for a plant processing an average of 100,000 lb of raw (no ultrafiltration or demineralization) acid whey daily. The effects of the cost of lactase

Table 2
TUBULAR REACTORS CONTAINING PARTICULATE CATALYSTS FOR LACTOSE HYDROLYSIS

			Catalyst characteristics			Reactor studies						
Enzyme source	Support	Immobilization method	Binding efficiency	Initial activity (IU/gm)	Half-life	Type	Capacity	Substrate(s)	Conversion	Temp.	Comments	Ref.
A. niger	Phenolic resin (Duolite® S-30) 10—50mesh	Adsorption + cross-linking with GA	75%	200—500 (45°C, pH4)	>4 wks. (45°C)	Fixed bed	30—300 mℓ/hr	Synthetic de-prot. whey	80%—96%	40—50°C	(1) Performance characteristics retained on 500× scale up. (2) Microbial contamination cited as formidable problem. Used dilute H_2O_2 and stored columns in cold to minimize microbial growth.	40, 41
Yeast	Phenolic resin Duolite® A-7	Covalent with triazinyl chloride	18%	—	>8 days	Fixed bed	SV = 3 (?)	Milk	100%	5°C		42
S. lactis	Comminuted hide collagen	Cross-linking with glutaraldehyde	—	—	>5 days (25°C)	Fixed bed	18L/hr	Sweet whey	85%	~25°C	(1) Plugging due to bacterial growth was a major obstacle even at low temperature despite attempts to sanitize column.	4
A. niger	CPG 120/200 mesh	Covalent	—	89	>165 days (55°C)	Fixed bed	1—5mℓ/min	Reconstituted acid whey	30—98%	55°C		3
S. lactis	CPG 120/200 mesh	Covalent	—	4.1	—	—	—	—	—	—		3
(Dairyland, Inc.)	CPG 40/80 mesh	Covalent	—	—	<3 days	Fixed bed	~2mℓ/min	Cheddar whey	< 38%	40°C	(1) Conversion dropped from 38% to 13% in 3 days. (2) Microbial growth caused overwhelming operational difficulties.	43
Fungal (miles)	CPG-ZrO_2 40/80 mesh	Covalent	—	50—100 (60°C)	60 days (40°C)	Fixed bed	1—2mℓ/min	5% Lactose 10% Lactose	80% (E/F = 3000)[a]	40°C 80% (E/F = 6000) 40°C	(1) Attempts to use this catalyst in fluidized bed were unsuccessful. (2) Neutralized, demineralized hydrolyzate was fed to reactor containing immobilized glucose isomerase. Final product was 1% lactose, 45% galactose, 20% fructose, and 25% glucose.	44
Yeast (Kyowa Hakko Kogyo Co.)	CPG-ZrO_2 40/80 mesh	Covalent	—	400—1100 (60°C)	60 days (30°C)	—	—	—	—	—	(1) Atypical yeast lactase-low pH optimum	44

Table 2 (continued)
TUBULAR REACTORS CONTAINING PARTICULATE CATALYSTS FOR LACTOSE HYDROLYSIS

			Catalyst characteristics			Reactor studies						
Enzyme source	Support	Immobilization method	Binding efficiency	Initial activity (IU/gm)	Half-life	Type	Capacity	Substrate(s)	Conversion	Temp.	Comments	Ref.
A. niger (Wallerstein)	ZrO_2-CPG 40/80 mesh	Covalent	25—80%	250—800	38 days (40°C)	Column	~1—2 mℓ/min	Lactose soln's., acid whey (UF), sweet whey (UF)	60—70%	30—50°C	(1) Suggested these might be suitable for sweet whey ($t_{1/2}$ = 53.7 days at 40°C) and possibly milk. (2) Microbial contamination stated to be a major problem — particularly in determination of $t_{1/2}$. (3) use of 2% gluteraldehyde and quaternary amines to sanitize industrial reactors suggested.	45
	Titania 40/80 mesh	Covalent	20—60%	200—500	25—29 days (40°C)	Column	~1—2 mℓ/min	Lactose soln's., acid whey (UF), sweet whey (UF)	60—70%	30—50°C		
A. niger (Wallerstein)	ZrO_2-CPG 40/80 mesh and Titania 40/80 mesh	Covalent	30—50%	~1150[a]	7 days[b,c] (60°C) 60 days[b,c] (50°C)	Column	4" × 18"	Lactose soln's; UF, demineralized whey	Equiv. to 380 IU/g; performance same as laboratory scale reactor	38°C	(1) Columns sanitized by backflushing daily for ½ hr with dilute acetic acid (pH 4). (2) Cost calculation presented in text.	46
A. niger (Wallerstein)	Alumina (100—150 μm)	Adsorp. + cross-link with glutaraldehyde	~50%	~700 (50°C)	50—60 days (50°C)	Fluidized bed	0.5L/min (3" × 6')	Lactose soln's; raw whey, UF whey	70—85%	50°C	(1) Columns sanitized daily with Iosan. (2) Particulates and microbial growth never a problem. (3) See text for cost estimate.	47-53

[a] E/F is the ratio of total enzyme activity to substrate volume metric flow rate expressed as activity units · min/cc. It is also referred to as the normalized residence time.
[b] Support to which value applies is not clear from data presented in reference.
[c] For deproteinized, deionized acid whey.

Table 3

REACTORS (OTHER THAN STANDARD TUBULAR) FOR LACTOSE HYDROLYSIS

Enzyme source	Immobilization method and reactor configuration	Capacity	Temp.	Conv.	Half-life	Comments	Ref.
E. coli (Worthington)	Entrapment in collagen membrane; spiral wound modular reactor	200 mℓ (reactor volume)	37°C	+100 %	Stable for 8 months — intermittent operation	(1) Pure lactose solutions (6%) used (2) Recycle operation only (3) Stable binding efficiency of 30%	54, 55
—	Microencapsulation	—	—	—	"Very stable"	(1) No details given (2) Suggested for use in vivo and in vitro	56
E. coli	Covalent bonding to polyisocyanate molded onto magnetic stirring bar	Small beaker	37°C	—	—	(1) Bar used for 43, 15 min - assays over period of 85 days (stored in cold when not used); lost only 14% of initial activity	57
E. coli (Worthington)	Covalent bonding to inside of nylon tube	0.3 cm × 40m 15 mℓ/hr	22°C	90% (20 hr)	>/wk (4°C)	(1) Reactor operated recycle; 300 mℓ skim milk at 9.3 cm/sec gave 90% conversion in 20 hr (2) Might have use in analytical applications	58
E. coli (Worthington)	Entrapped in collodion membrane disks; 39 supported disks (6" diam) mounted on rotating shaft — each disk in one compartment — axial flow from one compartment to next	5 gal/day	3—5°C	85%	>24 months (3—5°C)	(1) Mechanically complex but can handle substrates which are viscous and/or contain particles (2) Configuration well suited for multienzyme systems	59
E. coli & *S. fragilis*	Entrapped in cellulosic or other polymer fiber; fiber bundle in tubular reactor, woven fabric, others	100 mℓ/hr	25°C	75%	~94 days (yeast; skim milk substrate); ~100 days (*E. coli*; skim milk)	(1) Very long half-life (2) Microbial contamination removed easily (3) Can obtain very high enzyme loadings and binding efficiencies (4) Diffusion and hydrodynamic problems remain to be investigated	60—62

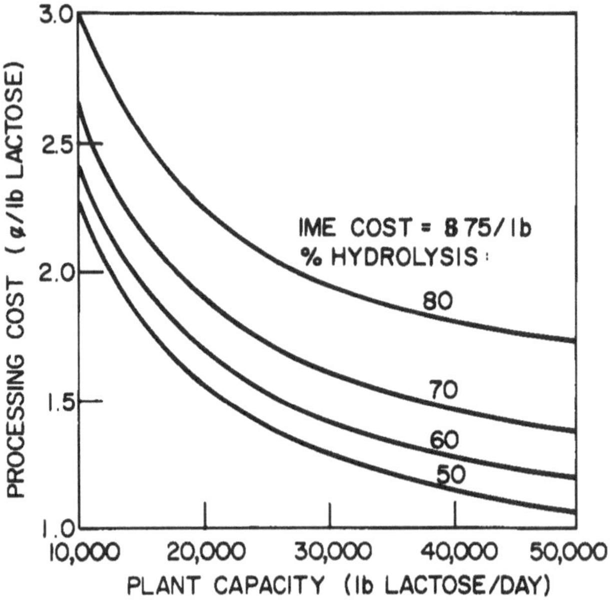

FIGURE 13. Effect of plant capacity on whey hydrolysis processing cost.[50]

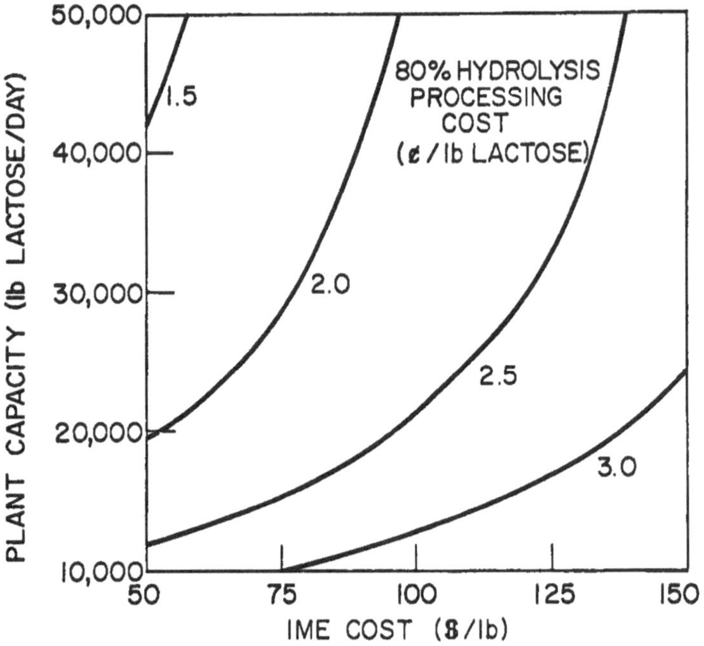

FIGURE 14. Whey hydrolysis processing cost as function of catalyst cost and plant capacity.[50]

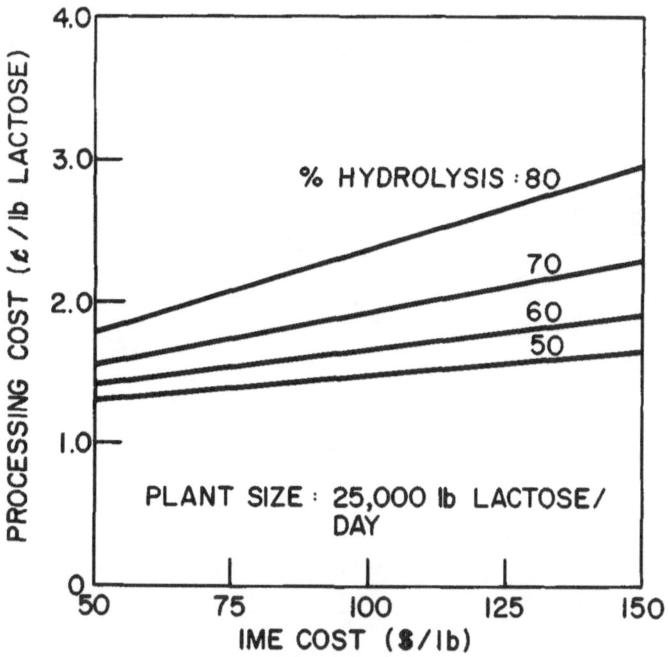

FIGURE 15. Dependence of whey hydrolysis processing cost on catalyst cost and conversion.[50]

and the useable lifetime of the immobilized enzyme catalyst on the cost hydrolysis are illustrated in Figure 16. The calculations, which are based on long-term experiments performed with 2.5-cm diameter columns processing whole raw whey, assume the use of a catalyst (*A. niger* lactase adsorbed on alumina and subsequently cross-linked with glutaraldehyde) having an average activity of 110 IU (37°C) and a half-life of 60 days in a reactor maintained at a temperature of 50°C. It was assumed also that the catalyst is replaced once every 60 days. Recent pilot plant experiments with whole whey and ultrafiltered whey indicate that the cost calculations were overly conservative[52, 53] and that 4 to 5 ¢/lb of lactose is a more realistic cost estimate. Also, whole whey or ultrafiltrate can be processed at the same cost. The costs of demineralization and concentration would be the same as those noted for the Corning process. Further work is required to determine if demineralization increases the half-life of the catalyst to such an extent that the cost is warranted.

Calculations of return on investment are so strongly dependent on individual circumstances, proposed use(s) of the hydrolyzate, and other market conditions that generalizations would be too highly speculative for presentation here. Coughlin and others have presented several specific economic case studies. It appears that there is still a great deal of commercial interest in large-scale immobilized enzyme lactose hydrolysis processes for both domestic and foreign use but that uncertainties in the sugar market have caused considerable hesitation. It is our expectation that these uncertainties will soon be overcome and that development of large-scale processes will proceed rapidly in the near future.

VII. REGULATORY ASPECTS

Enzymes are governed by the U.S. Food Drug and Cosmetic Act under the category of "food additives" in any instance where they become a component of food or are

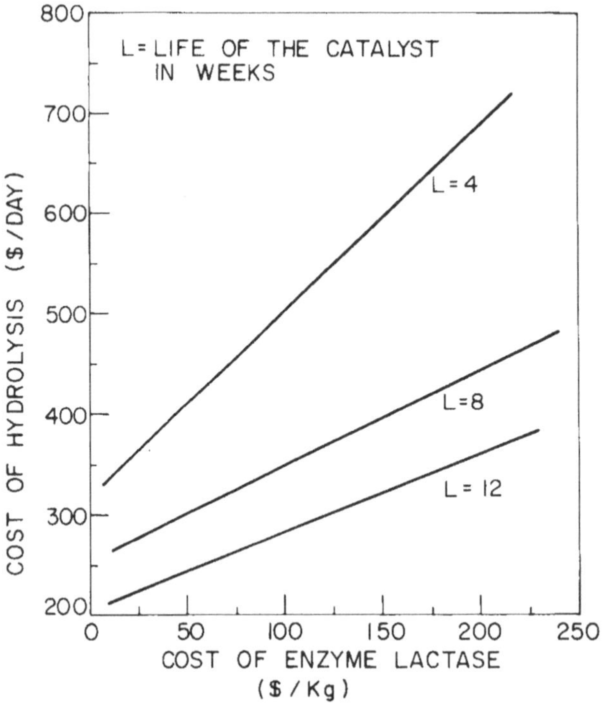

FIGURE 16. Cost of hydrolysis as a function of enzyme cost and useful life of IME.[51]

used as processing aids to affect the characteristics of food. The Act requires the review and regulation of such "food additives" by the Food and Drug Administration (FDA) unless the ingredient has been sanctioned specifically for such use in food by the U.S. Department of Agriculture (USDA) or FDA prior to 1958 *or* unless the ingredient is "generally recognized as safe" (GRAS). A number of microbial carbohydrase enzymes have been generally recognized as safe (GRAS) or given prior sanction for use in foods. As of June 1977,[64] the FDA was of the guarded opinion that the lactase enzymes derived from *A. niger* or from *S. fragilis* are GRAS but were reviewing the GRAS status of these enzymes on FDA initiative and also reviewing a petition seeking GRAS status for a lactase derived from *Saccharomyces lactis*, an enzyme which in 1968 could not be concluded to be GRAS based on "common use in food." GRAS status for an old substance must be based upon information that the substance was in common use in food or food manufacturing in the U.S. prior to 1958; for a new substance, GRAS status can be sought by a petition to FDA which contains adequate safety data for food additive approval, with simultaneous publication of such safety data so it can be generally reviewed, considered, and commented upon by the community of food-safety scientists. FDA is currently reviewing (both on its own initiative and in response to petitions) all available information on a large number of enzymes (including lactases) in order to assess, reconsider, and ultimately affirm (when the data justify such affirmation) the GRAS status of those enzymes.

The regulatory status of enzyme supports is less clear. Because the supports themselves presumably would not affect the characteristics of the food streams being processed it may be sufficient to provide incontrovertible evidence and safeguards to insure that the supports are 100% retained in the reactors and therefore would not become a component of the food stream. It seems that a safer approach however might be to utilize only those supports materials which already have or which could be admitted to the GRAS status, while at the same time taking all necessary precautions to prevent them from entering the food streams. Although further consideration of such matters is beyond the scope of the present paper they will be important concerns of manufacturers and processors engaged in the application of lactase for food processing.

ACKNOWLEDGMENT

Financial support for the preparation of this paper was provided, in part, by the National Science Foundation (RANN) through Grant No. AER72-03546 A06.

REFERENCES

1. **Woychik, J. H. and Holsinger, V. H.,** *Enzymes in Food and Beverage Processing, ACS Symposium Series No. 47, American Chemical Society, Washington, D.C., 1977, 67.*
2. **Hasselberger, F. X., Allen, B. R., Paruchuri, E. K., Charles, M., and Coughlin, R. W.,** Immobilized enzymes. Lactose bonded to stainless steel and other dense carriers for use in fluidized bed reactors, *Biochem. Biophys. Res. Commun.*, 57, 1054, 1974.
3. **Wierzbicki, L. E., Edwards, V. H., and Kosikowski, F. V.,** Hydrolysis of lactose in acid whey using β-galactosidase immobilized on porous glass particles: Preparation and characterization of a reusable catalyst for the production of low-lactose dairy products, *Biotech. Bioeng.*, 16, 397, 1974.
4. **Woychik, J. H. and Wondolowski, M. V.,** Lactose hydrolysis in milk and milk products by bound fungal β-galactosidase, *J. Milk Food Technol.*, 36, 31, 1973.
5. **Olson, A. C. and Stanley, W. L.,** Lactose and other enzymes bound to a phenol-formaldohyde resin with glutaraldehyde, *J. Agri. Food Chem.*, 21, 440, 1973.
6. **Hustad, G. O., Richardson, T. and Olson, N. F.,** Immobilization of β-galactosidase on an insoluble carrier with a poly-isocyanate polymer. I. Preparation and properties, *J. Dairy Sci.*, 56, 1111, 1973.
7. **Coughlin, R. W., Charles, M., Allen, B. R., Paruchuri, E. K., and Hasselberger, F. X.,** *AIChE Symp. Ser.*, No. 144, 70, 199 (1974).
8. **Woychik, J. H., Wondolowski, M. V., and Dahl, K. J.,** Preparation and application of immobilized β-galactosidase of *Saccharomyces lactis* in *Immobilized Enzymes in Food and Microbial Processes*, Olson, A. C. and Cooney, C. L., Eds., Plenum Press, New York, 1974.
9. **Coughlin, R. W. and Charles, M.,** unpublished work.
10. **Rosensweig, N. S.,** *J. Dairy Sci.*, 52, 585, 1969.
11. **Paige, D. M., Bayless, T. M., Ferry, G. C., and Graham, G. C.,** Lactose malabsorption and milk rejection in Negro children, *Johns Hopkins Med. J.*, 129, 163, 1971.
12. **Paige, D. M., Bayless, T. M., Huang, S., and Wexler, R.,** in *Physiological Effects of Food Carbohydrates*, ACS Symposium Series No. 15, American Chemical Society, Washington, D.C., 1975, 191.
13. **Gyuricsek, D. M. and Thompson, M. P.,** *Cult. Dairy Prod. J.*, 11, 12, Aug. 1976.
14. **Thompson, M. P. and Brower, D. P.,** *Cult. Dairy Prod. J.*, 11, 22, Feb. 1976.
15. **Holsinger, V. H. and Roberts, N. E.,** *Dairy and Ice Cream Field* 159(3), 30, 1976.
16. **Guy, E. J., Tamsma, A., Kontson, A., and Holsinger, V. H.,** Lactose-treated milk provides base to develop products for lactose-intolerant populations, *Food Prod. Develop.*, 8, 50, 1974.
17. **Holsinger, V. H., Posati, L. P., and DeVilbiss, E. D.,** Whey beverages: a review, *J. Dairy Sci.*, 57(8), 849, 1974.
18. **Guy, E. J.,** Paper D7, Abstract 71st Ann. Meeting Am. Dairy Sci. Ass., North Carolina State University, Raleigh, 1976.
19. **Atekonis, J. J.,** Paper D: ERRL Publication #3779, Proc. of the Whey Products Conf., Whey Products Institute, Chicago, No. 7, 1972.
20. **Webb, B. J.,** *J. Dairy Sci.*, 49, 1310, 1966.
21. **Guy, E. J., Vettel, H. E., and Pallansch, M. J.,** *J. Dairy Sci.*, 49, 1156, 1966.
22. **Loewenstein, M., Reddy, M. B., White, C. H., Speck, S. J., and Lunsford, T. A.,** Using cottage cheese whey fractions or their derivatives in ice cream, *Food Prod. Develop.*, 91, 1975.

23. Whittier, E. O. and Webb, B. H., *Byproducts From Milk*, Reinhold, New York, 1950, 223.
24. O'Leary, U. S., Green, R., Sullivan, B. C., and Holsinger, V. H., Alcohol production by selected yeast strains in lactose-hydrolyzed acid whey, *Biotechnol. Bioeng.*, 19, 1019, 1977.
25. Browne, H. H., *IEC (News Ed.*, 19, 1272, 1941.
26. Rogosa, M., Fermentation of lactose by yeasts, *J. Biol. Chem.*, 175, 314, 1948.
27. Charles, M., unpublished results.
28. Roland, J. F. and Alm, W. L., Wine fermentation using membrane processed hydrolyzed whey, *Biotechnol. Bioeng.*, 17, 1443, 1975.
29. Charles, M. and Radjai, M. K., paper presented at the First International Congress on Engineering and Food, Boston, August 1976.
30. Charles, M. and Radjai, M. K., paper presented at the 168th National ACS Meeting, San Francisco, August 1976.
31. Lucas, A. J., U.S. Patent 3,558,328, 1971.
32. Stimpson, E. G., U.S. Patent 2,668,765, 1954.
33. Guy, E. J., Ice cream manufacture with dairy products treated with lactose enzyme isolated from *Saccharomyces lactis*, *J. Dairy Sci.*, 56, 627, 1973.
34. Engel, W. G., *Food Proc.*, July, 1973, 32.
35. Stimpson, E. G., U.S. Patent 2,681,858, 1954.
36. Morgan, E. R., U.S. Patent 2,715,601, 1955.
37. Myers, R. P. and Stimpson, E. G., U.S. Patent 2,762,749, 1956.
38. Ohmiya, K., Ohashi, H., Kobayashi, T., and Shimizu, S., *Appl. and Env. Micro.*, 33, 137, 1977.
39. Charles, M., unpublished results.
40. Olson, A. C. and Stanley, W. L., The use of tannic acid and phenol-formaldehyde resins with glutaraldehyde to immobilize enzymes in *Immobilized Enzymes in Food and Microbial Processes*, Olson, A. C. and Cooney, C. L., Eds., Plenum Press, New York, 1974, 51.
41. Olson, A. C. and Stanley, W. L., Immobilization of enzymes on phenol-formaldehyde resins, in *Enzyme Engineering*, Vol. 2, Pye, E. K. and Wingard, L. B., Eds., Plenum Press, New York, 1974, 91.
42. Samejima, H. and Kimura, K., Immobilized enzymes using resinous carriers, in *Enzyme Engineering*, Vol. 2, Pye, E. K. and Wingard, L. B., Eds., Plenum Press, New York, 1974, 131.
43. Okos, E. S. and Harper, W. J., Activity and stability of β-galactosidase immobilized on porous glass, *J. Food Sci.*, 39, 88, 1974.
44. Weetall, H. H., Havewala, N. B., Pitcher, W. H., Jr., Detar, C. C., Vann, W. P., and Yaverbaum, S., The preparation of immobilized lactose and its use in the enzymatic hydrolysis of acid whey, *Biotechnol. Bioeng.*, 16, 295, 1974.
45. Weetall, H. H., Havewala, N. B., Pitcher, W. H., Jr., Detar, C. C., Vann, W. P., and Yaverbaum, S., The preparation of immobilized lactase and its use in the enzymatic hydrolysis of acid whey, *Biotechnol. Bioeng.*, 16, 689, 1974.
46. Ford, J. R. and Pitcher, W. H., Jr., Immobilized lactose systems, paper presented at Whey Products Conference, Rosemont, IL, Sept. 18, 1974.
47. Paruchuri, E. K., Ph.D. Dissertation, Lehigh University, Bethlehem, Pennsylvania, 1975.
48. Charles, M., Coughlin, R. W., Allen, B. R., Paruchuri, E. K., and Hasselberger, F. X., Lactose immobilized on stainless steel and other dense metal and metal oxide supports in *Immobilized Biochemicals and Affinity Chromatography*, Dunlap, R. B., Ed., Plenum Press, New York, 1974.
49. Coughlin, R. W., U.S. Patent 3,928,143, 1975.
50. Coughlin, R. W. and Charles, M., U.S. Patent 4,016,293, 1977.
51. Coughlin, R. W. and Charles, M., paper presented at the 5th North American Meeting of the Catalysis Society, Pittsburgh, April, 1977.
52. Charles, M. and Coughlin, R. W., Fluidized Bed Reactors for Immobilized Enzymes, paper presented at NSF/RANN Grantees Conference, University of Virginia, Charlottesville, May, 1976.
53. Charles, M. and Coughlin, R. W., unpublished results.
54. Eskamani, A., Gilbert, S. G., Leeder, J. G., and Vieth, W. R., paper presented at 75th National AIChE Meeting, Detroit, June 3-6, 1973.
55. Bernath, F. R. and Vieth, W. R., Collagen as a carrier for enzymes: materials science and process engineering aspects of enzyme engineering in *Immobilized Enzymes in Food and Microbial Processes*, Olson, A. C., and Cooney, C. L., Eds., Plenum Press, New York, 1974, 157.
56. Chang, T. M. S., Effects of different routes of in vivo administration of microencapsulated enzymes in *Enzyme Engineering*, Vol. 2, Pye, E. K. and Wingard, L. B., Eds., Plenum Press, New York, 1974, 419.

57. **Richardson, T. and Olson, N. F.**, Immobilized enzymes in milk systems in *Immobilized Enzymes in Food and Microbial Processes*, Olson, A. C. and Cooney, C. L., Eds., Plenum Press New York, 1974, 19.
58. **Ngo, T. T., Narinesingh, D., and Laidler, K. J.**, Hydrolysis of D-galactosidase in an open tubular lactose reactor, *Biotechnol. Bioeng.*, 18, 119, 1976.
59. **Worthington, C. C.**, The use of membrane-bound enzymes in an immobilized enzyme reactor in *Immobilized Biochemicals and Affinity Chromatography*, Dunlap, R. B. Ed., Plenum Press, New York, 1974, 235.
60. **Dinelli, D.**, Fibre-entrapped enzymes, *Process Biochem.*, 7, 9, 1972.
61. **Dinelli, D. S., Marconi, W., and Morisi, F.**, Fiber-entrapped enzymes in *Immobilized Enzymes, Antigens, Antibodies and Peptides*, Weetall, H. H., Ed., Marcel Dekker, New York, 1975, chap. 5.
62. **Dinelli, D. S.**, U.S. Patent 3,715,277, 1974.
63. **Gomaa, H., Charles, M., and Coughlin, R. W.**, submitted to *Biotechnol. Bioeng.*, 1979.
64. **Miles, C. I.**, paper presented at the 37th Annual Meeting of the Institute of Food Technologists, Philadelphia, June 6, 1977.

Chapter 7

IMMOBILIZED PROTEASES — POTENTIAL APPLICATION

Howard H. Weetall

TABLE OF CONTENTS

I. Introduction 176

II. Food Industries 176
  A. Dairy Industry 176
  B. Soya Hydrolysis 179
  C. Beer Chillproofing 179

III. Enzyme Inactivation with Immobilized Proteases 181

IV. Potential New Application of Immobilized Proteases 181
  A. Synthesis 181
  B. Biological Switches 181

V. Summary 182

References 182

I. INTRODUCTION

The application of immobilized enzymes to the food and beverage industry has long been recognized as a major area of endeavor. Enzymes such as glucose isomerase, glucoamylase, and lactase have been studied in depth and their applications extensively published.[1-9]

Within the food and drug industry there is also interest in the hydrolysis of protein. The reasons for this interest are the changes hydrolysis can cause in the characteristics of the treated protein. These changes include solubilization, texturization, and increased digestibility. Crude animal proteases have been used for several years for the partial hydrolysis of casein and soybean protein.

For two main reasons, little progress has been made in the use of immobilized proteases for industrial-type processes. First, the economics of immobilized proteolytic process are, in general, similar to existing soluble processes. Second, many bacterial proteases which could be effectively utilized in these processes are not considered GRAS (Generally Regarded As Safe) by the U.S. Government. Getting approval to use non-GRAS enzymes could be time consuming and expensive.

In spite of the many legal and financial problems associated with the application of immobilized protease to the food and beverage industry, a great deal of research has been carried out.

The potential value in using immobilized enzymes instead of soluble enzymes has been a powerful driving force. Taylor et al.[10] have listed many of the advantages offered by immobilized enzymes. These are summarized in Table 1.

Proteases have been immobilized on a wide variety of carriers. Since methods of immobilization and the characterization of various carriers has been extensively reviewed,[11-13] these are not the subject of this review.

There have been numerous reports on research applications of immobilized proteases for industrial-type areas of interest. However, as of this moment, none of the applications has been commercialized. The greatest potential for immobilized proteases is in the food and beverage industries.

II. FOOD INDUSTRIES

The industries capable of employing immobilized enzyme technology include: dairying, brewing, canning, and baby food processing.

A. Dairy Industry

It is feasible to develop an immobilized enzyme system for the continuous clotting of milk.[10,14-22] The clotting process involves an enzymatic primary step followed by a nonenzymatic secondary step. The enzymatic step, like most enzymatic reactions, is affected by temperature; if the reaction temperature decreases 10°C, the reaction rate is halved. On the other hand, the nonenzymic portion of the process is affected 15-fold for each 10°C change in temperature. This difference allows for the lowering of temperature sufficiently to permit the enzymic reaction to occur while inhibiting the nonenzymic portion of the clotting mechanism. Warming of the enzymatically treated milk would then cause immediate clotting.

Several proteolytic enzymes have been used experimentally to clot milk (Table 2). Although a direct comparison of the performance of these many derivatives is difficult, Taylor et al.[10] suggest that, based on the literature to date, pepsin is the best immobilized protease for coagulating milk. Although inactive at the pH values of milk, pepsin is quite active, very stable, and works quite well after immobilization, particularly on negatively charged supports.

Table 1
ADVANTAGES OF IMMOBILIZED ENZYMES FOR FOOD PROCESSING

1. Enzyme reusability
2. Ease of terminating enzyme reaction
3. More precise control
4. Less apparent product inhibition
5. Greater pH and thermal stability
6. No contamination due to added enzyme
7. Potential for use of non-GRAS enzymes
8. Continuous operation
9. Greater flexibility in reactor design

Table 2
IMMOBILIZED ENZYMES USED FOR CLOTTING MILK

Enzyme	Support	Method of immobilization	Ref.
Chymotrypsin	Porous glass	Covalent	15
Chymotrypsin	DEAE-cellulose	Ion exchange	16
Chymotrypsin	DEAE-cellulose	Ion exchange	17
Chymotrypsin	Amberlite® GC400I	Ion exchange	17
Muco michei protease	Porous glass	Covalent	15
Papain	CM-cellulose	Covalent	18
Papain	Polyacrylamide	Covalent	18
Pepsin	Porous glass	Covalent	15
Pepsin	Porous glass	Covalent	19
Pepsin	ZrO_2-coated glass	Covalent	20
Pepsin	ZrO_2-coated glass	Covalent	21
Pepsin	EMA resin	Covalent	22
Pepsin	Porous glass	Covalent	23
Rennin	Porous glass	Covalent	15
Rennin	CM-cellulose	Covalent	16
Rennet	Porous glass	Covalent	15
Rennet	Porous glass	Covalent	19
Trypsin	EMA-resin	Covalent	22

An examination of the literature indicates that, with immobilized pepsin, relatively slow inactivation rates were observed during continuous application. High flow rates could be achieved and high enzyme activity was possible.

A summary of the available data on milk coagulation with immobilized proteases is given in Table 3. It is obvious from this table that major differences were observed with different enzymes and carriers.

The most successful system reported in the literature appears to be that described by Cheryen et al.[14,18,19] In this system, pepsin was covalently coupled to zirconia coated porous glass particles which had been precoated with bovine serum albumin. The reactor was operated as a fluidized bed at low temperature. Coagulation of the milk was accompanied by acidification and warming. However, the active lifetime of the catalyst was not more than 2 days.

Using immobilized enzymes for milk clotting is an ideal method, since no enzyme contaminates the curd. Specific amounts of the proper enzymes could be added to the curd to ripen the cheese.

Other applications of proteolytic enzymes in the dairy industry include treatment of

Table 3
SUMMARY OF SOME MILK COAGULATION STUDIES USING IMMOBILIZED PROTEASES

Enzyme	Support	Activity	Rate of inactivation	Contact time	Initial clotting time	Reactor temp.	Ref.
Chymotrypsin	Agarose	Low	—	80 mℓ/day	115 min	4°C	16
Papain	CM-cellulose + polyacrylamide	Low	Rapid	3—4 mℓ/hr	5 min	4°C	18
Chymotrypsin	DEAE-cellulose	Low	—	15 min	30 sec	4°C	17
Pepsin	Porous glass	High	Gradual	1 min	Immediate	10°C	15
Rennet	Porous glass	High	Gradual	1 min	Immediate	10°C	15

milk with trypsin to increase shelf life. In these studies, Shipe et al.[23] treated fresh milk with trypsin covalently coupled to porous glass. The treated milk developed oxidized flavors much slower than did untreated milk. If the milk was overexposed to the enzyme, a bitter flavor developed. Table 4 gives some of the results of these experiments.

B. Soya Hydrolysis

The use of extracts from soya beans in the food industry is of ever increasing importance for meat substitutes and protein enrichment, particularly in soft drinks and in infant formulas.

It is of interest to hydrolyse soya protein to increase acid solubility but not change its bland flavor. To accomplish this, it is necessary to decrease the size of the protein molecules without releasing a large percentage of free amino acids which can cause "off" flavors. This was accomplished by Detar et al.[24] using a microbial protease covalently coupled to zirconia-coated controlled pore glass. Hydrolysis was carried out in a continuous manner using a batch-type system with continuous pH monitoring and maintenance. Operational half-life of the derivative using a soya extract was 5.06 days. The final product was found, after acidification, to contain 70 to 72% of the total protein in solubilized form and 84% of the total solids. Only 21% of the total soluble protein was converted to free amino acids. Figure 1 gives the results of a typical experiment. Using a total of 980 mg of enzyme, maximum hydrolysis was achieved in less than two hours with 70 g of soya protein. The quantity of protein hydrolysable during one 5-day half-life is considerable.

C. Beer Chillproofing

Fermented malt beverages such as beer and ales are produced by fermentation of worts obtained from mashes of barley malt and various grains. After fermentation, the beer is processed through several operations such as cold storage, carbonation, and filtration, to clarify the product. When beer is subjected to cold storage, a haze or turbidity forms in the beer as a consequence of high-molecular-weight complex formation of proteins with phenols and tannins. The complexes have low solubility at low temperature, precipitating from solution. Although these complexes do not affect the quality of the product, they do decrease the aesthetic appearance of the product when poured into a glass.

Chillproofing beer is a process by which a proteolytic enzyme is introduced into the beer after fermentation. The partial hydrolysis of the high-molecular-weight proteins prevents the haze formation upon cooling of the beer. During the pasteurization process, the enzymatic activity of the added protease is accelerated, accomplishing this task. Residual enzyme activity remains after pasteurization.

The most commonly used enzyme for chillproofing is papain. In the normal chillproofing process, 1 lb of chillproofing enzyme is used for each 100 to 200 bbl of beer. The cost of the treatment averages 5 ¢/bbl.

The use of an immobilized protease could decrease the cost of processing the beer and eliminate the problems, both technical and political, associated with the residual enzyme activity in beer.

The most complete study in this area was carried out by Wildi and Boyce.[25] These workers used an ethylene-malic anhydride copolymer of papain. Chillproofing was experimentally carried out by adding 75.7 g of immobilized papain to 100 bbl of wort and allowing it to react for 24 hr. After fermentation, the derivative was retained in the decanted beer for an additional week at 3°C followed by pasteurization and bottling. The polymer-enzyme was recovered and reused. In addition, these workers added the enzyme-polymer at other stages of the beer process with acceptable results. How-

Table 4
EFFECT OF IMMOBILIZED TRYPSIN ON THE OXIDATIVE STABILITY OF MILK

	Storage			
	After 3 days		After 6 days	
	TBA[a] values	Oxidized flavor[b]	Oxidized flavor	TBA values
Treated	0.028	0.0	0.032	0.0
	0.025	0.0	0.022	0.5
	0.027	0.0	0.024	0.0
	0.042	0.3	0.045	1.0
Untreated	0.042	2.0	0.044	2.5
	0.044	1.5	0.052	2.0
	0.046	1.3	0.053	0.7
	0.070	3.0	0.081	3.0
	0.092	2.1	0.092	2.5

[a] TBA values represent the optical densities observed after addition of 2-thiobarbitunic acid solution to a milk filtrate.

[b] Oxidized flavors were determined by a panel and scored on a basis of 0 to 4 with 4 representing very strong flavor.

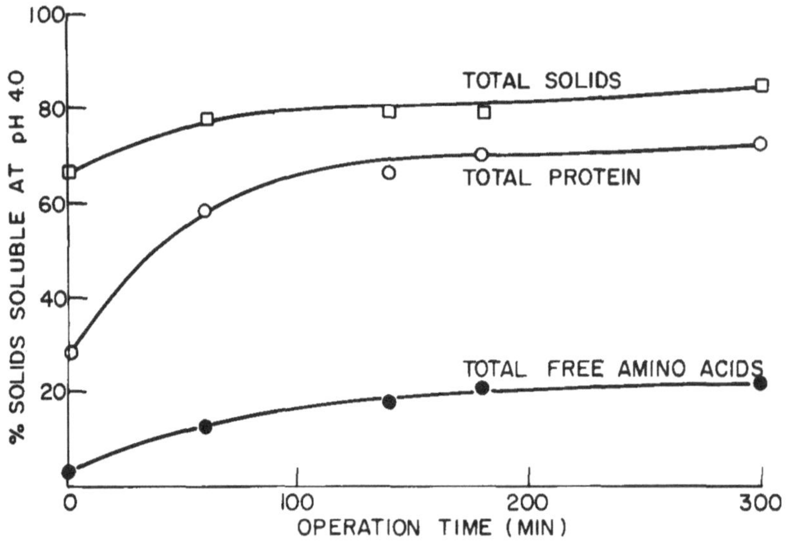

FIGURE 1. Hydrolysis of soya extract by immobilized protease.

ever, in all cases, long periods of time (7 days or more) were required for chillproofing with the enzyme-polymer.

In our own laboratories, we attempted to chillproof beer on a continuous basis using a packed-bed approach. Although the system worked successfully for up to 24 hr, it was unsuccessful for longer periods of time. As operation continued, the observed pressure drop across the column bed rose considerably until it reached unacceptable values. Back-flushing and repacking did not help. Since the particles used in these experiments were inorganic in nature and noncompressable, we assumed the increasing pressure drop was due to clogging or adsorption of material onto the particles. Prefiltering the beer through celite beds and other fine filter materials was to no avail. Examination of the particles indicated that copious quantities of protein-like material were indeed adsorbed to the particles.

It is obvious from the description of both these experiments that chillproofing with immobilized proteases is indeed possible but not yet truly feasible.

III. ENZYME INACTIVATION WITH IMMOBILIZED PROTEASES

There are several instances in enzymatic processing when one may want to stop an enzymatic reaction but cannot add exogenous inhibitors, denaturants, or heat. In the processing of fruit juices, for example, it could be of advantage to inactivate without heat the added enzymes used for classification.

Proteolytic enzymes covalently coupled to porous glass have been used for the inactivations of several enzymes.[26] *Streptomyces griseus* protease in a plug-flow reactor was used to inactivate glucose oxidase and pectin methylesterase. The data indicates that inactivation of the soluble enzymes did occur on passage through the immobilized protease column.

IV. POTENTIAL NEW APPLICATION OF IMMOBILIZED PROTEASES

A. Synthesis

One of the potential uses of immobilized proteases is synthesis. It is well known that soluble proteases can be induced to produce esters by adding the proper organic solvents at concentrations which do not precipitate or inactivate the enzyme.[27,28]

Recently,[29,30] it has been found that trypsin immobilized on inorganic supports can hydrolyze peptide bonds in concentrations of organic solvents as high as 95% (Table 3). It was also shown that peptide synthesis could be achieved with immobilized trypsin, leucine amino peptidase, and carboxypeptidase in 90% acetone, 80% ethanol, and 90% propanol, respectively.[30]

It is possible that in the not too distant future immobilized enzymes will be used for the synthesis of new peptides and protein-type molecules containing the essential amino acids necessary for health and growth. These synthetic proteins may become constituents of foodstuffs low in essential amino acids thus upgrading them for human consumption while producing products of excellent quality and texture.

B. Biological Switches

The physiology of all plants and animals contain mechanisms for turning reactions on and off. All of these switches are enzyme operated although they may be controlled by other factors either within or outside the organism. The biological switches, while of great theoretical interest, could also be of commercial value. Recent reports on light-sensitive soluble proteases[31] and immobilized proteases[32,33] bring to mind the possibility of silverless photography. Recent work in the Soviet Union indicates that this may already be a reality.

Inactivation of proteolytic enzymes in gels and on nylon fibers by applying tension[34] or pressure[35] to the derivative has been reported. Such experiments offer information on new ways of starting and stopping enzyme reactions other than separation of substrate and enzyme or through inhibition by adding reagents or denaturants. Such techniques could permit extremely close control of enzyme reactors far beyond that which is possible today.

V. SUMMARY

This review has attempted in a general way to explore a few of the possible applications for immobilized proteases. As yet, none have been commercialized. Commercialization will only come when the economics of a process necessitate the introduction of an immobilized enzyme process.

This review has also, in a small way, attempted to give the reader a small peek into the future application of immobilized enzymes in areas which heretofore have not been considered.

REFERENCES

1. **Vieth, W. R. and Venkatasubramanian, K.**, Process engineering of glucose isomerization by collagen-immobilized whole microbial cells, *Methods Enzymol.*, 44, 768, 1976.
2. **Weetall, H. H., Vann, W. P., Pitcher, W. H., Jr., Lee, D. D., Lee, Y. Y., and Tasao, G. T.**, Scale-up studies on immobilized, purified/glucoamylase, covalently coupled to porous ceramic support, *Methods Enzymol.*, 44, 776, 1976.
3. **Pitcher, W. H., Jr., Ford, J. R., and Weetall, H. H.**, The preparation, characterization, and scale-up of a lactase system immobilized to inorganic supports for the hydrolysis of acid whey, *Methods Enzymol.*, 44, 792, 1976.
4. **Paulsen, P. B. and Zitton, L.**, Continuous production of high-fructose syrup by cross-linked cell homogenates containing glucose isomerase, *Methods Enzymol.*, 44, 809, 1976.
5. **Takasaki, Y.**, Sugar isomerizing enzyme. Production and utilization of glucose isomerase from Streptomyces species, *Agric. Biol. Chem.*, 30, 1247, 1966.
6. **Takasaki, Y., Kosugi, A., and Kanbayoshi, A.**, Streptomyces glucose isomerase in *Fermentation Advances*, Perlman, D., Ed., Academic Press, New York, 1969.
7. **Schnyder, B. J.**, Continuous isomerization of glucose to fructose on a commercial basis, *Staerke*, 26, 409, 1974.
8. **Thompson, K. N., Johnson, R. A., and Lloyd, N. E.**, U.S. Patent 3,788,945, 1974.
9. **Weetall, H. H. and Havewala, N.**, Continuous production of dextrose from cornstarch. A study of reactor parameters necessary for commercial application, in *Enzyme Engineering*, Wingard, L. B., Jr., Ed., Interscience, New York, 1972, 241.
10. **Taylor, M. J., Richardson, T., and Olson, N. F.**, Coagulation of milk with immobilized proteases: a review, *J. Milk Food Technol.*, 36, 864, 1976.
11. **Colowick, S. P. and Kaplan, N. O., Eds.**, *Methods Enzymol.*, 34, 1974.
12. **Colowick, S. P. and Kaplan, N. O., Eds.**, *Methods Enzymol.*, 44, 1976.
13. **Bohak, Z. and Sharon, N.**, *Biotechnical Applications of Proteins and Enzymes*, Academic Press, New York, 1977.
14. **Cheryan, M., Van Wyk, P. J., Olson, N. F., and Richardson, T.**, Continuous coagulation of milk using immobilized enzymes in a fluidized-bed reactor, *Biotech. Bioeng.*, 17, 585, 1975.
15. **Green, M. L. and Crutchfield, G.**, Studies on the preparation of water-insoluble derivatives of rennin and chymotrypsin and their use in the hydrolysis of casein and the clotting of milk, *Biochem. J.*, 115, 183, 1969.
16. **Janusauskaite, V. B., Kozlov, L. V., and Antonov, V. K.**, Milk-coagulating and proteinase activity of modified alpha-chymotrypsin adsorbed on anion exchangers, *Prikl. Biokhim. Microbiol.*, 10, 410, 1974.
17. **Cooke, R. D. and Caygill, J. C.**, Possible utilization of plant proteases in cheesemaking, *Trop. Sci.*, 16, 149, 1974.
18. **Cheryan, M., Van Wyk, P. J., Olson, N. F., and Richardson, T.**, Secondary phase and mechanism of enzymic milk coagulation, *J. Dairy Sci.*, 58, 477, 1975.
19. **Cheryan, M., Van Wyk, P. J., Olson, N. F., and Richardson, T.**, Stability characteristics of pepsin-immobilized on protein-coated glass used for continuous milk coagulation, *Biotech. Bioeng.*, 18, 273, 1976.
20. **Ferrier, L. K., Richardson, T., Olson, N. F., and Hicks, C. L.**, Characteristics of insoluble pepsin used in a continuous milk-clotting system, *J. Dairy Sci.*, 55, 726, 1972.

21. **Goldstein, L.**, A new polyamine carrier for the immobilization of proteins. Water insoluble derivatives of pepsin and trypsin, *Biochim. Biophys. Acta*, 327, 132, 1973.
22. **Hicks, C. L., Ferrier, L. K., Olson, N. F., and Richardson, T.**, Immobilized pepsin treatment of skim milk and skim milk fractions, *J. Dairy Sci.*, 58, 19, 1975.
23. **Shipe, W. F., Senyk, G., and Weetall, H. H.**, Inhibition of oxidized flavor development in milk by immobilized trypsin, *J. Dairy Sci.*, 55, 647, 1972.
24. **Deter, C. C., Mason, R. D., and Weetall, H. H.**, Protein hydrolysis with immobilized enzymes, *Biotech. Bioeng.*, 17, 451, 1975.
25. **Wildi, B. S. and Boyce, D. C.**, U.S. Patent 3,597,219, 1971.
26. **Bliss, F. M. and Hultin, H. O.**, Enzyme inactivation by an immobilized protease in a plug flow reactor, *J. Food Sci.*, 42, 425, 1977.
27. **del Costillo, L. M. and Casteneda-Agullo, M.**, *Natl. Cancer Inst. Monogr.*, 27, 141, 1976.
28. **Amis, E. S.**, *Solvent Effects on Reaction Rates and Mechanisms*, Academic Press, New York, 1966.
29. **Weetall, H. H. and Vann, W. P.**, Studies on immobilized trypsin in high concentrations of organic solvents, *Biotech. Bioeng.*, 18, 105, 1976.
30. **Vann, W. P. and Weetall, H. H.**, Random peptide synthesis using immobilized enzymes in high concentrations of organic solvents, *J. Solid Phase Biochem.*, 1, 297, 1976.
31. **Berezin, I. V., Varfolomeyev, S. D., Klibanov, A. M., and Martinek, K.**, Light and ultrasonic regulation of alpha-chymotrypsin catalytic activity. Proflavine as a light- and sound-sensitive competitive inhibitor, *FEBS Lett.*, 39, 329, 1974.
32. **Nakomoto, Y., Karube, I., Terowaki, S., and Suzuki, S.**, Photocontrol of lactate dehydrogenase-spiropyran collagen membrane, *J. Solid Phase Biochem.*, 1, 143, 1976.
33. **Karube, I., Ishimori, Y., and Suzuki, S.**, Photocontrolled binding of cytochrome c to immobilized spiropyran, *J. Solid Phase, Biochem.*, 2, 9, 1977.
34. **Berezin, I. V., Klibanov, A. M., Samokhin, G. P., and Martinek, K.**, Mechanochemistry of immobilized enzymes: a new approach to studies in fundamental enzymology. I. Regulation by mechanical means of the catalytic properties of enzymes affected attached to polymer fibers, *Methods Enzymol.*, 44, 558, 1976.
35. **Berezin, I. V., Klibanov, A. M., Goldmacher, V. S., and Martinek, K.**, Mechanochemistry of immobilized enzymes: a new approach to studies in fundamental enzymology. II. Regulation of the rate of enzymic protein-protein interaction in polyacrylamide gel, *Methods Enzymol.*, 44, 571, 1976.

Chapter 8

APPLICATION AND POTENTIAL OF OTHER ENZYMES IN FOOD PROCESSING: AMINOACYLASE, ASPARTASE, FUMARASE, GLUCOSE OXIDASE-CATALASE, SULFHYDRYL OXIDASE, AND CONTROLLED RELEASE ENZYMES

Bhavender P. Sharma and Ralph A. Messing

TABLE OF CONTENTS

I. Introduction 186

II. Immobilized Enzymes and Microbes in Amino Acid Production 186
- A. Immobilized Aminoacylase 187
- B. Enzyme Reactor Studies 188
- C. Industrial Reactor System 189
- D. Immobilized Microbes for Amino Acid Production 190
  - 1. L-Aspartic Acid Production 191
  - 2. L-Malic Acid Production 194
- E. The Economics of Amino Acid Production with Immobilized Microbes 196

III. Applications and Potential of Immobilized Glucose Oxidase-Catalase 196
- A. Preservation of Canned Foods 199
- B. Preservation of Wrapped Foods 199
- C. Gluconic Acid Production 199
- D. Cold Pasteurization in Cheese Manufacture 200
- E. Analytical Applications 200

IV. Potential Application of Immobilized Sulfhydryl Oxidase for Flavor Modification 200

V. Controlled Release Enzymes and Conclusion 204

Acknowledgments 206

References 206

I. INTRODUCTION

Food processes are a function of a society at its particular developmental phase. Early man employed simple procedures to prepare his food. Those processes included the harvesting of fruits and vegetables, the removal of hides, and sectioning of the meat from animals. As the human community became more complex, the preparation of food included a cooking process which was accomplished either by an open fire or in smoldering pits. With the growth and spread of civilization, further processes were included for the preservation of food. These include salting, smoking, etc. With the advent of modern society and the further spread of civilization, requirements for long periods of storage and shipment around the world evolved more complex processing including canning and freezing as modes of preservation. During this evolution in food processing, the basic foods were identical to those of primitive man.

The population growth of the current era has initiated some concern with respect to the traditional food supply of the future. These concerns induced studies on additional preservation techniques such as those described under Glucose Oxidase-Catalase section and the Immobilized Sulfhydryl section. The potential shortage of the traditional food supplies has also spurred efforts in the direction of evaluating processes for converting required nutrients from nonconventional sources. Examples of these studies are reflected in the Immobilized Amino Acylase section and Amino Acid Production section.

Additional evidence of the search for new food sources may be noted in the massive efforts devoted to single cell protein.

The demand for sugars markedly increase as nations develop. The availability of farm land is reduced as the demand for manufacturing facilities is increased in those developing areas. Further stress is placed upon the land by the housing developments associated with manufacturing areas. The increase in industrialization throughout the world will ultimately lead to a shortage of land availability for the production of sugar from the traditional plant sources. This trend has encouraged efforts versed towards the more complete utilization of food crops and the identification of marine crops for sugar production. These efforts are exemplified by those programs throughout the world that have been addressed to the problem of converting cellulose to glucose by employing the enzymes characterized by the term "cellulase".

An area that is periphery to food processing is that of food, stain, or soil removal. This impinges upon the home processing of food as well as the industrial processing in that both the apparel of the food handler and the vessels in which the food is processed are affected.

II. IMMOBILIZED ENZYMES AND MICROBES IN AMINO ACID PRODUCTION

The first industrial application in the world of an immobilized enzyme system was demonstrated in 1969 when the Tanabe Seiyaku Co. Ltd. of Osaka, Japan brought its immobilized amino-acylase system on stream. Further details concerning this process will be discussed later in this chapter.

Amino acids, the building blocks of proteins, are an important part of human and animal nutrition. However, only the L-form can be metabolized and used for protein synthesis in the body and is thus the desirable form for food and medical applications. In Japan alone, the production of amino acids was a $300 million business in 1976.[1] The prices range from about $16/kg for L-aspartic acid to about $215/kg for L-proline and isoleucine.[2]

Amino acids can be produced by isolation from protein hydrolyzate, by fermentation, or by chemical synthesis. It is in the last option that immobilized aminoacylase is used. An extrapolation of the fermentation concept is the use of immobilized microorganisms. A few examples are mentioned following the discussion of immobilized aminoacylase.

A. Immobilized Aminoacylase

Chemical synthesis yields an optically inactive racemic mixture of the L- and D- isomers of amino acids. There are several methods for the resolution of this mixture.[3,4] These methods include the use of physical or chemical separations, as well as enzymatic procedures for resolution. Enzymes, such as aminoacylase, amino acid oxidases, or decarboxylases, can be employed for their specificity to attack preferentially one of the two forms. An effective resolution method, discussed in this section, is based on the hydrolysis of one isomer by immobilized aminoacylase.

$$\underset{\text{Acyl-DL-amino acid}}{\begin{array}{c}\text{R}-\text{CHCOOH}\\ |\\ \text{NHCOR}'\end{array}} + H_2O \xrightarrow{\text{Aminoacylase}} \underset{\text{L-amino acid}}{\begin{array}{c}\text{R}-\text{CHCOOH}\\ |\\ \text{NH}_2\end{array}} + \underset{\text{acyl-D-amino acid}}{\begin{array}{c}\text{R}-\text{CHCOOH}\\ |\\ \text{NHCOR}'\end{array}} \tag{1}$$

The products of the reaction, the L-amino acid, and the acyl-D-amino acid, can be easily separated due to the difference in their solubilities. The acyl-D-amino acid can be racemized and reused.

Tanabe Seiyaku Co. had used soluble aminoacylase for this optical resolution from 1954 until 1969 when they switched over to the immobilized aminoacylase process. Various aspects of the immobilized aminoacylase process have been well documented by Chibata and co-workers.[5-7]

Aminoacylase is obtained from *Aspergillus oryzae* cells grown on a wheat bran and rice hull based medium. Chibata and co-workers at the Tanabe Seiyaku Co. examined the immobilization of the aminoacylase by some 40 different carrier-immobilization-method combinations.[8] They selected the following as the most promising alternatives:

1. Entrapment in polyacrylamide
2. Ionic binding to diethylaminoethyl (DEAE)-sephadex
3. Covalent binding to iodoacetyl cellulose

Using ease of preparation, cost of immobilization per unit amount of product, operational stability, and ease of reactor activity regeneration as the criteria,[6] they determined the DEAE-sephadex to be the preferred method.

The aminoacylase polyacrylamide entrapment studies at Tanabe Seiyaku employed chemical polymerization. Potassium persulfate and *N-N'*-methylene-bis-acrylamide were chosen as the polymerization initiator and accelerator respectively. Entrapment of the enzyme on polyacrylamide and other polymers by irradiation, with a gamma cell at a dosage of 720 to 900 krd/hr, has been investigated.[9, 10] Aminoacylase has also been immobilized on mycelial pellets by Suzuki and co-workers by using albumin and glutaraldehyde.[11] A comparison of these methods with the DEAE-sephadex immobilization will probably evolve through these studies from Japan.

The properties of the free and the DEAE-sephadex immobilized aminoacylase with acetyl-DL-methionine as the substrate are summarized in Table 1. Certain concentra-

Table 1
PROPERTIES OF FREE AND DEAE-SEPHADEX IMMOBILIZED AMINOACYLASE[6]

Property	Free aminoacylase[a]	Aminoacylase immobilized to DEAE-sephadex by ionic binding
Optimum pH	7.5—8.0	7.0
Optimum temp. (°C)	60	72
Activation energy[b] (kcal/mol)	6.9	7.0
Optimum Co^{2+} (m*M*)[b]	0.5	0.5
K_m(m*M*)[b]	5.7	8.7
V_{max} (mol/hr)[b]	1.52	3.33
Heat stability (%)[c]		
60°C, 10 min	62.5	100
70°C, 10 min	12.5	87.5
Operation stability (half-life[d])	—	65 days at 50°C

[a] Data for acetyl-DL-methionine.
[b] All assays done at 37°C and pH 7.0.
[c] Remaining activity.
[d] The time required for 50% of the enzyme activity to be lost.

tions of protein denaturing agents, such as guanidine hydrochloride and urea, almost doubled the activity of the immobilized preparation, a phenomenon attributed to conformational changes during the binding to the DEAE-sephadex carrier.[12] Low concentrations of Co^{2+} were found to enhance the activity as well as thermal stability of the immobilized enzyme preparation. Organic solvents such as methanol, ethanol, n-propanol, t-butanol, and acetone activated the enzyme preparation in the presence of cobalt.[12] A similar effect was seen for the soluble enzyme. The protein denaturing agent sodium dodecylsulfate (SDS) and the solvent glycerol adversely affected both the free and the immobilized enzyme.

B. Enzyme Reactor Studies[12,13]

In the presence of an excess amount of the substrate, the Michaelis-Menten relationship (Chapter 2) can be written as follows:

$$-\frac{dS}{dt} = kE_0 \qquad (2)$$

where S = substrate concentration; t = time; k = specific constant expressing the rate of decomposition of the substrate-enzyme complex; and E_o = enzyme concentration.

Kinetic studies on the DEAE-sephadex aminoacylase system have demonstrated that the following expression, obtained by integration from the above equation, adequately fits the rate conversion data:

$$\frac{(S_0 - S)}{(S_0)} = \frac{kE_0}{(S_0)(V_s)} \qquad (3)$$

where S_o = inlet substrate concentration; S = outlet substrate concentration; V_s = space velocity = 1/residence time.

A straight line plot of the fractional conversion $(S_o-S)/(S_o)$ vs. the residence time was obtained for two different enzyme concentrations. The enzyme column activity, expressed as the specific constant k, was found to be the same regardless of whether the column was run in the upflow or the downflow mode.

Laboratory scale reactors, ranging in height to diameter ratio of 1.5 to 21.4, were used to study the effect of reactor dimensions. Mass transfer limitations were apparently absent as the reaction rate, expressed as k, was independent of the height to diameter ratio. The pressure drop across the enzyme reactor was found to be directly proportional to the column length and the flow rate. The relationship conformed with the Kozeny-Carman equation.

C. Industrial Reactor System

Based on the laboratory studies, a scaled-up version of the DEAE-sephadex-amino acylase system was developed at the Tanabe Seiyaku Co. The 1000-ℓ reactor at their Tokyo plant is shown in Figure 1. The reactor could be operated with automation (Figure 2). A highly simplified schematic diagram of the process is depicted in Figure 3. The excellent stability of the immobilized enzyme in a laboratory scale reactor may be noted from the data included in Figure 4 and from the reported half-life of about 65 days at 50°C. It was found that the column activity could be easily brought back to the initial level by charging the reactor with additional aminoacylase. No pretreatments were required for the reloading process.

The 1000-ℓ column reactors were operated at residence times of 0.5 hr to 1 hr depending on the L-amino acid being produced. As shown in the schematic in Figure 3, the effluent from the enzyme reactor is evaporated. The separated crude L-amino acid is collected by centrifugation and recrystallized from water. Yields of 91 to 94% have been obtained with the scheme. The carrier DEAE-sephadex has reportedly been used for 7 years without physical decomposition.[2]

The overall operating cost of the immobilized aminoacylase-based process is said to be about 60% of the batch process it replaced. Some of the classically cited advantages observed are noted below:

1. Better product yields are obtained due to the elimination of a purification step required in the soluble enzyme process.
2. Lower labor costs are achieved because the system is operated by automatic controls.
3. Almost an order of magnitude reduction in enzyme cost has been noted.

Although most of the data released by the Tanabe Seiyaku Co. have been for L-methionine production, the processes for L-alanine, L-phenylalanine, L-tryptophan, and L-valine have been in operation at that company. Optical resolutions in the commercial production of other amino acids, such as L-DOPA, D-phenylglycine, and L-hydroxyphenylglycine, are also in the planning phases.[2] The immobilized aminoacylase system has not been made available for licensing.[15]

Recently, the French company, Rhone Poulenc, has obtained a patent on a process using aminoacylase immobilized on TiO_2, silica, or alumina supports coated with basic cross-linked polymers.[16]

Other immobilized acylase systems for optical resolution of racemic mixtures of amino acids have also been tested.[17-19] Researchers at Snamprogetti in Italy have tested a pilot scale system on an acylase enzyme entrapped in cellulose triacetate fibers.[17,18]

FIGURE 1. Immobilized aminoacylase column reactors at Tanabe Seiyaku Co., Tokyo, Japan.

Laboratory scale studies with amino acylase immobilized in porous ceramics (SiO_2, SiO_2-Al_2O_3, and TiO_2-Al_2O_3) have also been conducted at Corning Glass Works.[19]

Immobilized D- and L- amino acid oxidases have also been examined as candidate systems for processing racemic mixtures of amino acids.[20,21] While their utility in commercial optical resolution is rather limited, immobilized D- and L- amino oxidase systems do work well for quantitative measurement of small amounts of L- and D- amino acids.[20,22]

D. Immobilized Microbes for Amino Acid Production

In order to complete the discussion of the subject of L-amino acid production, the use of immobilized "whole cells" for this purpose should be noted briefly. The Tanabe Seiyaku Co. has used two such processes on an industrial scale — one for L-aspartic acid production (since 1973) and the other for L-malic acid production (since 1974). The application, developed by Chibata and co-workers, is the first industrial use of immobilized microbes on a world-wide basis. Typically, under the concept utilized at Tanabe Seiyaku, the cells are entrapped in a polyacrylamide gel lattice. The cells are often treated to enhance the activity of the enzyme system in question. The entrapped cells are then used in a reactor much like an immobilized enzyme. Two examples are summarized in Table 2.

FIGURE 2. Control panel for continuous production of L-amino acids using immobilized aminoacylase at Tanabe Seiyaku Co., Tokyo, Japan.

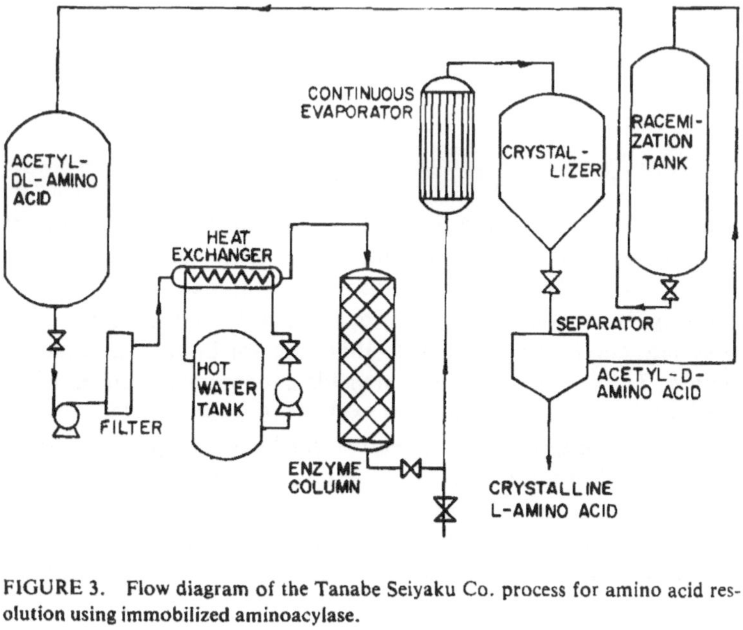

FIGURE 3. Flow diagram of the Tanabe Seiyaku Co. process for amino acid resolution using immobilized aminoacylase.

1. L-Aspartic Acid Production[6,7,23,24]

L-aspartic acid, used both as a food additive and for medical purposes, is produced by employing the aspartase enzyme in *Escherichia coli* cells entrapped in polyacrylamide.

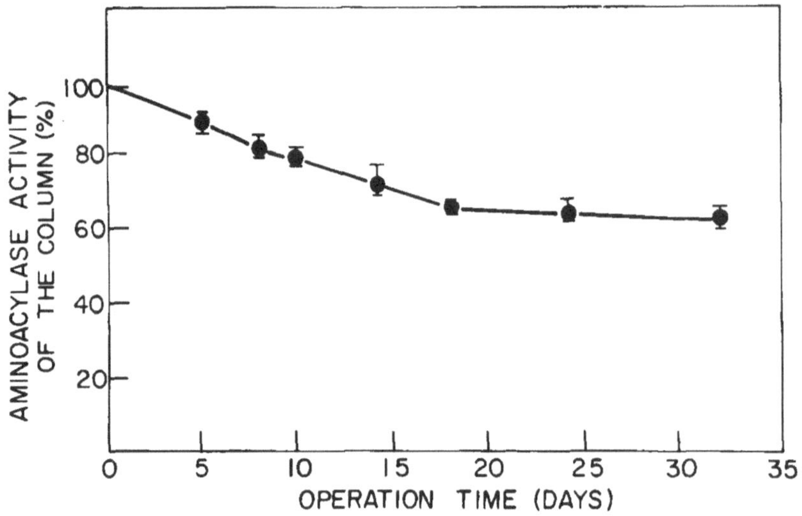

FIGURE 4. Operational stability of reactor containing aminoacylase immobilized onto DEAE-sephadex. A solution of 0.2 *M* acetyl-DL-methionine (pH 7.0, containing 5×10^{-4}M Co^{2+}) was applied to four DEAE-sephadex-aminoacylase columns (2.5 × 45 cm) by the method of upward flow for 32 days at 50°C at residence times of 0.2 to 1 hr. (From Tosa, T., Mori, T., Fuse, N., and Chibata, I., *Agric. Biol. Chem.*, 33, 1047, 1969. With permission.)

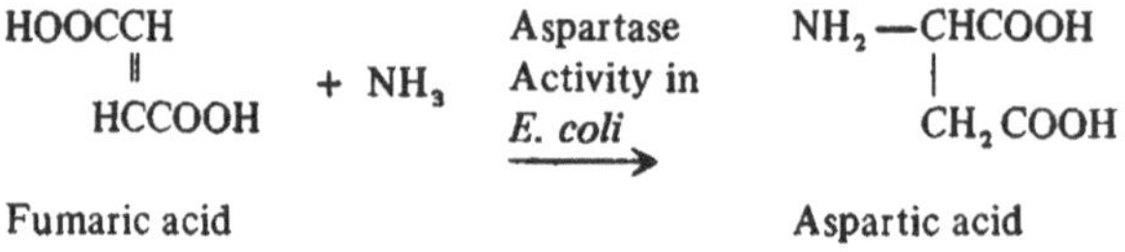

(4)

In practice, fumaric acid is dissolved in a 25% ammonia solution and the ammonium fumarate passed through the reactor containing immobilized *E. coli*.[24] The *E. coli* cells are incubated in a 1 m*M* Mg^{++} solution following entrapment. This improves the aspartase activity roughly tenfold, purportedly due to autolysis of cells in the gel lattice.[25]

The reactor system employed is similar to that used for immobilized amino-acylase, with one exception. The exothermic reaction yields a temperature difference across the reactor (6°C for a 1 *M* ammonium fumarate substrate solution). The enzyme column used (Figure 5) is, therefore, designed to remove the heat. The pressure drop observed for the entrapped *E. coli* reactor is about a factor of 7 lower than an equivalent sized DEAE-sephadex amino acylase reactor.

The activation energy for the reaction by immobilized aspartase in *E. coli* (Figure 6) is calculated to be 12.4 kcal/mol. The L-aspartic acid formation reaction has been found to follow zero order kinetics in a bench scale reactor treating 0.1 to 1 *M* ammonium fumarate solutions. Diffusion has been shown not to be a problem.[26] Stability data with laboratory reactors are impressive (Figure 7). Half-lives observed at 37°, 42°, and 45°C are 120, 46, and 21 days, respectively.[24] Another experiment at 37°C yielded a half-life of about 106 days.[23,26]

Since 1976, the Kyowa Hakko Co. of Japan has also been producing aspartic acid from fumaric acid and ammonia.[1]

Table 2
IMMOBILIZED MICROBES: PROPERTIES AND USE IN AMINO ACID PRODUCTION[23]

	L-Aspartic acid with *Escherichia coli*		L-Malic acid with *Brevibacterium ammoniagenes*	
Property	Free cells	Immobilized cells	Free cells	Immobilized cells
Optimum pH	10.5	8.5	7.5[a]	7.0
Optimum temperature °C	50	50	50	50
Heat stability conditions	50°C, 30 min	50°C, 30 min	55°C, 60 min	55°C, 60 min
Relative remaining activity (%)	49	58	63	65
Production conditions				
Substrate	—	1 *M* Ammonium fumarate	—	1 *M* Sodium fumarase
pH, temperature	—	8.5, 37°C	—	7.0, 37°C
Residence time	—	1.25 hr	—	4.35 hr
Product yield	—	95%	—	70%
Half-life	—	120 days at 37°C	—	52.5 days at 37°C

[a] Optimum pH for soluble fumarase.

FIGURE 5. Enzyme column reactor at Tanabe Seiyaku Co., Japan, using immobilized *Escherichia coli* cells for continuous production of L-aspartic acid.

2. L-*Malic Acid Production*[23,27]

L-malic acid (not to be confused with maleic acid) can be produced from fumaric acid and ammonia using the fumarase

$$\underset{\text{Fumaric acid}}{\begin{matrix} HOOC-CH \\ \| \\ HC-COOH \end{matrix}} + NH_3 \underset{}{\overset{\text{Fumarase}}{\rightleftharpoons}} \underset{\text{L-malic acid}}{\begin{matrix} HO-CH-COOH \\ | \\ CH_2COOH \end{matrix}} \qquad (5)$$

The equilibrium conversion is reported to be about 82%. The Tanabe Seiyaku Co., in a process also developed by Chibata and co-workers, produces L-malic acid by passing sodium fumarate through a reactor containing *Brevibacterium ammoniagenes* cells entrapped in polyacrylamide gel. The formation of succinic acid by the cells, an undesirable side reaction, can be suppressed by incubation of immobilized cells in a 1 *M* sodium fumarate solution containing 0.3% bile extract at 37°C for 20 hr.[23] In commercial production, the inexpensive bile powder is used as the source of bile acid. The treatment also enhances the fumarase activity of the *B. ammoniagenes* cells.

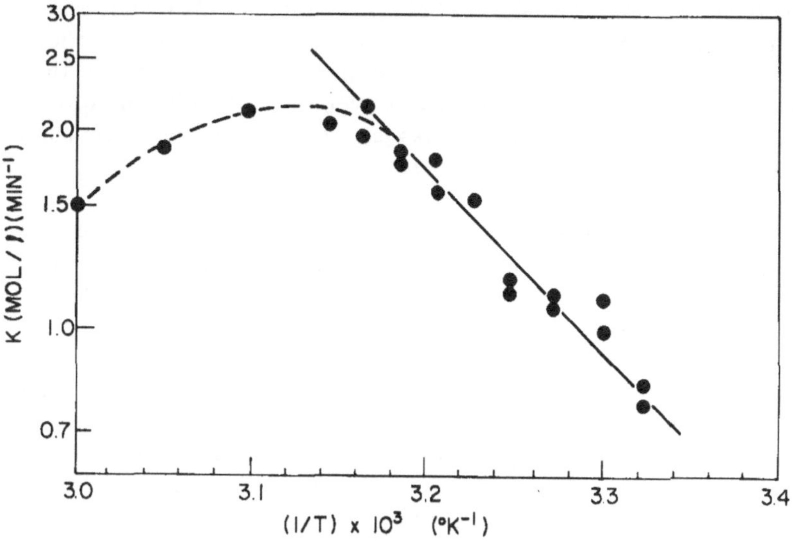

FIGURE 6. Arrhenius plot for aspartase activity in immobilized *E. coli* cells. A solution of 1 *M* ammonium fumarate (pH. 8.5) containing 1 m*M* Mg^{2+} was applied to the immobilized cell column at the indicated temperature. L-aspartic acid in effluent was measured. (From Sato, T., Mori, T., Tosa, T., Chibata, I., Furui, M., Yamashita, K., and Sumi, A., *Biotechnol. Bioeng.*, 17, 1797, 1975. With permission.)

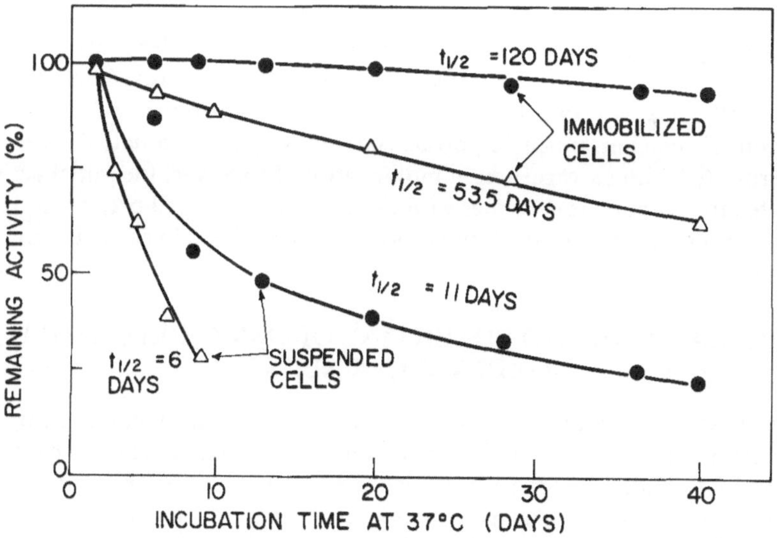

FIGURE 7. Stability of immobilized *E. coli* and *B. ammoniagenes* cells. Stability of suspended cells is also shown for comparison. In the case of *Escherichia coli*, intact cells and immobilized cells stood in 1 *M* ammonium fumarate (pH 8.5, containing 1 m*M* Mg^{2+}) at 37°C. In the case of *Brevibacterium ammoniagenes*, bile extract-treated cells and bile extract-treated immobilized cells stood in 1 *M* sodium fumarate (pH 7.0) at 37°C. At the indicated intervals, their remaining enzyme activities were determined, t½ indicates half-life of enzyme activity. (•) *E. coli* aspartase activity. (Δ) *B. ammoniagenes* fumarase activity. (From Chibata, I. and Tosa, T., *Advances in Applied Microbiology*, Vol. 22, Perlman, D., Ed., Academic Press, New York, 1977, 1. With permission.)

The reactor system used is similar to that for immobilized *E. coli.* It was found that unlike the aspartase-in-*E. coli* system, the fumarase activity of upper section of the *B. ammoniagenes* reactor decayed faster than that in the lower reaction. The flow through the reactor was from top to bottom. It was proposed that adsorption of poisonous substances, leakage of the enzyme, or leakage of the enzyme stabilizers could explain the occurrence.[27] The apparent half-life of the system, as shown in Figure 7, is about 54 days.

The Kyowa Hakko Co. was reported to have planned in 1978 to convert to a continuous, immobilized fumarase enzyme based process at its Bofu, Japan plant. The annual production of L-malic acid by Kyowa Hakko is estimated at about $1.6 million.[2]

G. The Economics of Amino Acid Production with Immobilized Microbes

Reports from Tanabe Seiyaku indicate their satisfaction with the economics of the immobilized microbe systems. The overall production cost of L-aspartic acid, using continuous processes based on entrapped *E. coli* cells, is about 60% of the conventional batch process. The major saving is in the cost of the enzyme, which is reduced by a factor of 7. There is, in addition, some reduction in the labor cost as a result of automation. The *B. ammoniagenes* based process for L-malic acid production is also said to compare favorably with the conventional production procedures from fruit juice, fermentation, or chemical synthesis followed by optical resolution of the DL-mixture.

An improvement over cell entrapment in polyacrylamide has been reported recently by Chibata and co-workers.[14] The new method of immobilization involves mixing a cell suspension with carrageenan, a polysaccharide used as a food additive, at 45 to 60°C, cooling to 10°C and then soaking the resulting gel in a 0.3*M* KCl solution. Granules, of any size or shape, can be obtained from the resulting gel. The stability of the immobilized cells can be improved significantly by a cross-linking treatment with glutaraldehyde and hexamethylenediamine. In contrast to the half-life of 120 days obtained with *E. coli* entrapped in polyacrylamide (Figure 7), a half-life of 686 days has been projected with carrageenan immobilization. Moreover, the initial activity has been reported to be about 11% higher with the carrageenan based preparation. Results with fumarase activity in *Brevibacterium flavum*, immobilized in carrageenan, are also encouraging.

III. APPLICATIONS AND POTENTIAL OF IMMOBILIZED GLUCOSE OXIDASE-CATALASE

Glucose oxidase and catalase will be covered together in this section, partly because they are often used together. In addition, almost all commercial glucose oxidase preparations contain catalase. This isn't all bad since the catalase serves to enhance the stability of an immobilized glucose oxidase preparation.[28] The commercial preparations for glucose oxidase also contain other enzymes such as invertase, amylase, maltase, and cellulase. These may either enhance or diminish the effectiveness of the immobilized enzyme preparation.

Glucose oxidase is one of the faster growing enzymes in the commercial market. While the overall enzyme market in the U.S. is expected to increase by about 38% from 1977 to 1985, the growth for glucose oxidase is anticipated to be about 62%. The projected U.S. market for 1985 is around $1.3 million.[29] The main reason is the diversity of applications in which glucose oxidase is involved.[30,32] The enzyme catalase also has many applications.[30,32] These are summarized in Tables 3 and 4, respectively. The principles behind these applications are best understood by noting the reactions catalyzed by the glucose oxidase-catalase system.[33a, 33b]

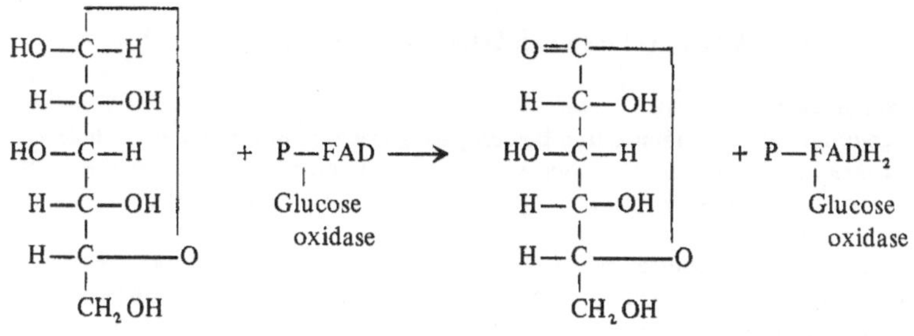

β-D-Glucose β-D-Gluconolactone

(6)

$$\underset{\text{Glucose oxidase}}{P\text{—}FADH_2} + O_2 \longrightarrow \underset{\text{Glucose oxidase}}{P\text{—}FAD} + H_2O_2 \quad (7)$$

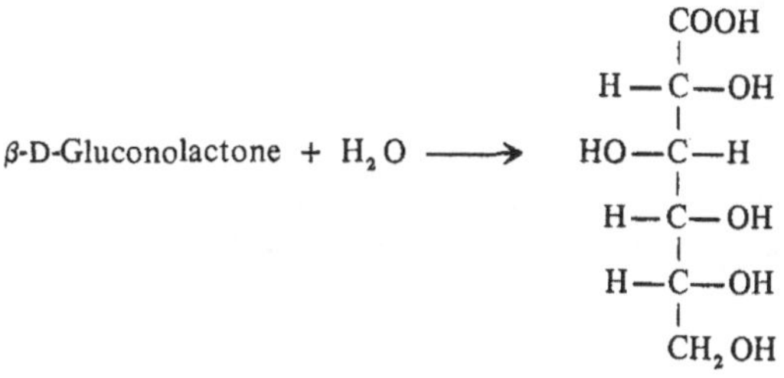

D-Gluconic Acid

(8)

$$2\,H_2O_2 \xrightarrow{\text{catalase}} 2\,H_2O + O_2 \quad (9)$$

NET REACTION

$$2\,C_6H_{12}O_6 + O_2 \xrightarrow{\text{Glucose oxidase-catalase}} 2\,C_6H_{12}O_7$$

Gluconic acid

(10)

Thus, the applications of glucose oxidase have to be directed at the removal of glucose or oxygen, or at the formation of gluconic acid or hydrogen peroxide.[31] The utility of catalase lies in its ability to decompose hydrogen peroxide with the production of oxygen. Under certain conditions, it can also participate in the oxidation of compounds like ethanol, methanol, and nitrate.[32,30] The immobilized glucose oxidase-catalase system has been reviewed recently.[33b-35] Some significant characteristics of the two enzymes are summarized in Table 5 for reference.

However, despite the widespread use of soluble glucose oxidase and catalase, the immobilized enzyme system has, to date, no commercially significant applications in the food industry. The major problem in using the immobilized glucose oxidase-catalase system is its lack of stability.[28,36-41] The half-lives of the immobilized enzyme system are so short that any potential economic advantages in using it are wiped out. The major reason is the enzyme inactivation due to hydrogen peroxide. Both glucose oxidase and catalase are deactivated by accumulated hydrogen peroxide, the product of

Table 3
APPLICATIONS OF GLUCOSE OXIDASE[31,32]

Removal of glucose
For color control and stabilization by desugaring egg albumin, egg yolks, whole egg, dried meat, and potatoes. Development of a brown color and off odors due to the Maillard reaction between glucose and proteins can be prevented by glucose removal.
Removal of oxygen
For flavor and color control by deoxygenation of carbonated soft drinks like orange, lemon, grapefruit, and citrus concentrates. Also for beer, precooked shrimp, foods in sealed containers, mayonnaise and salad dressings, white wine. The shelf life of foods where oxygen participates in reactions leading to color, odor change can be improved by deoxygenation. An example is the development of "sun starch flavor" in citrus beverages due to reactions involving light and oxygen.
Formation of gluconic acid
For reduction of glucose content in corn syrup and invert sugar.
Other applications
Glucose and mixed sugar analysis.
Galactose tolerance test.
Protection against oxidation of dry packaged foods such as whole milk powder, roasted coffee, and freeze dried products.

Table 4
APPLICATIONS OF CATALASE[30,32]

All applications of glucose oxidase where removal of the H_2O_2 formed or oxygen formation is desired, e.g., egg desugaring.
Stabilization of starter cultures in dairy industry.
Pasteurization of eggs.
Removal of residual H_2O_2
Oxidation of ethanol, methanol
Production of porosity in foam rubber, plastics, and insulating materials
Destruction of salmonella and staphylococci in foods such as eggs and starchy preparations
Bleaching and dying of textiles
Preservation of milk
Cold pasteurization of milk in production of swiss, cheddar, colby, washed, and granular cheese.

the former and the substrate for the latter. As is the case with many immobilized enzyme systems, mass transfer problems for the immobilized glucose oxidase-catalase system are especially significant.[36] Not only is the diffusion of the toxic H_2O_2 a factor, the diffusion of glucose and oxygen has to be considered also. The stability of the combined glucose oxidase-catalase system is somewhat better than that of immobilized glucose oxidase alone. Relatively stable preparation of glucose oxidase-catalase immobilized in 350 Å pore diameter titania have been achieved.[42a] Immobilization of glucose oxidase on manganese oxide (Mn_2O_3), capable of catalytic decomposition of H_2O_2, has also been tried.[42b] An increase in the ratio of the catalase to glucose oxidase in the immobilized glucose oxidase-catalase system has been suggested as a means of improving the system stability.[37]

Table 5
SOME CHARACTERISTICS OF GLUCOSE OXIDASE AND CATALASE

Property	Glucose oxidase	Catalase
Other names	β-D-Glucopyranose, aerodehydrogenase, notatin, microcide, p-FAD, corylophyline	Caperase, equilase, optilase
Common sources	*Aspergillus niger, Penicillium amagaskinense, Penicillium vitale*	Animal liver and kidney, *Micrococcus lysodeikticus, Aspergillus niger*
Some commercial names	Orazyme, fermocozyme, oxyBan, DeeO	Fermcolase, Takamine® catalase L
Molecular weight	140,000—190,000	225,000—250,000
Effective pH, temperature	4.5—7.0; 30°C—60°C	Beef liver: 6.5—7.5, 5—45°C; *A. niger*: 2.0—7.0
Food acceptability	[a]	[a]
Price (9/1978)	$60/lb DeeO (1500 units/g)	$60/*l* Takamine® catalase L (100 keil units/m*l*)

[a] Permitted as an optional ingredient without limitation in soda water, dried whole eggs, dried egg white. "Generally Recognized as Safe (GRAS)" application for glucose oxidase derived from *Aspergillus niger*, catalase from bovine liver, under consideration.[72] Catalase derived from *Micrococcus lysodeikticus* by a pure culture fermentation considered safe for H_2O_2 removal in cheese manufacture.[73]

A few applications aimed at using stabilized glucose oxidase or catalase or both are summarized below.

A. Preservation of Canned Foods

Thermal stabilization of glucose oxidase-catalase has been accomplished by coating the enzyme powder with methyl cellulose.[31,43,46] The enzymes are rapidly destroyed at temperatures of 45° to 60°C and above. Scott and co-workers at Searle and Co. stabilized the glucose oxidase-catalase system to withstand these high temperatures. The enzyme preparation was dried to a tablet form (at a temperature below 45°C) or on the top surface of a container. The enzyme is then coated with a hot water insoluble coating-dry methyl cellulose. The methyl cellulose coating, stable in the hot food material, releases the enzymes upon cooling. The enzymes can then perform their function in the food, e.g., that of oxygen consumption. Data from an application to an acid pH, hot filled food, indicate the initial residual activity to be 100% (Table 6).

B. Preservation of Wrapped Foods

Another example of a controlled release of the glucose oxidase-catalase system, also the work of Scott and co-workers, is the use of food wrappers coated with the enzymes. Some cheeses tend to form a brown ring as oxygen diffuses into the cheese during storage. Plastic wrappers coated with a glucose oxidase-catalase preparation were used to store loaves of processed cheese in a test.[31,47] The cheese loaves were protected from the entry of oxygen, and the resulting brown ring, by the enzymes for 10 weeks at 100°F and 90% humidity.

C. Gluconic Acid Production

Gluconic acid, estimated to be worth $6 to 7 million annually on a world-wide basis, is produced both by biosynthesis and chemical synthesis. It has low toxicity, an ability to form soluble complexes with metal ions, and characteristics appropriate for cleaning operations because it is relatively noncorrosive. In addition to its employment in a

Table 6
THERMAL STABILIZATION OF GLUCOSE OXIDASE-CATALASE BY COATING WITH METHYL CELLULOSE[31]

	Percent residual activity[a]			
	Control		Tablets	
Time (min)	Glucose oxidase	Catalase	Glucose oxidase	Catalase
0	100	100	100	100
3	0	43	94	94
6	0	15	91	90
9	0	0	84	87
12	0	0	81	82
15	0	0	78	73

[a] Methyl cellulose-coated tablets containing glucose oxidase were added to tomato juice at pH 4.4 and held at 87°C for the indicated times.

variety of industries, its use in the food industry includes that as an acidulant in cheesemaking and in washing and cleaning operations in the soft drink and brewing industries.

As can be seen from reactions (6) to (9), the enzyme glucose oxidase can be used for production of gluconic acid from glucose. A process based on immobilized glucose oxidase is reportedly being evaluated at the pilot plant level in Japan.[2] Although additional information is unavailable, it is anticipated that the immobilized glucose oxidase would be applied to the production of crystalline fructose. The glucose in high fructose corn syrup (Chapter 5) can be oxidized with glucose oxidase. The gluconate formed can be precipitated or ion exchanged, leaving the fructose in solution. The concept, based on soluble glucose oxidase, has been used in the U.S. on an industrial level.[48]

D. Cold Pasteurization in Cheese Manufacture

In the manufacture of cheddar type and swiss cheeses, milk is cold sterilized with dilute hydrogen peroxide. Soluble catalase is used to destroy the excess hydrogen peroxide after the pasteurization step. Immobilized catalase has been studied[49-52] for this application. The major problem encountered has been the very short half-life of the immobilized enzyme system.

E. Analytical Applications

One of the first examples of an analytical application of an immobilized enzyme is immobilized glucose oxidase. Measurement of glucose is an obvious application. Several researchers have successfully built such "enzyme electrodes" for research purposes.[22] Instruments marketed by the Radiometer Instrument Co. (Copenhagen, Denmark), The Leeds and Northrup Co. (North Wales, Pa.), and the Yellow Springs Instruments Co. (Yellow Springs, Ohio) are commercial examples.

IV. POTENTIAL APPLICATION OF IMMOBILIZED SULFHYDRYL OXIDASE FOR FLAVOR MODIFICATION

Included in the potential improvements for the $30 billion milk industry are those directed at the distribution, marketing, and storage of the 128 billion lb of fluid milk. One of those innovations is explored in this section.

From the standpoints of public health and economics of milk distribution, virtually all the "beverage" milk in the U.S. and many other countries is pasteurized. Pasteurization makes milk safer for human consumption as well as improves the shelf life. However, it also causes some changes in the milk constituents and milk characteristics. The extent of these changes depends upon the time and temperature used in the pasteurization or heat treatment. Figure 8 provides an overview of these changes. Some commonly practiced heat treatment schedules and temperatures are listed in Table 7. These are consistent with the U.S. Public Health Service criteria for applying the term "pasteurization".[53,57] Higher temperatures, i.e., temperatures above 175°F (79°C), offer some advantages. Those advantages include:

1. More efficient destruction of microorganisms
2. Longer shelf life, provided the milk is not contaminated following pasteurization

The emergence of dairy cooperatives and other large milk processors and of the complex marketing and distribution system in the U.S.[58] have resulted in relatively wide geographical areas covered by single milk processing plants; hence the desirability of longer shelf life. This has stimulated interest in asceptic or near-asceptic processing and packaging at milk. Ultra-high temperature (UHT) pasteurization, introduced in 1948,[54] offers the potential of long shelf lives at ambient temperatures. Packaged UHT milk is currently served to passengers on some international flights for use with tea/coffee. Generally, UHT is taken to mean a 2 sec or less holding time at temperatures ranging from 200°F (93°C) to 270°F (132°C).[55,57] Unfortunately, high temperature treatments such as the UHT impart a "cooked" flavor to milk.[53] Instability of milk proteins and the development of "oxidized" and/or "stale" flavor during storage also are initiated by UHT processing.[61,62]

The time of appearance and the extent of the stale flavor intensity depend upon the storage temperature and the dissolved oxygen level in the milk. Of the volatile flavor compounds formed during storage, diacetyl and, to a lesser extent, n-hexanal and propanal appear to be most significant in contributing to the flavor of UHT-treated milk.[61]

The "cooked" flavor impairment can be minimized by cooling the milk very rapidly after heating in the UHT treatment scheme.[56] The cooked flavor in sterilized milk usually peaks by itself during storage in 2 to 6 weeks.[60] Heat denaturation of milk proteins during UHT processing exposes and activates sulfhydryl groups.[71] It has been suggested that a direct cause and effect relationship exists between the exposure of these sulfhydryl groups and the development of the cooked flavor.[63,64]

An enzyme capable of oxidizing sulfhydryl groups, isolated in Germany, has been investigated by Swaisgood and co-workers at North Carolina State University. Results of a study from Cornell University, presented in Table 8, support the cause and effect relationship between the sulfhydryl level and the cooked flavor. As seen in Table 8, the sulfhydryl oxidase enzyme treatment seems to have reduced the strong cooked flavor to a relatively low level. The stoichiometry of the reaction has been suggested as follows:[63]

$$2\,RSH + O_2 \xrightarrow{\text{sulfhydryl oxidase}} RSSR + H_2O_2 \qquad (11)$$

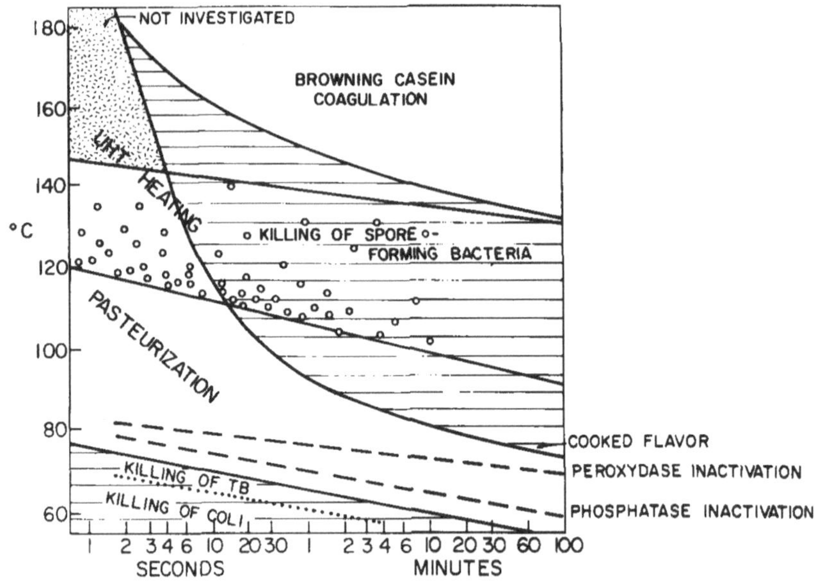

FIGURE 8. Heat treatments and their effect on milk characteristics. (Blohm, G., et al., German Dairy Industry, XVII International Dairy Congress, Munich, 1966. Taken from Hall, C. W. and Trout, G. M., *Milk Pasteurization*, AVI Publishing, Westport, Conn., 1968, chap. 2. With permission.)

Table 7
DURATIONS AND TEMPERATURES COMMONLY USED IN HEAT TREATMENT OF MILK[53]

Temp. °F	Temp. °C	Time
143 and/or 161	62 and/or 72	30 min and/or 15 sec, respectively
175—195	79—91	15 sec or less
200—212	93—100	Momentarily
Above 225	above 107	Momentarily up to 30 min

Thus, molecular oxygen serves as the electron acceptor. The enzyme activity, according to one method, is measured by the rate of oxygen consumption. Another method of the enzymatic activity measurement is based on the disappearance of the sulfhydryl groups.[65]

Details of the isolation, characterization, and immobilization studies at North Carolina State have been documented.[63, 65-71] A few key findings are noted below.

1. The enzyme can be isolated from bovine milk by treatment with rennin, followed by ammonium sulfate precipitation and several dissolution and centrifugation steps.
2. The presence of iron in the enzyme, apparently bound reversibly, is required for enzyme activity.
3. Optimum enzymatic activity is at 35°C and pH 7.

Table 8
ELIMINATION OF COOKED FLAVOR IN MILK BY SULFHYDRYL OXIDASE[64]

Sample	Sulfhydryl concentration[a]	Cooked flavor rating[b]
Control	0.18	3.4
1% enzyme solution	0.09	1.6
2% enzyme solution	0.05	1.1

[a] Expressed as optical density according to the method by Ellman.[65]
[b] Cooked flavor intensity scale: 1 = slight, to 4 = very strong cooked flavor.

4. The enzyme oxidizes sulfhydryl groups in small peptides such as reduced glutathione (GSH) as well as in proteins such as ribonuclease, chymotrypsinogen, and heated milk proteins.
5. Sulfhydryl oxidase can be immobilized on nonporous glass, in controlled-pore glass, and porous silica derivatized to provide aminopropyl or succinamidopropyl surfaces. Immobilization of the enzyme can be accomplished by chemical coupling with glutaraldehyde, carboxyl group activation with carbodiimide, and carbohydrate residue activation with periodate.
6. Sulfhydryl oxidase has a tendency to form large molecular aggregates, and so do the denatured milk proteins which the enzyme would have to treat for flavor modification.
7. Nonporous glass beads are as effective a support as controlled pore glass beads of 500 Å and 3000 Å average pore diameters. The comparison is based on covalent attachment of sulfhydryl oxidase to carbodiimide activated succinamidopropyl glass.

Stability data for immobilized sulfhydryl oxidase have been obtained with small laboratory scale reactors.[66] The reactors were operated intermittently for 5-hr periods. They were then washed and stored between operating periods in either a buffer, or a substrate, or an antimicrobial agent. For a reactor operated at 35°C with 0.8 m*M* reduced glutathione as the substrate and stored in 10% ethanol, a half-life of about 61 days was observed. However, the decline in the immobilized enzyme activity in reactors operated with heat treated milk (90°C for 30 min) was more rapid. Half-lives were in the range of 1 week at 23°C and 4 to 5 days at 35°C. The stability data at 23°C are shown in Figure 9. More recent results suggest that the enzyme activity can be improved by cross-linking with glutaraldehyde following immobilization.[70]

The studies at North Carolina State University signify the technical feasibility of using immobilized sulfhydryl oxidase for the elimination of the cooked flavor in UHT-processed milk. This, of course, holds the promise of storage of milk without refrigeration. Additional studies have refined the reactor sterilization and asceptic operation procedures. The current plans include the evaluation of the immobilized enzyme system in a 1 to 1.5-*ℓ* reactor in a small pilot-plant with a UHT sterilizer capable of processing about 50 *ℓ* of milk per hr.[60]

Some factors that require examination, if an immobilized sulfhydryl oxidase system were to be used on an industrial scale, are as follows:

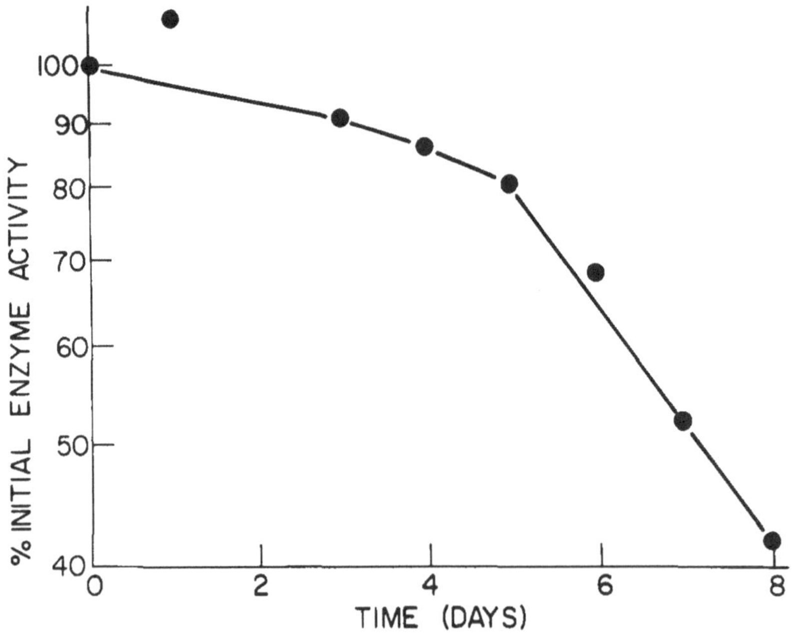

FIGURE 9. Stability of immobilized sulfhydryl oxidase. The immobilized enzyme reactor was operated daily for 5 hr at 23°C using heated skim milk as substrate. (From Swaisgood, H. E. and Horton, H. R., Utilization of sulfhydryl oxidase for the treatment of ultra-high temperature sterilized milk, semi-annual report, NSF Grant No. AER-73-07877 AO3, 1976.)

1. The cost effectiveness of using the immobilized enzyme treatment
2. The effect of the enzyme treatment on the stability of the UHT processed milk with respect to acquiring the oxidized flavor (methods for controlling the stale flavor in UHT treated milk during long term storage will have to be developed)
3. The significance of other problems, such as "gelation" due to lipid oxidation on long term storage (9 to 12 months)[60,71]
4. The marketing and customer acceptability of milk designed for ambient temperature storage

V. CONTROLLED RELEASE ENZYMES AND CONCLUSION

The commonly accepted definition of immobilized enzyme is that of an enzyme molecule permanently fixed to a carrier or to other enzyme molecules by cross-linking. In order to demonstrate the utility and versatility of immobilization techniques, that definition must be expanded to include enzymes that are fixed or immobilized on a temporary basis which will be released at an appropriate time or under appropriate conditions for a reaction or a specific purpose. The objectives of this approach may be to concentrate an enzyme in a particular region, or to stabilize an enzyme for storage under adverse conditions, or to release an enzyme under controlled conditions over a period of time such that the total quantity of enzyme is not rapidly destroyed in solution under adverse conditions and has an opportunity to react with the substrate, or for retaining the enzyme proximally to the substrate until the reaction is completed.

Mandels[74] pioneered a program for converting cellulosic residues and wastes to glucose with the enzyme system produced by the fungus *Trichoderma viride.* The free soluble enzyme system after addition to the reactor containing the cellulose forms a tight complex with that substrate and thus becomes immobilized. In this case the cellulose is both the carrier and the substrate for the enzyme. Since the enzyme is immobilized, it is readily retained within the reactor and concentrated on the surface of the substrate. When the cellulose is consumed by the conversion to glucose, the enzyme is released into solution and is subsequently reimmobilized by the addition of fresh cellulose. Recently, Mandels' group[75] has reported improvements in the production of cellulase with a new mutant strain of *Trichoderma reesi* MCG 77. Further refinements for the production of this enzyme system have been elucidated.[76] Another source of a cellulase system from the anaerobe *Clostridium thermocellum* ATCC 27405 has been extensively evaluated by the M.I.T. group.[77,78] This enzyme system appears to be very promising as an energy efficient catalyst for the continuous conversion of cellulose to glucose.

A combination of proteases and amylases were added to laundry detergents during the period of 1967 through the very early 1970s. The addition of these enzymes were proven to be rather effective against protein stains. The U.S. detergent manufacturers chose to remove these enzymes from their products because of reported allergic responses. Detergent enzymes are widely used throughout Europe at the present time. One of the problems encountered in the application of these enzymes was the limited shelf life due to the enzymatic decay in the presence of humidity and the high pH of the detergent. By bonding the enzymes to insoluble inorganic particles such as bentonite prior to addition to the dry detergents the enzyme shelf life was markedly improved.[79-81] The enzymes were readily released from the surface of the inorganic particles at elevated temperatures, 50 to 60°C, under the alkaline pH conditions, pH 9 to 11, of the detergents in solution. The bonding of the enzymes to these minute inorganic particles not only increased the enzyme storage stability in the dry detergent but also extended the effective life of the enzyme in solution under very adverse conditions of high temperature and pH by the gradual release of the enzyme from the surface over an extended period of time. This same approach was applied to cleaning cookware which had been utilized to prepare macaroni and cheese, eggs, etc.[82] The immobilized enzymes were added to dishwashing detergent which, in fact, is even more alkaline, pH 11 to 13, than the laundry detergent. The exposed ware was simply soaked with water at temperatures of 40 to 60°C for periods of less than 1 hr and the vessels were then simply rinsed and lightly wiped with a damp wash cloth. All cooking residues were removed without additional abrasives or scrubbing. Under these conditions, the inorganic bound enzymes were found to be far more effective than the controls which contained an equivalent level of free enzymes in the dry detergent or no enzyme addition to the detergent. Detergents without enzymes are very effective in removing oils and fats from surfaces; however, denatured proteins and charred carbohydrates resist detergents that are free of abrasives. Enzymes permanently immobilized would not be effective in washing applications due to the fact that they would not be able to gain access to the fixed stains or residues. Although this technology is not currently applied to commercial applications, it does demonstrate the effectiveness of controlled release immobilized enzymes for food applications.

The enzymatic conversions of fats, oils, and steroid structures in foods by immobilized enzymes appear to be a difficult task. The problems encountered are generally due either to the instability of the enzyme or to the fact that the enzymes must be delivered from a hydrophillic to a hydrophobic medium. In addition, it may be that the lipid is in a solid form. The fact that one may be dealing with both a liquid and a solid medium rules out the possibility of using a permanently immobilized enzyme for gaining stability. The alternative then appears to be controlled release enzymes for these applications. Modern society has become more diet conscious in recent years and with the implication that certain fatty substances may be adverse to health, the removal or conversion of these molecules will probably become more important with time. Enzymatic conversions may either facilitate their removal or reduce their adverse affect upon health.

Although the immobilized enzyme approach to these problems has not been seriously considered, there may be technology available at the current time which is applicable to the conversion of fatty substances. Gregoriadis[83,84] has described a procedure for the entrapment of enzymes in assemblages of phospholipids and other lipids that is termed liposomes. These liposome entrapped enzymes are not only insoluble in water but also appear to be very stable. It seems reasonable that these liposomes may be utilized to entrap enzymes that are capable of converting lipid or steriod structures, transporting the enzymes through the aqueous phase, delivering them to the hydrophobic phase where the membrane structure of the liposome may be dissolved and thus liberate the enzyme for reaction. A similar technique for encapsulation which employs a high molecular-weight paraffin and a surfactant has been described[85] as liquid membrane enzyme encapsulation. It would appear as though the liquid membrane approach to the controlled release of the enzyme in the hydrophobic environment would apply in the same manner as does the liposomes.

One may readily predict that the future of food processing will be more complex as the requirements and the pressures are applied due to the expanding population. It is not reasonable to believe that immobilized enzyme technology may be brought to bear on all of the future problems that arise in this industry; however, this solid phase technology has, to date, proven to be rather versatile and many of the future food processing problems may very well be resolved through this approach.

ACKNOWLEDGMENTS

We would like to thank Ms. Carol Allowatt and Ms. Nancy Foster for their assistance in preparation of this manuscript. We are also thankful to Dr. Ichiro Chibata for the photographs of the Tanabe Seiyaku Systems.

REFERENCES

1. **Yamada, K.**, Recent advances in industrial fermentation in Japan, *Biotechnol. Bioeng.*, 19, 1563, 1977.
2. Immobilized Enzymes and Cells in Japan, Special Survey Report, The Int. Technical Information Institute, Tokyo, Japan, 1978.
3. **Dinelli, D., Marconi, W., Cecere, F., Galli, G., and Morisi, F.**, A new method for the production of optically active aminoacids, in *Enzyme Engineering*, Vol. 3, Pye, E. K. and Weetall, H. H., Eds., Plenum Press, New York, 1978, 477.
4. **Johnson, J. C.**, *Amino Acids Technology*, Chem. Technol. Rev. No. 108, Noyes Data Corp., Park Ridge, N.J., 1978.
5. **Chibata, I., Tosa, T., Sato, T., and Mori, T.**, Production of L-amino acids by aminoacylase adsorbed on DEAE-sephadex, in *Methods in Enzymology*, Vol. 44, Mosbach, K., Ed., Colowick, S. P., and Kaplan, N. O., Series Eds., Academic Press, New York, 1976, 746.
6. **Chibata, I. and Tosa, T.**, Industrial applications of immobilized enzymes and immobilized microbial cells, in *Applied Biochemistry and Bioengineering*, Vol. 1, Wingard, L. B., Katchalski-Katzir, E., and Goldstein, L., Eds., Academic Press, New York, 1976, 329.

7. **Chibata, I., Tosa, T., Sato, T., Mori, T., and Yamamoto, K.,** Applications of immobilized enzymes and immobilized microbial cells for L-malic acid production, in *Immobilized Enzyme Technology: Research and Applications,* Weetall, H. H. and Suzuki, S., Eds., Plenum Press, New York, 1975, 111.
8. **Chibata, I., Tosa, T., Sato, T., Mori, T., and Matsuo, Y.,** Preparation and industrial application of immobilized aminoacylases, *Proc. IV IFS: Fermentation Technology Today,* Society for Fermentation Technology, Tokyo, 1972, 383.
9. **Kawashima, K. and Umeda, K.,** Immobilization of enzymes by radiopolymerization of acrylamide, *Biotechnol. Bioeng.,* 16, 609, 1974.
10. **Kawashima, K. and Umeda, K.,** Immobilization of enzymes by radiocopolymerization of monomers, *Biotechnol. Bioeng.,* 17, 599, 1975.
11. **Hirano, K., Karube, I., and Suzuki, S.,** Aminoacylase pellets, *Biotechnol. Bioeng.,* 19, 311, 1977.
12. **Tosa, T., Mori, T., and Chibata, I.,** Studies on continuous enzyme reactions. VII. Activation of water-insoluble aminoacylase by protein denaturing agents, *Enzymologia,* 40, 49, 1971.
13. **Tosa, T., Mori, T., Fuse, N., and Chibata, I.,** Studies on continuous enzyme reactions. V. Kinetics and industrial application of aminoacylase column for continuous optical resolution of acyl-DL-amino acids, *Agric. Biol. Chem.,* 33, 1047, 1969.
14. **Chibata, I., Tosa, T., Sato, T., Yamamoto, K., Takata, I., and Nishida, Y.,** New method for immobilization of microbial cells and its industrial application, in *Enzyme Engineering,* Vol. 4, Broun, G. B., Manecke, G., and Wingard, L. B., Jr., Eds., Plenum Press, New York, 1978, 335.
15. **Sweigart, R. D.,** Approach of the Enzyme User to Enzyme Developments and Markets, presented at Conf. on Enzyme Economics, Chicago, Ill., May 31 to June 2, 1978.
16. Rhone Poulenc Patent BE 855 051, in "Heartcut", Pilato, L. A. and Reichle, W. T., *Chem. Technol.,* 8, 309, 1978.
17. **Bartoli, F., Bianchi, G. E., and Zaccardelli, D.,** Production of L-tryptophan, in *Enzyme Engineering,* Vol. 4, Broun, G. B., Manecke, G., and Wingard, L. B., Jr. Eds., Plenum Press, New York, 1978, 279.
18. **Dinelli, D., Marconi, W., and Morisi, F.,** Fiber-entrapped enzymes in *Methods in Enzymology,* Vol. 44, Mosbach, K., Ed., Colowick, S. P. and Kaplan, N. O., Series Eds., Academic Press, New York, 1976, 227.
19. **Weetall, H. H. and Detar, C. C.,** Studies on aminoacylase immobilized on porous ceramic carriers, *Biotechnol. Bioeng.,* 16, 1537, 1974.
20. **Tosa, T., Sano, R., and Chibata, J.,** Immobilized D-amino acid oxidase, *Agric. Biol. Chem.,* 38, 1529, 1974.
21. **Weetall, H. H. and Baum, G.,** Preparation and characterization of insolubilized L-amino acid oxidase, *Biotechnol. Bioeng.,* 12, 399, 1970.
22. **Guilbault, G. C.,** Enzyme electrodes and solid surface fluorescence methods, in *Methods in Enzymology,* Vol. 44, Mosbach, K., Ed., Colowick, S. P. and Kaplan, N. O., Series Eds., Academic Press, New York, 1976, 589.
23. **Chibata, I. and Tosa, T.,** Transformation of organic compounds by immobilized microbial cells, in *Advances in Applied Microbiology,* Vol. 22, Perlman, D., Ed., Academic Press, New York, 1977, 1.
24. **Chibata, I., Tosa, T., and Sato, T.,** Production of L-aspartic acid by microbial cells entrapped in polyacrylamide gels, in *Methods in Enzymology,* Vol. 44, Mosbach, K., Ed., Colowick, S. P. and Kaplan, N. O., Series Eds., Academic Press, N. Y., 1976, 73.
25. **Chibata, I., Tosa, T., and Sato, T.,** Immobilized aspartase-containing microbial cells: preparation and enzymatic properties, *Appl. Microbiol.,* 27, 1974, 878.
26. **Sato, T., Mori, T., Tosa, T., Chibata, I., Furui, M., Yamashita, K., and Sumi, A.,** Engineering analysis of continuous production of L-aspartic acid by immobilized *Escherichia coli* cells in fixed beds, *Biotechnol. Bioeng.,* 17, 1797, 1975.
27. **Yamamoto, K., Tosa, T., Yamashita, K., and Chibata, I.,** Kinetics and decay of fumarase activity of immobilized *Brevibacterium ammoniagenes* cells for continuous production of L-malic acid, *Biotechnol. Bioeng.,* 19, 1101, 1977.
28. **Messing, R. A.,** Immobilized glucose oxidase and catalase in controlled pore titania, in *Immobilized Enzymes in Food and Microbial Processes,* Olson, A. C. and Cooney, C. L., Eds., Plenum Press, New York, 1974, 149.
29. **Wolnak, B.,** Present Status of the Enzyme Industry in the U.S. presented at Conf. on Enzyme Economics, Chicago, Ill., May 31 to June 2, 1978.
30. **Scott, D.,** Oxidoreductases, in *Enzyme in Food Processing,* 2nd ed., Reed, G., Ed., Academic Press, New York, 1975, 222.
31. **Scott, D.,** Oxidoreductases, in *Enzymes in Food Processing,* 2nd ed., Reed, G., Ed., Academic Press, New York, chap. 19.

32. Enzymes from Miles, Technical Information, Oxidoreductases and Lipases, Miles Laboratories, Inc., Elkhart, Ind., 1978.
33a. Underkoffer, L. A., Properties and Applications of the Fungal Enzyme Glucose Oxidase, in Proc. of the Int. Symp. on Enzyme Chemistry, Tokyo and Kyoto, 1957.
33b. Bouin, J. C., Attallah, M. T., and Hultin, H. O., The glucose oxidase-catalase system, in *Methods in Enzymology,* Vol. 44, Mosbach, K., Ed., Colowick, S. P. and Kaplan, N. O., Series Eds., Academic Press, New York, 1976, 478.
34. Bouin, J. C., Dudgeon, P. H., and Hultin, H. O., Effect of enzyme ratio and pH on the efficiency of an immobilized dual catalyst of glucose oxidase and catalase, *J. Food Sci.,* 41, 886, 1976.
35. Greenfield, P. F. and Laurence, R. L., Characterization of glucose oxidase and catalase on inorganic supports, *J. Food Sci.,* 40, 906, 1975.
36. Lilly, M. D., Cheetham, P. S. J., and Dunnill, P., Mass transfer problems with immobilized oxidases, in *Enzyme Engineering,* Vol. 3, Pye, E. K. and Weetall, H. H., Eds., Plenum Press, New York, 1978, 73.
37. Buchholz, K. and Jaworek, D., Kinetics and stability of immobilized glucose oxidase and catalase, in *Enzyme Engineering,* Vol. 3, Pye, E. K. and Weetall, H. H., Eds., Plenum Press, New York, 1978, 139.
38. Kittrell, J. R., Laurence, R. I., and Hultin, H. O., Studies of a Catalytic Reactor Using an Immobilized Multi-enzyme System, in Enzyme Technology Grantees-Users Conf., University of Pennsylvania, May 1975, Pye, E. K., Ed., NTIS PB 265 548, NSF/RA-760032, National Technical Information Service, Springfield, Va., 1976, 34.
39. Krishnaswamy, S. and Kittrell, J. R., Deactivation studies of immobilized glucose oxidase, *Biotechnol. Bioeng.,* 20, 821, 1978.
40. Buchholz, K. and Godelmann, B., Macrokinetics and operational stability of immobilized glucose oxidase and catalase, *Biotechnol. Bioeng.,* 20, 1220, 1978.
41. Prenosil, J. E., Immobilized glucose-oxidase/catalase: deactivation in a differential reactor, in *Enzyme Engineering,* Vol. 4, Broun, G. B., Manecke, G., and Wingard, L. B., Jr., Eds., Plenum Press, New York, 1978, 99.
42a. Messing, R. A., Simultaneously immobilized glucose oxidase and catalase in controlled pore titania, *Biotechnol. Bioeng.,* 16, 897, 1974.
42b. Duvnjak, Z. and Lilly, M. D., The immobilization of glucose oxidase to manganese oxide, *Biotechnol. Bioeng.,* 18, 737, 1976.
43. Scott, D., U.S. Patent 3,006,815, 1961.
44a. Scott, D., U.S. Patent 3,160,508, 1964.
44b. Scott, D., U.S. Patent 3,162,537, 1964.
45. Scott, D., U.S. Patent 3,193,393, 1965.
46. Weiland, H., *Enzymes in Food Processing and Products,* Food Processing Review No. 23, Noyes Data Corporation, Park Ridge, N. J., 1972, 255.
47. Sarett, B. L. and Scott, D., U.S. Patent 2,765,233, 1956.
48. Hamilton, B. K., Colton, C. K., and Cooney, C. L., Glucose isomerase: a case study of enzyme process technology, in *Immobilized Enzymes in Food and Microbial Processes,* Olson, A. C. and Cooney, C. L., Eds., Plenum Press, New York, 1974, 118.
49. Ferrier, L. K., Richardson, T., and Olson, N. F., Crystalline catalase insolubilized with glutaraldehyde, *Enzymologia,* 42, 273, 1972.
50. O'Neill, S. P., Inactivation of immobilized catalase by hydrogen peroxide in continuous reactors, *Biotechnol. Bioeng.,* 14, 201, 1972.
51. Balcom, J., Foulkes, P., Olson, N. F., and Richardson, T., Immobilized catalase, *Process Biochem.,* 6, 42, 1971.
52. Richardson, T. and Olson, N. F., Immobilized enzymes in milk systems, in *Immobilized Enzymes in Food and Microbial Processes,* Olson, A. C. and Cooney, C. L., Eds., Plenum Press, New York, 197, 19.
53. Hall, C. W. and Trout, G. M., *Milk Pasteurization,* AVI Publishing, Westport, Conn., 1968, chap. 2.
54. Hall, C. W. and Trout, G. M., *Milk Pasteurization,* AVI Publishing, Westport, Conn., 1968, chap. 1.
55. Hall, C. W. and Trout, G. M., *Milk Pasteurization,* AVI Publishing, Westport, Conn., 1968, 49.
56. Hall, C. W. and Trout, G. M., *Milk Pasteurization,* AVI Publishing, Westport, Conn., 1968, 50.
57. Henderson, J. L., *The Fluid-Milk Industry,* 3rd ed., AVI Publishing, Westport, Conn., 1971, chap. 13.
58. Henderson, J. L., *The Fluid-Milk Industry,* 3rd ed., AVI Publishing, Westport, Conn., 1971, chap. 17.

59. **Henderson, J. L.,** *The Fluid-Milk Industry,* 3rd ed., AVI Publishing, Westport, Conn., 1971, 334.
60. **Swaisgood, H. E.,** personal communication, 1978.
61. **Jeon, I. J., Thomas, E. L., and Reineccius, G. A.,** Production of volatile flavor compounds in ultrahigh-temperature processed milk during aseptic storage, *J. Agric. Food Chem.,* 26, 1183, 1978.
62. **Shipe, W. F., Senyk, G., and Weetall, H. H.,** Inhibition of oxidized flavor development in milk by immobilized trypsin, *J. Dairy Sci.,* 55, 647, 1972.
63. **Swaisgood, H. E. and Horton, H. R.,** Characteristics of soluble and immobilized sulfhydryl oxidase, in Enzyme Technology Grantees-Users Conf., University of Pennsylvania, May, 1975, Pye, E. K., Ed., NTIS PB 265548 NSF/RA-760032, 1976, 94.
64. **Shipe, W. F.,** Enzymatic modification of milk flavor, in *Enzymes in Food and Beverage Processing,* Ory, R. L. and St. Angelo, A. J. Eds., ACS Symp. Ser. #47, American Chemical Society, Washington, D.C., 1977, 57.
65. **Horton, H. R. and Swaisgood, H. E.,** Immobilization as a means of investigating the acquisition of tertiary structure in chymotrypsinogen, in *Methods in Enzymology,* Vol. 44, Mosbach, K., Ed., Colowick, S. P. and Kaplan, N. O., Series Eds., Academic Press, New York, 1976, 522.
66. **Swaisgood, H. E. and Horton, H. R.,** Utilization of Sulfhydryl Oxidase for the Treatment of Ultra-High Temperature Sterilized Milk, Semi-Annual Report, NSF Grant No. AER 73-07877 A03, Department of Food Science and Technology, North Carolina State University, Raleigh, 1976.
67. **Janolino, V. G. and Swaisgood, H. E.,** Isolation and characterization of sulfhydryl oxidase from bovine milk, *J. Biol. Chem.,* 250, 2532, 1975.
68. **Patrick, P. S. and Swaisgood, H. E.,** Sulfhydryl and disulfide groups in skim milk as affected by direct ultra-high-temperature heating and subsequent storage, *J. Dairy Sci.,* 59, 594, 1976.
69. **Janolino, V. G. and Swaisgood, H. E.,** Effect of support pore size on activity of immobilized sulfhydryl oxidase, *J. Dairy Sci.,* 61, 393, 1978.
70. **Swaisgood, H. E.,** Utilization of Immobilized Sulfhydryl Oxidase for Flavor Modification, presented at the 8th Northeast Regional Meeting of the American Chemical Society, Boston, Mass., June 1978.
71. **Swaisgood, H. E., Janolino, V. G., and Horton, H. R.,** Immobilized sulfhydryl oxidase, *AIChE Symp., Ser.,* 74, 25, 1978.
72. **Reed, G.,** Health and legal aspects of the use of enzymes, in *Enzymes in Food Processing,* 2nd ed., Reed, G., Ed., Academic Press, New York, 1975, 549.
73. Congressional Federal Register, 173.135, 426, 1977.
74. **Mandels, M., Kostlick, J., and Parizek, R.,** The use of adsorbed cellulase in the continuous conversion of cellulose to glucose, *J. Polym. Sci.,* 36, 445, 1971.
75. **Gallo, B. J., Andreotti, R., Roche, C., and Mandels, M.,** Cellulase Production by a New Mutant Strain of *Trichoderma versei* MCG 77, Abstr. Symp. on Biotechnology in Energy Production and Conservation, Gatlinburg, Tenn., Oak Ridge National Laboratory, Oak Ridge, Tenn., 1978, 18.
76. **Ryu, D., Andreotti, R., Mandels, M., Gallo, B., and Reese, E. T.,** Studies on Quantitative Physiology of *Trichoderma versei* with Two-Stage Continuous Culture for Cellulase Production, Abstracts 176th American Chemical Society Natl. Meeting, Paper #MICR 024, American Chemical Society, Washington, D.C., 1978.
77. **Garcia-Martinez, D. V., Shinmyo, A., and Demain, A. L.,** Improvements in Cellulase Production by *Clostridium thermocellum,* Abstracts 176th American Chemical Society Natl. Meeting, Paper #MICR 025, 1978.
78. **Cooney, C. L. and Wang, D. I. C.,** Simultaneous Cellulose Production by Cellulolytic Anaerobic Bacteria, Abstr. Symp. on Biotechnology in Energy Production and Conservation, Gatlinburg, Tenn., 1978, 19.
79. **Messing, R. A., Fischer, D. J., and Hutchins, J. R., III,** A stabilized protease complex for heavy duty detergents, *Deterg. Spec.,* 22, 1969.
80. **Messing, R. A.,** U.S. Patent 3,666,627, 1972.
81. **Messing, R. A.,** U.S. Patent 3,802,997, 1974.
82. **Messing, R. A.,** unpublished.
83. **Gregoriadis, G.,** Enzyme entrapment in liposomes, in *Methods in Enzymology,* Vol. 44, Mosbach, K., Ed., Colowick, S. P. and Kaplan, N. O., Series Eds., Academic Press, New York, 1976, 218.
84. **Gregoriadis, G.,** Medical applications of liposome-entrapped enzymes, in *Methods in Enzymology,* Vol. 44, Mosbach, K., Ed., Colowick, S. P. and Kaplan, N. O., Series Eds., 1976, 698.
85. **Li, N. N., Mohan, R. R., and Brusca, D. R.,** U.S. Patent 3,740,315, 1973.

INDEX

A

Acetals, 68
Acetone, 181
Achromobacter liquidum, 13
Acid anhydride, 103
Acid clay, 123
Acid cleaner, 49
Acid hydrolysis, 83
Acidic methylation, 7
Acidulation, 159
Acrolein, 67
Acrylamide, 120, 125
  polymerized, 136
Actinoplanes, 100
Activated charcoal, 123
Activation energy, 33, 129
Activity decay, 38
Activity loss, 41
  mechanism, 38
  type, 37
Acyl-D-amino acid, 187
Acylation, 42, 103, 104
Adsorbed enzyme, 56
Adsorbed protein, 3
Adsorption, 2—6, 8, 66, 88, 91, 94, 98, 117, 122, 123, 132, 133, 138, 181
Aerobacter aerogenes, 134
Aglycone residue, 135
L-Alanine, 189
Alcohol, 68, 154, 158
Alcohol dehydrogenase, 42, 46
Alcohol production, 160, 162
Aldehyde, 68
Aldehyde dehydrogenase, 46
Aldehyde reaction, 103
Aldose, 86
Alkylamine silica, 119
Alkylation, 103, 104
Alpha-amylase, 87, 92, 93, 107
  potential and use, 115—119, 131—135
Alpha-chymotrypsin, 32, 42
Alpha-galactosidase, 114, 115, 141—143
Alpha-galactoside, 115
Alternaria, 76
Aluminum hydroxide, 138
Ames Salmonella/microsome mutagenicity assay, 67
Ames test, 69
Amide, 68
Amine, 68
  lactase-collagen complex, 49
Aminidation, 103, 104
Amino acids, 186—196
DL-Amino acid, 56
L-Amino acid, 191
Amino acid oxidase, 187, 192
Aminoacylase, 2, 185—188
p-Aminobenzene sulfonylethyl cellulose, 125
Amino glass, 119
Aminostyrene, 135
Aminostyrene bead, 140
Ammonia, 192
Ammonium chloride, 76
Ammonium persulfate, 142
Ammonium phosphate, 90
Amylase, 196, 205
α-Amylase, 87, 92, 93, 107
  potential and use, 115—119, 131—135
β-Amylase, 17, 114, 115, 131, 132, 133
  coupling, 7
γ-Amylase, 119, 133
Amyloglucosidase adsorption, 5
Amylopectin, 83, 115, 131
Amylose, 83, 117, 128
Anhydride, 68
Anion exchange resin, 107
Anionic resin, 140
β-Anomer, 93, 94
Antibiotics, 63
Antimicrobials, 76
Aqueous chloroform, 130
Aqueous sugar solution, 155
D-Arabinose, 86
Aroma of food, 66
Aromatic, 68
Arrhenius plot, 32
  aspartase activity, 195
  immobilized glucoamylase, 34
Arrhenius relationship, 134
Arthrobacter, 90, 100
Ash, 84
Aspartase, 185
Aspartic acid, 192
L-Aspartic acid, 186
  production, 191
Asperigillis, 76, 127
  flavus, 75
  niger, 8, 38, 93, 130, 155, 164—166, 169
  oryzac, 8, 12
Attached enzyme, solid surface, 56
Azide, 68, 103
Azo linkage, 130

B

Bacillus
  cereus, 75
  subtilis, 41, 92, 129
Back-flushing, 180
Backmixing, 34, 35
Bacteria, see also Microbes, 75—77, 101, 102, 156
Bacterial protease, 176
Bang-bang-bang policy, 46
Barley, 134, 135, 179

Barley malt, 132
Bed height, pressure drop, 43
Beer chillproofing, 179
Belgium, definition of food additive, 65
Bentonite, 116, 138, 139
Beta-amylase, 17, 114, 115, 131—133
  coupling, 7
Beta-fructo furanoside, 115, 136
Beta-fructo pyranose, 108
Beta-galactosidase, 4, 11, 32, 38
Beta-glucose, 119
Beta-glycosyl, 92
Beverage industry, 55, 69—72
BHA, see Butylated hydroxyanisole
Bifunctional alkyl chain, 66
Binding agent, 118, 126
Biological switches, 181
Biotite, 141
Bisulfite, 107
Blanching, 58
Bodenstein number, 34
Borate, 106
Bovine serum albumin, 177
Brevibacterium
  ammoniagenes, 193—196
  flavum, 196
Brick, 139
Broth, 90
Buffer capacity, 97
Butanol, 133
Buttermilk production, 156
Butylated hydroxyanisole, 77

C

Calamity, legislative response to, 60
Canada, food additives, 65
Cancer, 60
Cancer Institute, 64
Candida utilis, 8
Canned food, 199
Canning plant, 58
Capital investment, long-term, 70
Carbamylation, 103, 104
Carbodiimide, 68, 103
Carbohydrases, see specific enzymes
Carbohydrate degradation, 84
Carboxymethylcellulose, 128
Carcinogen, 61
Carcinogenic potency, 69
Carrageenan, 196
Carrier, 117, 118, 121, 123, 133, 137, 138
  cost, 48
  inorganic, 119, 126, 139
  organic, 139
  silicate, 129
Carrier geometry, mass transfer, 48
Carrier particle, 42
Carrier regeneration, 5, 6
Casein, 176
Catalase, 41, 196
  application, 198
  characteristic, 199
  system, 46
Catalase inactivation, 39
Catalyst profile, 46
Catalytic service life, 106
Catechol, 3
Catechol-formaldehyde polymer, 3
Cation exchange resin, 88, 102
Cationic carrier, 140
Celite, 127
Cellulase, 186, 196
Cellulose, 118, 126, 139, 189
Cellulose anion exchanger, 56
Cellulose bead, 8, 128
Cellulose-bound enzyme, 134
Cellulose diacetate, 136
Cellulose triacetate, 17
Cellulose triacetate filter, 121
Cellulosics, 125
Charcoal, 138
Cheese, 156, 200, 205
Chemical agent, 68
Chemical process, 48
Chemical solvent, 68
Chemical synthesis, 196
Chitin granule, 105
Chloro-triazine, 68
Chromatographic separation, 97
Chymotrypsin, 177, 178
  immobilization, 17
α-Chymotrypsin, 32, 42
Cladosporium, 75, 76
Clostridium, 75
  thermocellum, 205
CM-cellulose, 139, 177
Coagulation, 103
Codex Alimentarius, 64
Cold pasteurization, 200
Collagen, 116, 118, 123, 126, 127, 138, 139, 167
Collodion membrane disk, 167
Color, food, 66
Commercial sterilization, 58
Contamination, microbial, 120
Continuous process plant, 69
Controlled release enzyme, 185, 204—206
Corn, 83, 87
  nutritive sweetener, 84
  wet milling process, 87, 90
Corn starch, 83
  hydrolysis, 93
Copolymerization, 10, 11, 66
Corn syrup, 66, 82, 83, 89
  high fructose, see High fructose corn syrup
Cost estimate of reactor system, 47
Cottage cheese, 156, 157
Coupling, 48
Covalent attachment, 2, 6, 39, 66, 101, 105, 118, 125, 126, 133, 139, 140, 167, 187
Cross-linking, 2, 10, 11, 67, 76, 126

agent, 102, 138
intramolecular, 42
Crystalline glucose, 83, 88
Crystallization, 107, 108
Cyanogen bromide, 8, 104, 127
Cyanogen bromide activation, 103, 135
Cyanogen halogenide, 68
Cyclohexyl isocyanide, 134

D

Deactivation energy, 37
DEAE, see Diethylamino-ethyl cellulose
DEAE-cellulose, 2, 8, 10, 56, 94, 102, 124, 140, 177
DEAE-Sephadex, 2, 17, 56, 124, 138, 187, 188
DEAE-Sephadex aminoacylase system, 188, 189
Decarboxylase, 187
DE dextrin, 125
Delaney clause, 61
Desorption, 3, 37, 42
Dextranase, 4
Dextransucrase, 37
Dextrin, 119, 122, 131
Dextrose solution, 88
Diacetyl, 201
Dialysis paper, 128
m-Diaminabenzene, 129
Diauxic lag, 159
Diazo compound, 68
Diazo coupling, 103, 129
Diazonium salt reactions, 105
Dicarbonyl cellulose, 104
Dicetyl phosphate, 122
Diethylamino-ethyl cellulose, 56, 91
Diethylaminoethyl-Sephadex, 187
Difluorobenzene, 68
Digestibility, 66
Diisocyanate, 8
Diisocyanatediazo, 8
Diisocyanateglutaraldehyde, 8
Dimethyladipimidate hydrochloride, 142
p-Dimethylaminopropionitrile, 142
L-DOPA, 189
Dry milling, 89
Duolite®, 2, 3

E

ECD-Sepharose, 28, 117
Economic constraint, 69—70
EDA, see Ethylenediamine
EDTA, 116
Efficiency improvement, 34, 47
Egg, 198, 205
Electric field, 36
Electrodeposition, 138
Electrolyte activator, 49
Elixir sulfanilamide, 59
Elution chromatography, 88
Endohydrolase, 115
Endomyces glucoamylase, 124
Entrapment, 10, 17, 66, 116, 120, 121, 132, 133, 136, 137, 167, 187
Enzacryl®, 116
EPA, 59
Epoxy compound, 68
Equipment cost, 47
Erythrocyte, 118
Escherichia coli, 12, 86, 129, 167, 191—193, 195, 196
Ester, 68
Ethanol, 159, 181
Ether, 68
Ether alcohol, 68
Ethyl cellulose, 136
Ethylenediamine, effect on enzyme loading, 7, 9
Exohydrolase, 92
Exothermic reaction, 192

F

FAO/WHO, 64, 65
Fat, 84
FDA, 59, 63—65, 75
Federal Register, 62
Federation of American Societies, for Experimental Biology, 65
Feed, 48
Feed temperature profile, 99
Feed treatment cost, 48
Fermentation, 90, 159, 187, 196
Fermentor broth, 101
Fiber entrapment, 10
Fixed bed, 35, 39
Flavor modification, sulfhydryl oxidase in, 200—206
Flow rate, 6, 36, 189
Fluid food flavor, 73
Fluidized bed, 35, 142
Food additive, 62—65, 158, 169, 170
Food and Drug Administration, see FDA
Food poisoning, 75
Food processing
activity loss, 37—42
applications of immobilized enzymes, 66—69
amino acids, 186—196
controlled release, 204—206
glucose oxidase-catalase, 196—200
sulfhydryl oxidase, 200—204
cost estimates, 47—48
design considerations, 48—49
economic constraints, 69—72
governmental regulations, 59—66
heat transfer, 37
mass transfer
backmixing, 34—35

dispersion, 34—35
electrostatic effects, 35—36
external, 25—28
internal, 28—34
transient behavior of reactors, 36—37
microbial control of reactors, 75—77
multienzyme systems, 45—47
parameters affecting the industry, 57—59
pressure drop, 42—43
proteases, 176—181
reaction kinetics, 20—25
reactor behavior, 20—25
reactor operating strategy, 43—45
reactor types, 16—20
technological factors, 72—74
temperature effects, 37
Formaldehyde, 143
Fractionation, 97, 107
France, 65
Free enzyme, 93
Free glucoamylase, 120
β-Fructo pyranose, 108
Fructose, 56, 66, 81, 82, 86, 88, 89, 93, 94, 96, 99, 136
crystallization, 107—108
fractionation, 107
isomerization reaction from glucose, 105—107
D-Fructose, 82
β-Fructofuranoside, 115, 136
β-D-Fructofuranosyl, 135
Fruit juice, 196
L-Fucose, 86
Fumarase, 185
Fumaric acid, 192
Fungus, 77
Fusarium, 76

G

α-D-Galactopyranosyl, 141
Galactose, 154, 155, 159
D-Galactose, 154
α-Galactosidase, 114, 115, 141—143
β-Galactosidase, 4, 11, 32, 38
β-D-Galactosidase, 154
α-Galactoside, 115
Gamma-amylase, 119, 133
Gel, 11, 120, 123, 125, 133, 142, 181
acrylic, 136
lattice, 192
organic, 117, 121, 137
packed, 122
polyacrylamide, 141, 194
silica, 127
Germ, 89
Glass, 102, 127, 130, 139, 177
Glass bead, 203
D-Glucan, 83
4-Glucanohydrolase, 115
Glucoamylase, 8, 12, 34, 38, 87, 92, 107, 155, 164—166, 169
adsorbed, 124
immobilization, 7, 9, 47, 119—131
adsorption, 123
commercial use, 120
entrapment, 121
isolated, 93
native, 125
technology, 70
Gluconic acid, 198, 199
Glucose, 57, 119, 130, 155
controlled release enzymes, 204
fermentation of LH whey, 159
fractionation, 97, 107
in manufacture of HFCS, see High fructose corn syrup
isomerization reaction to fructose, 85—86, 93—96, 105—107
removal, 198
structural formula, 154
β-Glucose, 119
D-Glucose, 82, 83, 86, 87, 90, 93
Glucose isomerase
adsorption, 2—4
coupling, 8
heat transfer, 37
immobilization, 39—40
manufacture of HFCS, see High Fructose corn syrup
production of, 98—105
reactor, 17
Glucose oxidase, 46, 138, 197—199
Glucose oxidase-catalase, 185, 196, 198, 200
Glucose-1-phosphate, 86
Glucose-6-phosphate, 86
Glutaraldehyde, 8, 67, 77, 103, 105, 127, 129, 130, 135, 140, 196, 203
Glutaraldoxine, 67
Glutaric acid, 67
Gluten, 89
Glycocalyx, 76
Glycoside hydrolase, 135, 141
β-Glycosyl, 92
Governmental regulation, 59—61
Granular carbon, 87
Grape juice, 159
GRAS, 61—63, 65
Grinding, 89
GSH, see Reduced glutathione
Guanidino, 125
Guanidino cellulose, 8
Guoy-Chapman solution, 36

H

HACCP, see Hazard Analysis and Critical Control Point
Halogenoalkyl derivative, 68
Hazard Analysis and Critical Control Point, 75—77

Health hazard, 58
Heat transfer, 37
Heat treatment, 101, 202
Heterogeneous catalysis, 16
Hexamethylenediamine, 7, 196
Hexamethylphosphoramide, 41
n-Hexanal, 201
HFCS, see High fructose corn syrup
High fructose corn syrup, 66
  definition, 82
  manufacture using glucose isomerase
    concentration, 96—97
    crystallization, 107—108
    fractionation, 97—98, 107
    historical background, 83—87
    hydrolysis of starch, 92—93
    isomerization of glucose to fructose, 93—96
    isomerization reaction glucose ⇌ fructose, 105—106
    isomerization reactions to attain high ratios of fructose to glucose, 106—107
    hydrolysis of starch, 92—93
    isomerization of glucose to fructose, 93—96
    isomerization reaction glucose ⇌ fructose, 105—106
    isomerization reactions to attain high ratios of fructose to glucose, 106—107
    manufacture of starch, 89
    overall process, 87—89
    production of glucose isomerase, 99—100
    production of immobilized glucose isomerase, 89—92, 100—105
    refining, 96—97
HMDA, see Hexamethylene-diamine
Holland, enzyme regulations, 65
Hollow fiber, 12
Hornblende-invertase conjugate, 140
Hull, 89
Humic acid, 3
Hydrazine treatment, 7
Hydrochloric acid, 136
Hydrogen peroxide, 49, 197
Hydrolysis, 92, 119, 130, 154, 170
Hydrolyzed lactose syrup, 71—73
Hydrophilic substrate, 135
Hydrophobic carrier, 135
Hydrophobic interaction, 132
p-Hydroxybenzaldehyde, 3
Hydroxylapatite, 122
L-Hydroxyphenylglycine, 189
Hydroxysuccinimide ester, 103

I

IMES, see Immobilized enzyme systems
Immobilization, 8, 90
Immobilization method, 6
Immobilized enzyme systems, 75
Immobilized isomerase, definition, 82
Incidental additive, 66
Infant formula, 179
Insolubilizing enzyme, 67
Insolubilizing method, 66
Interfacial coacervation, 11
Interfacial polymerization, 11
Internal diffusion effect, 29
Intramolecular oxidoreductase, 82
Intraparticle diffusion, 131
Invertase, 8, 135—140, 196
Invert sugar, 85
Iodoacetyl cellulose, 187
Iodophor, 49
Ion exchange, 66, 123
Ionic binding, 187, 188
Iron filing, 11
γ-Irradiation, 120—136
Isocyanate, 68
Isoleucine, 186
Isomaltose, 119, 130
Isomerase, 82, 89
Isomerization reaction, 106
Isothermal operation, 44
Isotoic anhydride, 105

J

Japan, 100, 115
Jet cooker, 115
Joint Expert Committee, 64

K

Koalin, 116
Keto furanose, 93
Ketone, 68
Ketose, 86
Kinetic analysis, 24
Klebsiellba lactis, 12
Kluyveromyces fragilis, 159, 160
Kozeny-Carman equation, 189

L

Labor cost, 47
Lactase, 2, 3, 114
  applications to milk, 155—158
  applications to whey, 40, 158—163
  immobilization, 45, 47, 153, 164
  regulatory aspects, 169—171
  sources of, 155
  structural formula, 154—155
  technology of application, 71, 163—169
Lactose, 82, 155, 156, 159
Lactose-hydrolyzed, 155
Laundry detergent, 205
Lecithin, 122
Leuconostoc mesenteraides, 143

LH, see Lactose-hydrolyzed
LH acid whey, 161, 162
LH whey permeate, 162
α-Limit dextrin, 115
β-Limit dextrin, 132
Lineweaver-Burk plot, 33, 36
Linking agent, 125, 133
Liposome, 206
Liquid consumption, 73
Liquid drying, 11
Liquid food, 58
Liquor, 88
Lysing enzyme, 101
Lytic enzyme, 91

M

Magnetic recovery, 11
Maize dextrin, 131
L-Malic acid, 194
Maltase, 196
Maltodextrin, 131
Maltoheptoses, 92
Maltohexaoses, 92
Maltose, 82, 93, 122, 124, 128, 129, 131, 132
β-Maltose, 131
Maltotriose, 92, 131
Marine crop, 186
Mass transfer
  carrier geometry, 48
  coefficient, 26
  electrostatic effect, 35
  external, 25, 31
  internal, 28, 31
  limitation, 10, 25, 189
  rate, 25, 27
  testing, 28
Mechanical kneading, 108
Metal salt, 68
Methionine, 47
Methyl cellulose, 199, 200
N-N′-Methylene-bis-acrylamide, 187
N,N-Methylene-bisacrylamidel tetramethylethylenediamine copolymer, 116
γ-Methyl-polyglutamate, 17
Michaelis constant, 36, 120, 122, 128, 133—135
Michaelis-Menten
  equation, 21, 24, 27, 122
  kinetics, 19, 29—31, 36, 46
  rate expression, 20
  relationship, 188
Microbes
  contamination, 37, 38, 48
  control, 57, 75—77
  growth, 48
  in amino acid production, 186—196
Microcapsule, 31
Micrococcus lysodeikticus, 199
Microencapsulation, 10, 11, 167
Milk, 154—156, 163, 164, 176—178, 198, 201—203
Milk sugar, 154
Mold, 76
Monila, 75
Monomer concentration, 10
Morteriella vinacea, 141
Mother liquor, 86, 88
Muco michei, 177
Mucor, 76
Multienzyme system, 45
Multiple column system, 45
Muscovite, 141
Mutagenicity testing, 69
Mutarotase, 138
Mycotoxin, 63

N

National Academy of Sciences, 65
NDGA, see Nordihydroguararetic acid
Neoplastic hazard level, 67
Nigerose, 119
p-Nitrophenyl-α-D-galactoside, 142
Nitrous acid, 7, 118, 119, 126, 127, 129
Nonisomerase protein, 91
Nordihydroguararetic acid, 77
Nutritional need, 58
Nutritive, 82
Nutritive sweetener, 82, 84

O

Oligosaccharide, 125, 131
Organic solvent, 41
OSHA, 59
Oxidation, 198
Oxidative stability of milk, 180
Oxygen removal, 198

P

Papain, 178, 179
Particle crushing, 32
Pasteurization, 58, 179, 201
Patent, fermentation process, 99
Penicillin, 47
Penicillium, 75, 76
Performance improvement, 48
Periodate, 203
Permeate, 71
Pepsin, 176—178
Petrochemical feedstock, 129
o-Phenanthroline, 41
Phenol-formaldehyde, 140
Phenolic antioxidant, 77
D,L-Phenylalanine methylester, 32

L-Phenylalanine, 189
D-Phenylglycine, 189
Phenyl maltoside, 116
Phloroglucinol-formaldehyde polymer, 3
Phosgene derivative, 68
Phospholipid, 206
Pilot scale operation, 74
Plugging, 49
PNPG, see *p*-Nitrophenyl-α-D-gelactoside
Polyacrylamide, 105, 120, 187
Polyacrylamide gel, 116
Polyalcohol, 108
Polyaminostyrene, 135
Polymeric aldehydes, 104—105
Polymerization, 10, 11
Polysaccharide glycocalyx, 76
Pore diameter, 102, 129
Pore diffusion, 28, 32, 33, 49, 124, 130, 131
Porosity, 198
Porus alumina bead, 2
Porous alumina carrier, 3
Porous alumina sample, 5
Porous ceramic material, 102
Porous paper, 124
Porous silica bead, 136
Potassium persulfate, 187
Pressure drop, 42, 49
Product purity, 97
Productive capacity, 69
Productivity, 45, 57—59
L-Proline, 186
Propanol, 181, 201
Propyl gallate, 77
Proteases, 101, 182, 205
  inhibition, 41
  potential applications, 175
    biological switches, 181
    enzyme inactivation, 181
    food industries, 176—181
    synthesis, 181
Protein, 3, 41, 84, 134, 125, 128
  cotton, 127
  denaturing agent, 188
  hydrolysis, 176
  hydrolyzate, 187
  immobilization, 181
  milk, 201
  recovery, 71
  stain, 205
  synthesis, 186
  synthetic, 181
Protein loading, 4
Proteolytic enzyme, 181
Pseudomonas putida, 13
Pullulan, 134
Pullulanase, 114, 131—135
Pulp, 127
Pyroxene, 141

Q

Quality assurance, 75
Quaternary ammonium salt, 91

R

Raffinose, 115, 142
Rate equation, 74
Reaction, kinetics, 20
Reactions, 104—105, 128
Reactor
  activity loss, 37—42
  amino acid production, 188—190, 192, 194
  behavior, 20—25
  controlled release enzymes, 205
  cost estimates, 47—48
  design considerations, 48—49
  heat transfer, 37
  immobilization of glucoamylase, 119, 121, 124
  immobilization of glucose isomerase, 91, 100—102
  immobilization of invertase, 137
  immobilization of proteases, 181
  isomerization of glucose to fructose, 95, 97, 105—106
  kinetics, 20—25
  lactase production, 155, 164—167
  mass transfer
    backmixing, 34—35
    dispersion, 34—35
    electrostatic effects, 35—36
    external, 25—28
    internal, 28—34
    transient behavior, 36—37
  membrane sandwich, 12
  microbial control, 75—77
  milk production, 203
  multienzyme systems, 45—47
  operating strategy, 43—45
  pressure drop, 42—43
  temperature effects, 37
  types, 16-20
Synthesis, use of proteases, 181
Thiocarbamylation, 104
Reduced glutathione, 203
Regeneration, 47
Regulated food additive, 62
Regulatory aspect, 169
Rennet, 178
Rennin, 202
Research and development, 59
Resorcinol-formaldehyde polymer, 3
D-Ribulose, 86

S

Saccharification step, 87
Saccharomyces
  cerevisiae, 159, 163
  fragilis, 167
  lactis, 155, 163, 165, 170
  pastorianus, 136
Salicylaldehyde, 3
Salmonella, 75, 198
Salt, 129
Sand, 116, 139, 141
Sanitary condition, 72
Sanitizing agent, 49
Sanitizing system, 76
SDS, see Sodium dodecylsulfate
Shelf life, 201
Sherbet, 159
Shigella, 75
Silane-glutaraldehyde, 8, 42
Silica, 127, 130, 135
Silverless photography, 181
SMA, see Styrene-maleic anhydride
Sodium chloride concentration, 4
Sodium dodecylsulfate, 188
Soft drink, 179
Solubilized enzyme, 101
Sorption, 101, 103
Soya extract, 180
Soya hydrolysis, 179
Soya protein, 179
Soybean, 141, 142, 176
Soy flour, 141
Soy protein, 141
Soy sauce, 163
Starch, 83, 89, 134
  autoclaved, 87
  hydrolysis, 92
  liquified, 125
  manufacture, 89
  sweetener from, 85
Sterigmatocystis, 75
Streptomyces, 90, 93, 100, 101, 106
  griseus, 181
Styrene-maleic anhydride, 7
Sucrose, 82, 84, 85, 154
Sugar beet, 115, 143
Sugar utilization pattern, 160, 161
Sulfhydryl, 186, 201
Sulfhydryloxidase, 185, 200, 203, 204
Sulphonyl chloride, 103
Superoxide dismutase, 41
Supersaturation, 108
Surface attachment, 41
Sweden, definition of food additive, 65
Sweetener, 82
Sweet potato, 132, 133
Switzerland, enzyme regulation, 65
System design consideration, 48

T

Tannic acid, 103
Temperature
  effect, 37
  enzyme activity, 37
  operating, 48
  optimal policy, 46
Textile, 198
Texture of food, 66
Thermal denaturation, 37
Thermolysin, 7
Thiele moduli, 124
Thiocarbamylation, 103
Thioisocyante, 68
Thiol-disulfide, 104, 117
Thiophosgene, 134
Titanic complex, 125
Titanium salt, 140
Titanous complex, 125
Toxicologic evaluation of food additives, 64
Transglucosidase, 93
Trichoderma
  reesi, 205
  viride, 204
Trypsin, 181
L-Tryptophan, 189

U

Ultrafiltration membrane, 16
United Kingdom, definition of enzyme, 65
Urea, 116, 141
U.S. Department of Agriculture, 59, 64, 65
U.S. Public Health Service, 59
U.S. Treasury, 59

V

L-Valine, 189
Vinylating agent, 11

W

Wet milling, 89
Wet spun synthetic fiber, 17
Wheat, 135
Whey, 40, 72, 142, 154, 155, 158, 163, 164, 168, 169
Wine, 74, 159
Wrapped food, preservation, 199

X

Xanthan gum, 158, 162, 163
Xanthomonas compestris, 162, 163
Xylan, 99
Xylose, 99
Xylose isomerase, 82, 86, 89, 90
D-Xylulose, 82, 86

Y

Yeast, 75, 76, 137, 156, 159, 165
Yoghurt, 156, 157